Asset Analytics

Performance and Safety Management

Series Editors

Ajit Kumar Verma, Western Norway University of Applied Sciences, Haugesund, Norway

P. K. Kapur, Centre for Interdisciplinary Research, Amity University, Noida, India

Uday Kumar, Division of Operation and Maintenance Engineering, Luleå University of Technology, Luleå, Sweden

The main aim of this book series is to provide a floor for researchers, industries, asset managers, government policy makers and infrastructure operators to cooperate and collaborate among themselves to improve the performance and safety of the assets with maximum return on assets and improved utilization for the benefit of society and the environment.

Assets can be defined as any resource that will create value to the business. Assets include physical (railway, road, buildings, industrial etc.), human, and intangible assets (software, data etc.). The scope of the book series will be but not limited to:

- Optimization, modelling and analysis of assets
- Application of RAMS to the system of systems
- Interdisciplinary and multidisciplinary research to deal with sustainability issues
- Application of advanced analytics for improvement of systems
- Application of computational intelligence, IT and software systems for decisions
- Interdisciplinary approach to performance management
- Integrated approach to system efficiency and effectiveness
- Life cycle management of the assets
- Integrated risk, hazard, vulnerability analysis and assurance management
- Adaptability of the systems to the usage and environment
- Integration of data-information-knowledge for decision support
- Production rate enhancement with best practices
- Optimization of renewable and non-renewable energy resources

Review Process

The proposal for each volume follows multi-pronged review process. The first level of review (single blind) is conducted by the series editors who may or may not decide to enlist the help of editorial board members for a second level of review. Proposals may also undergo a third level of peer review (double blind) if recommended by the Series Editors.

The series follows Ethics Statement found in the Springer standard guidelines here. https://www.springer.com/us/authors-editors/journal-author/journal-author-helpdesk/before-you-start/before-you-start/1330#c14214

P. K. Kapur · Omar Alhazmi · Shakshi Singhal
Editors

Data-centric Approaches to Industrial Decisions

Technology, Digitisation and Business Decisions

Springer

Editors
P. K. Kapur
Amity Center for Interdisciplinary Research
Amity University
Noida, Uttar Pradesh, India

Omar Alhazmi
Department of Computer Science
Taibah University
Medina, Saudi Arabia

Shakshi Singhal
XLRI Xavier School of Management
Delhi NCR Campus
Jhajjar, Haryana, India

ISSN 2522-5162 ISSN 2522-5170 (electronic)
Asset Analytics
ISBN 978-981-96-7555-5 ISBN 978-981-96-7556-2 (eBook)
https://doi.org/10.1007/978-981-96-7556-2

This Springer imprint is published by the registered company Springer Nature Singapore Pte Ltd.
The registered company address is: 152 Beach Road, #21-01/04 Gateway East, Singapore 189721, Singapore

Contents

Chapter 1
Data Preprocessing and Fault Big Data Analysis for Open Source Software

Yoshinobu Tamura and Shigeru Yamada

Abstract Recently, many open source software have been used under several situation. Especially, several open source software are implemented under the network computing areas of edge computing and clouds. We propose the data preprocessing and fault big data analysis based on deep learning for open source software. In particular, we analyze the open source fault big data from various points of view. Also, the fault data sets are analyzed by using the frequency encoding and count encoding from the special characteristics of fault detection phenomena for open source software. Moreover, we discuss the characteristics of fault big data in open source software. Furthermore, several numerical examples for fault big data analysis in open source software are shown in this paper. Then, we focus on several open source software. We compare the difference of the deep learning and the stochastic models from the standpoint of fault estimation.

Keywords Data preprocessing · Deep learning · Fault big data · Open source software

Y. Tamura (✉)
Yamaguchi University, Yamaguchi, Japan
e-mail: tamuray@yamaguchi-u.ac.jp

S. Yamada
Tottori University, Emeritus Professor, Tottori, Japan
e-mail: bex2yama@muse.ocn.ne.jp

P. K. Kapur et al. (eds.), *Data-centric Approaches to Industrial Decisions*, Asset Analytics, https://doi.org/10.1007/978-981-96-7556-2_1

1.1 Introduction

Several methods of software reliability assessment have been proposed by several reserchers [1, 2]. Especially, several analysis methods for open source software (OSS) have been developed by several researchers [3, 4]. Moreover, our research group has proposed several methods based on deep learning by using the fault big data [5–7].

Our motivation of this paper is to confirm the trend of fault big data. Then, it is very important to confirm the trend of fault big data before the data preprocessing. We discuss the normality of fault big data. In particular, we analyze by using the discriminant analysis.

1.2 Data Analysis

First, we show a part of results data after the data preprocessing as follows:

	Opened	Changed	Reporter	Assignee	Product	Component	Status	Resolution	Hardware	OS	Severity	Version	Summary
1	17.2	0.0	53.0	105.0	19,017.0	2963.0	19,017.0	2504.0	2868.0	4150.0	3157.0	48.0	57.0
2	0.1	0.0	53.0	1745.0	19,017.0	2963.0	19,017.0	2504.0	14,781.0	14,331.0	3157.0	48.0	40.0
3	3.7	2259.5	53.0	65.0	19,017.0	451.0	19,017.0	1833.0	14,781.0	14,331.0	3157.0	48.0	34.0
4	10.2	0.0	53.0	1745.0	19,017.0	2963.0	19,017.0	429.0	14,781.0	14,331.0	3157.0	48.0	69.0
5	3.8	0.0	43.0	54.0	19,017.0	451.0	19,017.0	8819.0	14,781.0	14,331.0	3157.0	48.0	50.0
6	0.0	1228.3	43.0	105.0	19,017.0	2963.0	19,017.0	8819.0	14,781.0	14,331.0	3157.0	48.0	50.0
7	32.6	0.0	6.0	54.0	19,017.0	106.0	19,017.0	1907.0	14,781.0	14,331.0	3157.0	367.0	73.0
8	0.6	1027.4	6.0	55.0	19,017.0	106.0	19,017.0	8819.0	14,781.0	14,331.0	3157.0	367.0	96.0
9	9.6	0.0	1.0	1745.0	19,017.0	451.0	19,017.0	2504.0	14,781.0	14,331.0	3157.0	367.0	150.0
10	1.1	914.5	25.0	404.0	19,017.0	2963.0	19,017.0	1833.0	14,781.0	14,331.0	3157.0	48.0	55.0
11	5.9	0.0	43.0	54.0	19,017.0	451.0	19,017.0	8819.0	14,781.0	14,331.0	3157.0	367.0	62.0
12	0.0	0.0	43.0	54.0	19,017.0	418.0	19,017.0	8819.0	14,781.0	14,331.0	3157.0	367.0	60.0
13	0.0	0.0	43.0	54.0	19,017.0	1386.0	19,017.0	8819.0	14,781.0	14,331.0	3157.0	367.0	60.0
14	0.0	204.6	43.0	54.0	19,017.0	2963.0	19,017.0	8819.0	14,781.0	14,331.0	3157.0	367.0	58.0
15	0.0	0.0	43.0	54.0	19,017.0	106.0	19,017.0	8819.0	14,781.0	14,331.0	3157.0	367.0	61.0
16	4.5	1027.4	30.0	54.0	19,017.0	205.0	19,017.0	8819.0	14,781.0	14,331.0	3157.0	367.0	36.0
17	0.9	0.0	5.0	54.0	19,017.0	205.0	19,017.0	8819.0	14,781.0	14,331.0	3157.0	1424.0	35.0
18	3.5	1158.5	140.0	1745.0	19,017.0	2963.0	19,017.0	1833.0	14,781.0	14,331.0	3157.0	367.0	91.0
19	3.3	8.5	30.0	54.0	19,017.0	753.0	19,017.0	8819.0	14,781.0	14,331.0	3157.0	367.0	42.0
20	0.3	0.0	101.0	69.0	19,017.0	451.0	19,017.0	1633.0	14,781.0	14,331.0	3157.0	133.0	70.0
21	0.0	0.0	101.0	1745.0	19,017.0	418.0	19,017.0	1633.0	14,781.0	14,331.0	3157.0	133.0	35.0
22	0.0	0.0	101.0	1745.0	19,017.0	42.0	19,017.0	1907.0	14,781.0	14,331.0	3157.0	133.0	38.0

(continued)

(continued)

	Opened	Changed	Reporter	Assignee	Product	Component	Status	Resolution	Hardware	OS	Severity	Version	Summary
23	0.5	0.0	52.0	54.0	19,017.0	451.0	19,017.0	8819.0	14,781.0	14,331.0	3157.0	367.0	45.0
24	0.1	141.4	52.0	1745.0	19,017.0	418.0	19,017.0	429.0	14,781.0	14,331.0	3157.0	367.0	33.0
25	0.1	1027.7	20.0	404.0	19,017.0	753.0	19,017.0	8819.0	14,781.0	14,331.0	3157.0	367.0	56.0
26	0.0	0.0	101.0	105.0	19,017.0	2963.0	19,017.0	1907.0	14,781.0	14,331.0	3157.0	133.0	53.0
27	0.1	0.0	11.0	1745.0	19,017.0	418.0	19,017.0	2504.0	14,781.0	14,331.0	3157.0	367.0	31.0
28	4.7	939.5	54.0	1745.0	19,017.0	106.0	19,017.0	2504.0	14,781.0	14,331.0	3157.0	367.0	66.0
29	1.0	0.5	54.0	1745.0	19,017.0	106.0	19,017.0	2504.0	14,781.0	14,331.0	3157.0	367.0	70.0
30	2.1	0.0	33.0	404.0	19,017.0	753.0	19,017.0	8819.0	14,781.0	14,331.0	3157.0	367.0	82.0
31	7.0	1027.5	7.0	105.0	19,017.0	2963.0	19,017.0	8819.0	14,781.0	14,331.0	3157.0	367.0	29.0
32	1.1	0.0	8.0	7.0	19,017.0	753.0	19,017.0	8819.0	14,781.0	14,331.0	3157.0	1424.0	43.0
33	3.0	1181.9	108.0	69.0	19,017.0	451.0	19,017.0	8819.0	14,781.0	14,331.0	3157.0	367.0	78.0
34	8.0	0.0	160.0	404.0	19,017.0	753.0	19,017.0	8819.0	14,781.0	14,331.0	3157.0	133.0	59.0
35	0.2	1183.2	160.0	404.0	19,017.0	753.0	19,017.0	8819.0	14,781.0	14,331.0	3157.0	133.0	53.0
36	0.7	1231.1	160.0	96.0	19,017.0	2963.0	19,017.0	1907.0	14,781.0	14,331.0	3157.0	368.0	48.0
37	0.5	0.0	1.0	404.0	19,017.0	753.0	19,017.0	8819.0	14,781.0	14,331.0	3157.0	1424.0	74.0
38	0.6	1231.1	11.0	11.0	19,017.0	2963.0	19,017.0	8819.0	14,781.0	14,331.0	3157.0	1424.0	48.0
39	0.1	0.0	8.0	404.0	19,017.0	753.0	19,017.0	8819.0	14,781.0	14,331.0	3157.0	1424.0	47.0
40	0.0	0.0	160.0	404.0	19,017.0	753.0	19,017.0	8819.0	14,781.0	14,331.0	3157.0	133.0	64.0
41	0.9	2414.1	149.0	105.0	19,017.0	2963.0	19,017.0	8819.0	14,781.0	14,331.0	3157.0	133.0	41.0
42	2.7	0.0	54.0	55.0	19,017.0	106.0	19,017.0	8819.0	14,781.0	14,331.0	3157.0	367.0	64.0
43	1.1	1230.7	19.0	661.0	19,017.0	2963.0	19,017.0	8819.0	14,781.0	4150.0	3157.0	133.0	43.0
44	2.3	0.0	43.0	55.0	19,017.0	106.0	19,017.0	8819.0	14,781.0	14,331.0	3157.0	367.0	78.0

(continued)

(continued)

	Opened	Changed	Reporter	Assignee	Product	Component	Status	Resolution	Hardware	OS	Severity	Version	Summary
45	1.0	0.0	149.0	54.0	19,017.0	753.0	19,017.0	8819.0	14,781.0	14,331.0	3157.0	133.0	41.0
46	0.2	0.0	160.0	404.0	19,017.0	753.0	19,017.0	8819.0	14,781.0	14,331.0	3157.0	133.0	42.0
47	2.5	2362.8	147.0	1745.0	19,017.0	2963.0	19,017.0	2504.0	14,781.0	14,331.0	3157.0	367.0	87.0
48	1.2	0.0	45.0	404.0	19,017.0	753.0	19,017.0	8819.0	14,781.0	14,331.0	3157.0	367.0	80.0
49	0.1	1230.6	101.0	96.0	19,017.0	2963.0	19,017.0	8819.0	14,781.0	14,331.0	3157.0	133.0	33.0
50	0.0	0.0	101.0	55.0	19,017.0	106.0	19,017.0	8819.0	14,781.0	14,331.0	3157.0	133.0	36.0
51	3.0	0.0	34.0	404.0	19,017.0	753.0	19,017.0	8819.0	14,781.0	14,331.0	3157.0	1424.0	46.0
52	3.4	208.2	25.0	11.0	19,017.0	418.0	19,017.0	2504.0	14,781.0	14,331.0	3157.0	133.0	69.0
53	0.3	940.1	54.0	1745.0	19,017.0	753.0	19,017.0	1907.0	14,781.0	14,331.0	3157.0	367.0	48.0
54	1.2	0.0	54.0	55.0	19,017.0	106.0	19,017.0	8819.0	14,781.0	14,331.0	3157.0	367.0	44.0
55	0.0	0.0	54.0	55.0	19,017.0	106.0	19,017.0	1833.0	14,781.0	14,331.0	3157.0	367.0	42.0
56	0.0	0.0	54.0	55.0	19,017.0	106.0	19,017.0	1833.0	14,781.0	14,331.0	3157.0	367.0	88.0
57	0.0	0.6	54.0	55.0	19,017.0	106.0	19,017.0	1833.0	14,781.0	14,331.0	3157.0	367.0	39.0
58	0.0	0.0	54.0	55.0	19,017.0	106.0	19,017.0	1833.0	14,781.0	14,331.0	3157.0	367.0	112.0
59	0.0	0.3	54.0	55.0	19,017.0	106.0	19,017.0	1833.0	14,781.0	14,331.0	3157.0	367.0	47.0
60	0.0	0.0	54.0	55.0	19,017.0	106.0	19,017.0	1833.0	14,781.0	14,331.0	3157.0	367.0	53.0
61	0.0	0.5	54.0	55.0	19,017.0	106.0	19,017.0	1833.0	14,781.0	14,331.0	3157.0	367.0	139.0
62	0.0	0.0	54.0	55.0	19,017.0	106.0	19,017.0	1833.0	14,781.0	14,331.0	3157.0	367.0	81.0
63	0.0	0.0	54.0	55.0	19,017.0	106.0	19,017.0	1833.0	14,781.0	14,331.0	3157.0	367.0	70.0
64	0.0	0.0	54.0	55.0	19,017.0	106.0	19,017.0	1833.0	14,781.0	14,331.0	3157.0	367.0	68.0
65	0.0	0.1	54.0	55.0	19,017.0	106.0	19,017.0	1833.0	14,781.0	14,331.0	3157.0	367.0	56.0
66	0.0	0.0	54.0	55.0	19,017.0	106.0	19,017.0	1833.0	14,781.0	14,331.0	3157.0	367.0	58.0

(continued)

(continued)

	Opened	Changed	Reporter	Assignee	Product	Component	Status	Resolution	Hardware	OS	Severity	Version	Summary
67	2.5	0.0	43.0	54.0	19,017.0	451.0	19,017.0	8819.0	14,781.0	14,331.0	3157.0	367.0	62.0
68	0.8	0.2	149.0	695.0	19,017.0	1386.0	19,017.0	8819.0	14,781.0	14,331.0	3157.0	133.0	60.0
69	2.5	1231.0	54.0	211.0	19,017.0	2963.0	19,017.0	1907.0	14,781.0	14,331.0	3157.0	367.0	52.0
70	2.3	0.0	30.0	404.0	19,017.0	753.0	19,017.0	8819.0	14,781.0	14,331.0	3157.0	367.0	42.0
71	0.8	2378.8	25.0	71.0	19,017.0	2963.0	19,017.0	8819.0	14,781.0	14,331.0	3157.0	367.0	78.0
72	3.4	0.0	2.0	661.0	19,017.0	418.0	19,017.0	8819.0	14,781.0	14,331.0	3157.0	367.0	68.0
73	7.6	1230.4	46.0	105.0	19,017.0	2963.0	19,017.0	8819.0	14,781.0	14,331.0	3157.0	133.0	124.0
74	0.3	0.0	34.0	190.0	19,017.0	418.0	19,017.0	8819.0	14,781.0	14,331.0	3157.0	1424.0	36.0
75	1.1	0.0	11.0	1745.0	19,017.0	418.0	19,017.0	2504.0	14,781.0	14,331.0	3157.0	367.0	66.0
76	0.7	2170.6	24.0	211.0	19,017.0	2963.0	19,017.0	8819.0	14,781.0	14,331.0	3157.0	133.0	68.0
77	1.0	0.0	46.0	105.0	19,017.0	2963.0	19,017.0	8819.0	14,781.0	4150.0	3157.0	133.0	66.0
78	1.1	0.0	138.0	661.0	19,017.0	418.0	19,017.0	8819.0	14,781.0	14,331.0	3157.0	367.0	79.0
79	0.2	0.1	138.0	695.0	19,017.0	1386.0	19,017.0	2504.0	14,781.0	14,331.0	3157.0	368.0	67.0
80	0.3	0.2	20.0	69.0	19,017.0	451.0	19,017.0	8819.0	14,781.0	14,331.0	3157.0	367.0	36.0
81	0.6	1230.5	138.0	1745.0	19,017.0	2963.0	19,017.0	1633.0	14,781.0	14,331.0	3157.0	133.0	60.0
82	2.7	0.0	54.0	55.0	19,017.0	106.0	19,017.0	8819.0	14,781.0	14,331.0	3157.0	133.0	87.0
83	0.1	0.0	1.0	1745.0	19,017.0	753.0	19,017.0	1633.0	14,781.0	14,331.0	3157.0	1424.0	53.0
84	0.1	543.6	20.0	69.0	19,017.0	451.0	19,017.0	8819.0	14,781.0	14,331.0	3157.0	367.0	38.0
85	0.2	0.0	160.0	404.0	19,017.0	753.0	19,017.0	8819.0	14,781.0	14,331.0	3157.0	133.0	78.0
86	0.7	1118.1	54.0	55.0	19,017.0	106.0	19,017.0	2504.0	14,781.0	14,331.0	3157.0	133.0	23.0
87	1.1	1231.7	138.0	105.0	19,017.0	2963.0	19,017.0	8819.0	14,781.0	14,331.0	3157.0	133.0	68.0
88	0.9	0.0	138.0	105.0	19,017.0	2963.0	19,017.0	1633.0	14,781.0	14,331.0	3157.0	133.0	37.0

(continued)

(continued)

	Opened	Changed	Reporter	Assignee	Product	Component	Status	Resolution	Hardware	OS	Severity	Version	Summary
89	0.0	0.0	138.0	1745.0	19,017.0	2963.0	19,017.0	533.0	14,781.0	14,331.0	3157.0	133.0	39.0
90	2.7	0.0	54.0	1745.0	19,017.0	106.0	19,017.0	2504.0	14,781.0	14,331.0	3157.0	133.0	50.0
91	0.7	1230.4	149.0	695.0	19,017.0	1386.0	19,017.0	8819.0	14,781.0	14,331.0	3157.0	133.0	68.0
92	0.8	0.0	149.0	1745.0	19,017.0	344.0	19,017.0	1907.0	14,781.0	14,331.0	3157.0	133.0	41.0
93	1.4	989.1	2.0	1745.0	19,017.0	106.0	19,017.0	1833.0	14,781.0	14,331.0	3157.0	367.0	82.0
94	0.6	0.0	23.0	1745.0	19,017.0	753.0	19,017.0	2504.0	2868.0	14,331.0	3157.0	367.0	48.0
95	4.8	1138.1	84.0	190.0	19,017.0	418.0	19,017.0	8819.0	14,781.0	14,331.0	3157.0	133.0	54.0
96	0.2	0.5	147.0	1745.0	19,017.0	106.0	19,017.0	1907.0	14,781.0	14,331.0	3157.0	368.0	65.0
97	0.0	0.0	138.0	661.0	19,017.0	418.0	19,017.0	8819.0	14,781.0	14,331.0	3157.0	367.0	69.0
98	4.5	1231.0	54.0	1745.0	19,017.0	451.0	19,017.0	1907.0	14,781.0	14,331.0	3157.0	133.0	48.0
99	2.5	0.0	16.0	98.0	19,017.0	753.0	19,017.0	1907.0	14,781.0	14,331.0	3157.0	367.0	91.0
100	0.9	0.5	16.0	98.0	19,017.0	753.0	19,017.0	1907.0	14,781.0	14,331.0	3157.0	367.0	50.0
101	0.1	0.0	43.0	1745.0	19,017.0	753.0	19,017.0	429.0	14,781.0	14,331.0	3157.0	133.0	146.0
102	0.7	2323.2	1.0	211.0	19,017.0	2963.0	19,017.0	2504.0	14,781.0	14,331.0	3157.0	367.0	51.0
103	0.9	0.0	34.0	1745.0	19,017.0	418.0	19,017.0	533.0	14,781.0	14,331.0	3157.0	368.0	56.0
104	4.0	0.0	8.0	3.0	19,017.0	753.0	19,017.0	8819.0	14,781.0	14,331.0	3157.0	368.0	59.0
105	8.9	2089.6	34.0	1745.0	19,017.0	2963.0	19,017.0	2504.0	14,781.0	14,331.0	3157.0	368.0	52.0
106	0.0	0.0	34.0	1745.0	19,017.0	2963.0	19,017.0	1907.0	14,781.0	14,331.0	3157.0	368.0	52.0
107	0.1	0.0	52.0	404.0	19,017.0	753.0	19,017.0	8819.0	14,781.0	14,331.0	3157.0	368.0	51.0
108	11.3	1050.2	14.0	1745.0	19,017.0	205.0	19,017.0	2504.0	14,781.0	14,331.0	3157.0	367.0	64.0
109	4.0	0.0	43.0	404.0	19,017.0	753.0	19,017.0	8819.0	14,781.0	14,331.0	3157.0	133.0	63.0
110	2.9	0.0	52.0	404.0	19,017.0	753.0	19,017.0	8819.0	14,781.0	14,331.0	3157.0	368.0	29.0

(continued)

(continued)

	Opened	Changed	Reporter	Assignee	Product	Component	Status	Resolution	Hardware	OS	Severity	Version	Summary
111	1.4	0.0	43.0	1745.0	19,017.0	753.0	19,017.0	1907.0	14,781.0	14,331.0	3157.0	133.0	59.0
112	0.4	1071.9	54.0	13.0	19,017.0	106.0	19,017.0	1833.0	14,781.0	14,331.0	3157.0	368.0	26.0
113	1.1	0.0	52.0	190.0	19,017.0	1386.0	19,017.0	2504.0	14,781.0	14,331.0	3157.0	368.0	48.0
114	1.4	0.3	1.0	1745.0	19,017.0	451.0	19,017.0	1907.0	14,781.0	14,331.0	3157.0	133.0	63.0
115	2.3	1230.8	138.0	24.0	19,017.0	2963.0	19,017.0	8819.0	14,781.0	14,331.0	3157.0	133.0	45.0
116	2.5	0.0	8.0	695.0	19,017.0	1386.0	19,017.0	8819.0	14,781.0	14,331.0	3157.0	368.0	36.0
117	4.9	0.0	54.0	98.0	19,017.0	106.0	19,017.0	8819.0	14,781.0	14,331.0	3157.0	368.0	44.0
118	1.5	0.2	10.0	404.0	19,017.0	753.0	19,017.0	1633.0	14,781.0	14,331.0	3157.0	368.0	39.0
119	7.6	0.0	241.0	174.0	19,017.0	344.0	19,017.0	1907.0	14,781.0	14,331.0	3157.0	368.0	28.0
120	0.8	326.8	54.0	55.0	19,017.0	106.0	19,017.0	8819.0	14,781.0	14,331.0	3157.0	368.0	67.0
121	0.3	0.1	138.0	1745.0	19,017.0	106.0	19,017.0	1907.0	14,781.0	14,331.0	3157.0	368.0	41.0
122	2.8	0.1	138.0	1745.0	19,017.0	106.0	19,017.0	1907.0	14,781.0	14,331.0	3157.0	368.0	34.0
123	0.1	0.0	138.0	1745.0	19,017.0	753.0	19,017.0	1633.0	14,781.0	14,331.0	3157.0	368.0	60.0
124	1.7	213.7	138.0	71.0	19,017.0	389.0	19,017.0	8819.0	14,781.0	14,331.0	3157.0	368.0	34.0
125	0.0	0.0	138.0	71.0	19,017.0	42.0	19,017.0	8819.0	14,781.0	14,331.0	3157.0	368.0	70.0
126	0.0	0.0	138.0	71.0	19,017.0	389.0	19,017.0	8819.0	14,781.0	14,331.0	3157.0	368.0	53.0
127	0.2	112.9	138.0	1745.0	19,017.0	2067.0	19,017.0	1833.0	14,781.0	14,331.0	3157.0	368.0	65.0
128	0.2	0.0	10.0	12.0	19,017.0	473.0	19,017.0	8819.0	14,781.0	14,331.0	3157.0	368.0	42.0
129	1.5	1353.3	46.0	130.0	19,017.0	1386.0	19,017.0	8819.0	14,781.0	4150.0	3157.0	368.0	91.0
130	0.4	0.0	54.0	55.0	19,017.0	106.0	19,017.0	8819.0	14,781.0	14,331.0	3157.0	368.0	98.0
131	4.1	0.4	45.0	69.0	19,017.0	451.0	19,017.0	8819.0	14,781.0	14,331.0	3157.0	368.0	64.0
132	0.8	0.0	54.0	55.0	19,017.0	106.0	19,017.0	8819.0	14,781.0	14,331.0	3157.0	368.0	135.0

(continued)

(continued)

	Opened	Changed	Reporter	Assignee	Product	Component	Status	Resolution	Hardware	OS	Severity	Version	Summary
133	2.3	0.0	54.0	3.0	19,017.0	451.0	19,017.0	429.0	14,781.0	14,331.0	3157.0	368.0	34.0
134	0.8	0.0	46.0	404.0	19,017.0	753.0	19,017.0	1907.0	14,781.0	14,331.0	3157.0	368.0	55.0
135	3.0	1342.2	0.0	198.0	19,017.0	753.0	19,017.0	8819.0	14,781.0	14,331.0	3157.0	368.0	53.0
136	0.0	0.0	43.0	695.0	19,017.0	1386.0	19,017.0	8819.0	14,781.0	14,331.0	3157.0	368.0	40.0
137	0.0	0.0	43.0	10.0	19,017.0	670.0	19,017.0	8819.0	14,781.0	14,331.0	3157.0	368.0	38.0
138	2.4	0.0	160.0	71.0	19,017.0	42.0	19,017.0	8819.0	14,781.0	14,331.0	3157.0	368.0	48.0
139	5.1	1344.0	25.0	105.0	19,017.0	2963.0	19,017.0	1907.0	14,781.0	14,331.0	3157.0	368.0	56.0
140	1.6	0.0	17.0	190.0	19,017.0	418.0	19,017.0	628.0	14,781.0	14,331.0	3157.0	855.0	100.0
141	0.1	1230.7	1.0	1745.0	19,017.0	2963.0	19,017.0	1833.0	14,781.0	14,331.0	3157.0	368.0	32.0
142	4.9	0.0	46.0	98.0	19,017.0	753.0	19,017.0	1907.0	14,781.0	14,331.0	3157.0	368.0	79.0
143	2.6	132.9	4.0	404.0	19,017.0	753.0	19,017.0	1907.0	14,781.0	14,331.0	3157.0	1424.0	59.0
144	0.1	0.0	40.0	1745.0	19,017.0	753.0	19,017.0	429.0	14,781.0	14,331.0	3157.0	368.0	82.0
145	6.4	2168.2	122.0	54.0	19,017.0	2963.0	19,017.0	8819.0	14,781.0	14,331.0	3157.0	368.0	14.0
146	0.1	0.0	1.0	27.0	19,017.0	1386.0	19,017.0	1633.0	14,781.0	14,331.0	3157.0	368.0	78.0
147	6.2	0.0	0.0	1745.0	19,017.0	753.0	19,017.0	2504.0	2868.0	4150.0	3157.0	368.0	69.0
148	1.9	132.4	3.0	404.0	19,017.0	753.0	19,017.0	8819.0	14,781.0	530.0	3157.0	1424.0	98.0
149	5.9	2089.7	11.0	1745.0	19,017.0	2963.0	19,017.0	1633.0	14,781.0	14,331.0	3157.0	368.0	82.0
150	0.9	0.0	122.0	54.0	19,017.0	2963.0	19,017.0	115.0	14,781.0	14,331.0	3157.0	368.0	82.0
151	15.3	0.0	23.0	404.0	19,017.0	753.0	19,017.0	8819.0	14,781.0	14,331.0	3157.0	368.0	72.0
152	8.1	0.0	149.0	695.0	19,017.0	1386.0	19,017.0	8819.0	14,781.0	14,331.0	3157.0	855.0	77.0
153	3.8	0.0	9.0	1745.0	19,017.0	753.0	19,017.0	1633.0	14,781.0	14,331.0	3157.0	855.0	68.0
154	0.0	56.1	9.0	404.0	19,017.0	753.0	19,017.0	628.0	14,781.0	14,331.0	3157.0	855.0	79.0

(continued)

(continued)

	Opened	Changed	Reporter	Assignee	Product	Component	Status	Resolution	Hardware	OS	Severity	Version	Summary
155	6.3	0.0	24.0	1745.0	19,017.0	753.0	19,017.0	2504.0	14,781.0	14,331.0	3157.0	367.0	47.0
156	0.8	54.4	53.0	326.0	19,017.0	473.0	19,017.0	2504.0	14,781.0	14,331.0	3157.0	855.0	53.0
157	0.4	895.0	15.0	661.0	19,017.0	389.0	19,017.0	2504.0	14,781.0	14,331.0	3157.0	855.0	18.0
158	5.6	0.0	149.0	695.0	19,017.0	1386.0	19,017.0	8819.0	14,781.0	14,331.0	3157.0	368.0	41.0
159	1.1	0.0	2.0	404.0	19,017.0	753.0	19,017.0	1633.0	14,781.0	14,331.0	3157.0	1424.0	40.0
160	0.0	148.6	2.0	198.0	19,017.0	753.0	19,017.0	8819.0	14,781.0	14,331.0	3157.0	1424.0	109.0
161	7.7	802.8	149.0	695.0	19,017.0	1386.0	19,017.0	8819.0	14,781.0	14,331.0	3157.0	855.0	42.0
162	6.2	0.0	149.0	695.0	19,017.0	1386.0	19,017.0	8819.0	14,781.0	14,331.0	3157.0	855.0	53.0
163	5.0	1231.0	9.0	130.0	19,017.0	205.0	19,017.0	1633.0	14,781.0	14,331.0	3157.0	855.0	59.0
164	6.7	0.1	1.0	26.0	19,017.0	2067.0	19,017.0	8819.0	14,781.0	14,331.0	3157.0	368.0	87.0
165	3.2	0.0	9.0	174.0	19,017.0	344.0	19,017.0	1907.0	14,781.0	14,331.0	3157.0	855.0	46.0
166	4.7	0.0	43.0	190.0	19,017.0	418.0	19,017.0	8819.0	14,781.0	14,331.0	3157.0	368.0	82.0
167	7.2	1231.8	16.0	1745.0	19,017.0	2963.0	19,017.0	1633.0	14,781.0	4150.0	3157.0	855.0	46.0
168	9.2	0.0	45.0	198.0	19,017.0	753.0	19,017.0	8819.0	14,781.0	14,331.0	3157.0	855.0	63.0
169	0.0	0.3	134.0	695.0	19,017.0	1386.0	19,017.0	628.0	14,781.0	14,331.0	3157.0	855.0	73.0
170	0.0	1230.9	134.0	1745.0	19,017.0	2963.0	19,017.0	2504.0	14,781.0	14,331.0	3157.0	855.0	80.0
171	3.8	0.0	70.0	15.0	19,017.0	753.0	19,017.0	8819.0	14,781.0	14,331.0	3157.0	855.0	76.0
172	0.0	747.4	149.0	22.0	19,017.0	753.0	19,017.0	8819.0	14,781.0	14,331.0	3157.0	855.0	34.0
173	0.1	111.2	14.0	69.0	19,017.0	451.0	19,017.0	2504.0	14,781.0	14,331.0	3157.0	855.0	74.0
174	0.6	0.2	19.0	2.0	19,017.0	451.0	19,017.0	1633.0	14,781.0	14,331.0	3157.0	368.0	55.0
175	1.2	0.0	122.0	404.0	19,017.0	753.0	19,017.0	8819.0	14,781.0	14,331.0	3157.0	368.0	72.0
176	0.0	1230.9	122.0	26.0	19,017.0	106.0	19,017.0	8819.0	14,781.0	14,331.0	3157.0	368.0	87.0

(continued)

(continued)

	Opened	Changed	Reporter	Assignee	Product	Component	Status	Resolution	Hardware	OS	Severity	Version	Summary
177	0.1	0.9	122.0	15.0	19,017.0	753.0	19,017.0	8819.0	14,781.0	14,331.0	3157.0	368.0	76.0
178	5.8	0.0	138.0	1745.0	19,017.0	2067.0	19,017.0	1633.0	14,781.0	14,331.0	3157.0	855.0	47.0
179	0.1	0.0	128.0	695.0	19,017.0	1386.0	19,017.0	533.0	14,781.0	14,331.0	3157.0	855.0	95.0
180	8.0	489.0	134.0	190.0	19,017.0	418.0	19,017.0	8819.0	14,781.0	14,331.0	3157.0	855.0	61.0
181	2.3	0.0	43.0	82.0	19,017.0	473.0	19,017.0	8819.0	14,781.0	14,331.0	3157.0	855.0	108.0
182	3.4	262.2	52.0	22.0	19,017.0	328.0	19,017.0	1833.0	14,781.0	14,331.0	3157.0	855.0	21.0
183	1.1	1828.4	188.0	1745.0	19,017.0	753.0	19,017.0	2504.0	14,781.0	4150.0	3157.0	855.0	51.0
184	0.0	0.0	16.0	1745.0	19,017.0	451.0	19,017.0	2504.0	14,781.0	4150.0	3157.0	855.0	50.0
185	0.1	0.0	19.0	1745.0	19,017.0	1386.0	19,017.0	429.0	14,781.0	14,331.0	3157.0	855.0	41.0
186	2.0	0.0	134.0	695.0	19,017.0	1386.0	19,017.0	2504.0	1340.0	4150.0	3157.0	855.0	47.0
187	3.2	0.0	9.0	198.0	19,017.0	753.0	19,017.0	8819.0	14,781.0	14,331.0	3157.0	855.0	48.0
188	0.8	0.0	2.0	1745.0	19,017.0	753.0	19,017.0	2504.0	14,781.0	14,331.0	3157.0	1424.0	71.0
189	0.0	23.3	2.0	198.0	19,017.0	753.0	19,017.0	8819.0	14,781.0	14,331.0	3157.0	1424.0	32.0
190	1.5	0.0	40.0	22.0	19,017.0	753.0	19,017.0	2504.0	14,781.0	14,331.0	3157.0	855.0	36.0
191	5.1	285.5	8.0	198.0	19,017.0	328.0	19,017.0	1833.0	14,781.0	14,331.0	3157.0	855.0	61.0
192	1.7	483.4	70.0	71.0	19,017.0	42.0	19,017.0	8819.0	14,781.0	14,331.0	3157.0	855.0	40.0
193	6.9	1344.1	149.0	695.0	19,017.0	1386.0	19,017.0	8819.0	14,781.0	14,331.0	3157.0	855.0	87.0
194	1.1	0.0	53.0	326.0	19,017.0	473.0	19,017.0	8819.0	14,781.0	14,331.0	3157.0	855.0	24.0
195	4.0	857.9	0.0	82.0	19,017.0	473.0	19,017.0	8819.0	14,781.0	14,331.0	3157.0	855.0	71.0
196	0.7	0.0	138.0	22.0	19,017.0	753.0	19,017.0	8819.0	14,781.0	14,331.0	3157.0	1424.0	69.0
197	2.0	1230.9	1013.0	1745.0	19,017.0	2963.0	19,017.0	1633.0	14,781.0	14,331.0	3157.0	1424.0	41.0
198	0.0	0.1	1013.0	1745.0	19,017.0	2963.0	19,017.0	533.0	14,781.0	14,331.0	3157.0	1424.0	87.0

(continued)

(continued)

	Opened	Changed	Reporter	Assignee	Product	Component	Status	Resolution	Hardware	OS	Severity	Version	Summary
199	0.1	0.0	1013.0	1745.0	19,017.0	205.0	19,017.0	1633.0	14,781.0	14,331.0	3157.0	1424.0	42.0
200	0.1	0.5	1013.0	1745.0	19,017.0	205.0	19,017.0	1633.0	14,781.0	14,331.0	3157.0	1424.0	72.0
201	0.0	0.0	1013.0	1745.0	19,017.0	418.0	19,017.0	1633.0	14,781.0	14,331.0	3157.0	1424.0	31.0
202	0.0	0.0	1013.0	1745.0	19,017.0	418.0	19,017.0	1633.0	14,781.0	14,331.0	3157.0	1424.0	90.0
203	0.0	0.0	1013.0	1745.0	19,017.0	451.0	19,017.0	1633.0	14,781.0	14,331.0	3157.0	1424.0	101.0
204	0.0	0.3	1013.0	1745.0	19,017.0	451.0	19,017.0	1633.0	14,781.0	14,331.0	3157.0	1424.0	34.0
205	0.1	0.1	1013.0	203.0	19,017.0	2067.0	19,017.0	533.0	14,781.0	14,331.0	3157.0	1424.0	59.0
206	0.7	0.0	149.0	695.0	19,017.0	1386.0	19,017.0	8819.0	14,781.0	14,331.0	3157.0	855.0	64.0
207	6.0	0.0	43.0	326.0	19,017.0	473.0	19,017.0	1833.0	14,781.0	14,331.0	3157.0	855.0	190.0
208	0.0	0.0	43.0	82.0	19,017.0	473.0	19,017.0	8819.0	14,781.0	14,331.0	3157.0	855.0	82.0
209	0.2	215.5	43.0	82.0	19,017.0	473.0	19,017.0	1907.0	14,781.0	14,331.0	3157.0	855.0	81.0
210	7.0	464.4	1013.0	1745.0	19,017.0	389.0	19,017.0	1833.0	14,781.0	14,331.0	3157.0	1424.0	44.0
211	0.0	132.2	1013.0	1745.0	19,017.0	418.0	19,017.0	1633.0	14,781.0	14,331.0	3157.0	1424.0	46.0
212	0.5	0.0	0.0	1745.0	19,017.0	2067.0	19,017.0	8819.0	14,781.0	4150.0	3157.0	1424.0	37.0
213	2.9	0.5	0.0	1745.0	19,017.0	2067.0	19,017.0	2504.0	14,781.0	4150.0	3157.0	1424.0	42.0
214	0.9	0.1	1013.0	1745.0	19,017.0	1386.0	19,017.0	1633.0	14,781.0	14,331.0	3157.0	1424.0	66.0
215	0.0	0.0	1013.0	1745.0	19,017.0	418.0	19,017.0	1633.0	14,781.0	14,331.0	3157.0	1424.0	67.0
216	2.6	0.0	352.0	404.0	19,017.0	328.0	19,017.0	8819.0	14,781.0	14,331.0	3157.0	855.0	92.0
217	0.4	498.6	1013.0	1745.0	19,017.0	1386.0	19,017.0	1633.0	14,781.0	14,331.0	3157.0	1424.0	56.0
218	3.1	0.0	0.0	1745.0	19,017.0	2067.0	19,017.0	2504.0	14,781.0	4150.0	3157.0	1424.0	27.0
219	0.7	0.0	352.0	404.0	19,017.0	753.0	19,017.0	1833.0	14,781.0	14,331.0	3157.0	855.0	78.0
220	0.0	0.0	352.0	326.0	19,017.0	473.0	19,017.0	8819.0	14,781.0	14,331.0	3157.0	855.0	36.0

(continued)

(continued)

	Opened	Changed	Reporter	Assignee	Product	Component	Status	Resolution	Hardware	OS	Severity	Version	Summary
221	0.2	698.1	1013.0	1745.0	19,017.0	418.0	19,017.0	1633.0	14,781.0	14,331.0	3157.0	1424.0	64.0
222	0.4	0.0	2.0	190.0	19,017.0	1386.0	19,017.0	628.0	14,781.0	4150.0	3157.0	855.0	66.0
223	0.0	0.0	45.0	198.0	19,017.0	753.0	19,017.0	429.0	14,781.0	14,331.0	3157.0	855.0	57.0
224	1.2	0.0	43.0	326.0	19,017.0	473.0	19,017.0	1833.0	14,781.0	14,331.0	3157.0	855.0	86.0
225	7.8	596.4	128.0	695.0	19,017.0	1386.0	19,017.0	533.0	2868.0	4150.0	3157.0	855.0	32.0
226	0.7	0.3	11.0	695.0	19,017.0	1386.0	19,017.0	2504.0	2868.0	4150.0	3157.0	855.0	111.0
227	0.9	129.2	1013.0	69.0	19,017.0	451.0	19,017.0	2504.0	14,781.0	14,331.0	3157.0	2110.0	60.0
228	2.2	0.0	45.0	1745.0	19,017.0	753.0	19,017.0	1833.0	14,781.0	14,331.0	3157.0	855.0	69.0
229	0.8	168.7	23.0	1745.0	19,017.0	670.0	19,017.0	2504.0	14,781.0	14,331.0	3157.0	855.0	154.0
230	0.4	427.7	1013.0	1745.0	19,017.0	1386.0	19,017.0	1633.0	14,781.0	14,331.0	3157.0	1424.0	53.0
231	0.8	0.1	11.0	97.0	19,017.0	1386.0	19,017.0	1907.0	2868.0	4150.0	3157.0	855.0	53.0
232	0.3	0.0	5.0	22.0	19,017.0	753.0	19,017.0	8819.0	14,781.0	14,331.0	3157.0	855.0	54.0
233	0.2	101.9	10.0	8.0	19,017.0	473.0	19,017.0	1907.0	14,781.0	14,331.0	3157.0	855.0	67.0
234	4.4	1826.9	0.0	1745.0	19,017.0	2963.0	19,017.0	2504.0	14,781.0	14,331.0	3157.0	1424.0	26.0
235	4.5	0.0	43.0	1745.0	19,017.0	2067.0	19,017.0	1907.0	14,781.0	14,331.0	3157.0	855.0	56.0
236	2.7	0.0	23.0	1745.0	19,017.0	753.0	19,017.0	1833.0	14,781.0	14,331.0	3157.0	834.0	50.0
237	0.8	596.9	43.0	190.0	19,017.0	1386.0	19,017.0	2504.0	14,781.0	14,331.0	3157.0	855.0	86.0
238	0.4	197.0	1013.0	1745.0	19,017.0	218.0	19,017.0	533.0	14,781.0	14,331.0	3157.0	1424.0	83.0
239	0.7	1033.4	11.0	1745.0	19,017.0	2067.0	19,017.0	628.0	14,781.0	14,331.0	3157.0	855.0	80.0
240	0.9	0.0	228.0	25.0	19,017.0	753.0	19,017.0	2504.0	14,781.0	14,331.0	3157.0	1424.0	82.0
241	0.2	0.0	7.0	326.0	19,017.0	473.0	19,017.0	1907.0	14,781.0	14,331.0	3157.0	855.0	50.0
242	1.3	98.4	45.0	280.0	19,017.0	473.0	19,017.0	8819.0	14,781.0	14,331.0	3157.0	855.0	74.0

(continued)

(continued)

	Opened	Changed	Reporter	Assignee	Product	Component	Status	Resolution	Hardware	OS	Severity	Version	Summary
243	2.7	0.0	7.0	326.0	19,017.0	473.0	19,017.0	1907.0	14,781.0	14,331.0	3157.0	855.0	53.0
244	1.2	1827.3	23.0	1745.0	19,017.0	2963.0	19,017.0	1633.0	14,781.0	14,331.0	3157.0	834.0	77.0
245	0.1	0.0	52.0	404.0	19,017.0	753.0	19,017.0	8819.0	14,781.0	14,331.0	3157.0	855.0	48.0
246	0.1	0.0	122.0	82.0	19,017.0	473.0	19,017.0	1907.0	14,781.0	14,331.0	3157.0	855.0	92.0
247	0.0	474.7	45.0	82.0	19,017.0	473.0	19,017.0	8819.0	14,781.0	14,331.0	3157.0	855.0	91.0
248	0.0	0.0	45.0	280.0	19,017.0	473.0	19,017.0	8819.0	14,781.0	14,331.0	3157.0	855.0	69.0
249	0.2	697.8	15.0	1745.0	19,017.0	2067.0	19,017.0	1907.0	14,781.0	14,331.0	3157.0	855.0	87.0
250	9.1	0.5	1.0	130.0	19,017.0	418.0	19,017.0	628.0	14,781.0	14,331.0	3157.0	1424.0	67.0
251	5.4	488.8	2.0	190.0	19,017.0	418.0	19,017.0	8819.0	14,781.0	14,331.0	3157.0	855.0	59.0
252	0.2	0.0	2.0	1745.0	19,017.0	753.0	19,017.0	2504.0	14,781.0	14,331.0	3157.0	855.0	37.0
253	5.1	28.8	14.0	2.0	19,017.0	753.0	19,017.0	8819.0	14,781.0	14,331.0	3157.0	855.0	56.0
254	0.1	0.0	14.0	2.0	19,017.0	753.0	19,017.0	8819.0	14,781.0	14,331.0	3157.0	855.0	47.0
255	1.7	568.0	140.0	1745.0	19,017.0	451.0	19,017.0	1633.0	14,781.0	14,331.0	3157.0	855.0	81.0
256	0.0	0.0	140.0	103.0	19,017.0	451.0	19,017.0	8819.0	14,781.0	14,331.0	3157.0	855.0	73.0
257	0.4	0.4	0.0	1745.0	19,017.0	451.0	19,017.0	2504.0	2868.0	4150.0	3157.0	1424.0	54.0
258	0.6	0.0	52.0	326.0	19,017.0	473.0	19,017.0	8819.0	14,781.0	14,331.0	3157.0	855.0	56.0
259	5.2	697.7	34.0	1745.0	19,017.0	451.0	19,017.0	120.0	2868.0	4150.0	3157.0	834.0	63.0
260	0.6	0.1	15.0	1745.0	19,017.0	2067.0	19,017.0	1907.0	14,781.0	14,331.0	3157.0	855.0	41.0
261	0.9	0.5	140.0	1745.0	19,017.0	1386.0	19,017.0	533.0	14,781.0	14,331.0	3157.0	855.0	85.0
262	0.8	0.0	1013.0	1745.0	19,017.0	2067.0	19,017.0	533.0	14,781.0	14,331.0	3157.0	1424.0	44.0
263	3.4	1230.9	160.0	1745.0	19,017.0	2963.0	19,017.0	1907.0	14,781.0	14,331.0	3157.0	1424.0	65.0
264	1.3	0.0	160.0	1745.0	19,017.0	1386.0	19,017.0	1633.0	14,781.0	14,331.0	3157.0	855.0	55.0

(continued)

(continued)

	Opened	Changed	Reporter	Assignee	Product	Component	Status	Resolution	Hardware	OS	Severity	Version	Summary
265	0.8	0.0	45.0	326.0	19,017.0	473.0	19,017.0	8819.0	14,781.0	14,331.0	3157.0	855.0	31.0
266	0.0	0.0	45.0	404.0	19,017.0	328.0	19,017.0	8819.0	14,781.0	14,331.0	3157.0	855.0	25.0
267	6.2	0.0	1.0	1.0	19,017.0	33.0	19,017.0	8819.0	14,781.0	14,331.0	3157.0	855.0	71.0
268	0.0	5.8	45.0	280.0	19,017.0	473.0	19,017.0	1907.0	14,781.0	14,331.0	3157.0	855.0	57.0
269	0.9	691.9	8.0	188.0	19,017.0	2067.0	19,017.0	1833.0	14,781.0	14,331.0	3157.0	855.0	40.0
270	0.2	1231.0	15.0	96.0	19,017.0	2963.0	19,017.0	628.0	14,781.0	14,331.0	3157.0	855.0	40.0
271	0.6	0.0	3.0	132.0	19,017.0	344.0	19,017.0	8819.0	14,781.0	4150.0	3157.0	834.0	43.0
272	5.3	0.0	140.0	1745.0	19,017.0	1386.0	19,017.0	1633.0	14,781.0	14,331.0	3157.0	368.0	88.0
273	1.9	0.0	53.0	22.0	19,017.0	328.0	19,017.0	8819.0	14,781.0	14,331.0	3157.0	855.0	38.0
274	0.1	1868.5	34.0	8.0	19,017.0	473.0	19,017.0	8819.0	14,781.0	14,331.0	3157.0	42.0	44.0
275	0.7	0.0	53.0	326.0	19,017.0	473.0	19,017.0	8819.0	14,781.0	14,331.0	3157.0	855.0	32.0
276	3.9	698.0	21.0	156.0	19,017.0	2067.0	19,017.0	1833.0	14,781.0	14,331.0	3157.0	834.0	64.0
277	1.3	1230.9	101.0	1745.0	19,017.0	2067.0	19,017.0	1907.0	14,781.0	14,331.0	3157.0	855.0	74.0
278	0.0	0.0	101.0	23.0	19,017.0	670.0	19,017.0	8819.0	14,781.0	14,331.0	3157.0	834.0	74.0
279	0.0	0.0	101.0	96.0	19,017.0	2963.0	19,017.0	1633.0	14,781.0	14,331.0	3157.0	368.0	74.0
280	0.0	0.1	101.0	1745.0	19,017.0	389.0	19,017.0	1633.0	14,781.0	14,331.0	3157.0	368.0	74.0
281	0.0	0.0	101.0	190.0	19,017.0	1386.0	19,017.0	1633.0	14,781.0	14,331.0	3157.0	368.0	74.0
282	0.0	0.0	101.0	349.0	19,017.0	670.0	19,017.0	1633.0	14,781.0	14,331.0	3157.0	368.0	74.0
283	0.0	0.0	101.0	1745.0	19,017.0	106.0	19,017.0	1633.0	14,781.0	14,331.0	3157.0	368.0	74.0
284	0.7	0.0	45.0	326.0	19,017.0	473.0	19,017.0	1833.0	14,781.0	14,331.0	3157.0	855.0	75.0
285	0.0	0.0	45.0	326.0	19,017.0	473.0	19,017.0	1907.0	14,781.0	14,331.0	3157.0	855.0	75.0
286	0.5	0.0	43.0	326.0	19,017.0	473.0	19,017.0	1833.0	14,781.0	14,331.0	3157.0	855.0	100.0

(continued)

(continued)

	Opened	Changed	Reporter	Assignee	Product	Component	Status	Resolution	Hardware	OS	Severity	Version	Summary
287	4.1	0.0	37.0	82.0	19,017.0	473.0	19,017.0	8819.0	14,781.0	14,331.0	3157.0	855.0	39.0
288	0.3	698.1	2.0	97.0	19,017.0	1386.0	19,017.0	8819.0	14,781.0	14,331.0	3157.0	834.0	53.0
289	1.3	0.0	53.0	326.0	19,017.0	473.0	19,017.0	8819.0	14,781.0	14,331.0	3157.0	855.0	35.0
290	0.0	0.0	53.0	280.0	19,017.0	473.0	19,017.0	8819.0	14,781.0	14,331.0	3157.0	855.0	32.0
291	0.0	255.8	53.0	326.0	19,017.0	473.0	19,017.0	8819.0	14,781.0	14,331.0	3157.0	834.0	38.0
292	1.3	1673.0	1013.0	1745.0	19,017.0	2963.0	19,017.0	533.0	14,781.0	14,331.0	3157.0	1424.0	35.0
293	4.4	0.0	101.0	34.0	19,017.0	2963.0	19,017.0	8819.0	14,781.0	14,331.0	3157.0	855.0	53.0
294	1.8	0.0	0.0	114.0	19,017.0	753.0	19,017.0	1833.0	14,781.0	14,331.0	3157.0	855.0	54.0
295	3.2	588.1	114.0	155.0	19,017.0	753.0	19,017.0	1633.0	14,781.0	14,331.0	3157.0	855.0	62.0
296	1.0	0.0	5.0	404.0	19,017.0	328.0	19,017.0	1833.0	14,781.0	14,331.0	3157.0	855.0	75.0
297	4.0	596.3	45.0	1745.0	19,017.0	418.0	19,017.0	1907.0	14,781.0	14,331.0	3157.0	855.0	79.0
298	0.0	0.0	45.0	326.0	19,017.0	473.0	19,017.0	1907.0	14,781.0	14,331.0	3157.0	855.0	76.0
299	0.2	596.0	160.0	1745.0	19,017.0	2067.0	19,017.0	1633.0	14,781.0	14,331.0	3157.0	1424.0	21.0
300	0.0	1230.8	160.0	1745.0	19,017.0	2963.0	19,017.0	1633.0	14,781.0	14,331.0	3157.0	1424.0	33.0
301	0.7	0.0	40.0	198.0	19,017.0	753.0	19,017.0	8819.0	14,781.0	14,331.0	3157.0	834.0	91.0
302	0.3	0.0	37.0	404.0	19,017.0	328.0	19,017.0	8819.0	14,781.0	14,331.0	3157.0	248.0	51.0
303	3.0	0.0	3.0	404.0	19,017.0	328.0	19,017.0	2504.0	14,781.0	14,331.0	3157.0	834.0	59.0
304	0.7	596.4	228.0	1745.0	19,017.0	451.0	19,017.0	2504.0	14,781.0	14,331.0	3157.0	834.0	51.0
305	0.9	1231.0	228.0	1745.0	19,017.0	2963.0	19,017.0	2504.0	14,781.0	14,331.0	3157.0	834.0	71.0
306	0.2	0.0	114.0	198.0	19,017.0	753.0	19,017.0	429.0	14,781.0	14,331.0	3157.0	834.0	58.0
307	0.0	0.0	3.0	404.0	19,017.0	753.0	19,017.0	2504.0	14,781.0	4150.0	3157.0	834.0	62.0
308	0.2	596.4	20.0	132.0	19,017.0	344.0	19,017.0	8819.0	14,781.0	14,331.0	3157.0	834.0	57.0

(continued)

(continued)

	Opened	Changed	Reporter	Assignee	Product	Component	Status	Resolution	Hardware	OS	Severity	Version	Summary
309	0.4	0.0	1013.0	203.0	19,017.0	2067.0	19,017.0	533.0	14,781.0	14,331.0	3157.0	1424.0	56.0
310	0.1	0.0	138.0	1745.0	19,017.0	389.0	19,017.0	1907.0	14,781.0	14,331.0	3157.0	834.0	50.0
311	0.0	0.0	138.0	2.0	19,017.0	42.0	19,017.0	1833.0	14,781.0	14,331.0	3157.0	834.0	61.0
312	0.0	0.0	138.0	2.0	19,017.0	42.0	19,017.0	8819.0	14,781.0	14,331.0	3157.0	834.0	78.0
313	0.1	0.0	138.0	155.0	19,017.0	753.0	19,017.0	8819.0	14,781.0	14,331.0	3157.0	834.0	25.0
314	0.2	1828.3	84.0	71.0	19,017.0	389.0	19,017.0	8819.0	14,781.0	14,331.0	3157.0	834.0	79.0
315	0.0	0.0	84.0	404.0	19,017.0	328.0	19,017.0	8819.0	14,781.0	14,331.0	3157.0	834.0	86.0
316	6.9	0.0	45.0	404.0	19,017.0	753.0	19,017.0	1907.0	14,781.0	14,331.0	3157.0	834.0	45.0
317	0.0	0.0	45.0	1745.0	19,017.0	753.0	19,017.0	2504.0	14,781.0	14,331.0	3157.0	834.0	78.0
318	0.2	597.0	51.0	3.0	19,017.0	344.0	19,017.0	8819.0	14,781.0	14,331.0	3157.0	834.0	50.0
319	1.0	0.0	138.0	25.0	19,017.0	753.0	19,017.0	8819.0	14,781.0	14,331.0	3157.0	834.0	66.0
320	1.6	596.3	2.0	1745.0	19,017.0	1386.0	19,017.0	1907.0	2868.0	4150.0	3157.0	834.0	58.0
321	1.9	1231.1	1013.0	1745.0	19,017.0	2963.0	19,017.0	1907.0	14,781.0	14,331.0	3157.0	1424.0	51.0
322	6.2	0.0	134.0	1745.0	19,017.0	2963.0	19,017.0	533.0	14,781.0	14,331.0	3157.0	834.0	59.0
323	0.8	0.0	1013.0	1745.0	19,017.0	2963.0	19,017.0	533.0	14,781.0	14,331.0	3157.0	1424.0	56.0
324	0.1	0.0	0.0	1745.0	19,017.0	451.0	19,017.0	2504.0	2868.0	4150.0	3157.0	1424.0	98.0
325	0.2	0.0	45.0	326.0	19,017.0	473.0	19,017.0	1833.0	14,781.0	14,331.0	3157.0	855.0	57.0
326	0.2	1824.8	160.0	92.0	19,017.0	2963.0	19,017.0	1633.0	14,781.0	14,331.0	3157.0	834.0	68.0
327	0.5	0.0	1013.0	1745.0	19,017.0	2963.0	19,017.0	533.0	14,781.0	14,331.0	3157.0	1424.0	59.0
328	5.0	0.0	1013.0	203.0	19,017.0	2067.0	19,017.0	533.0	14,781.0	14,331.0	3157.0	1424.0	42.0
329	1.3	0.0	37.0	1745.0	19,017.0	344.0	19,017.0	1833.0	14,781.0	14,331.0	3157.0	834.0	53.0
330	1.1	596.7	228.0	1745.0	19,017.0	451.0	19,017.0	1633.0	14,781.0	14,331.0	3157.0	834.0	46.0

(continued)

(continued)

	Opened	Changed	Reporter	Assignee	Product	Component	Status	Resolution	Hardware	OS	Severity	Version	Summary
331	3.0	0.0	16.0	1745.0	19,017.0	42.0	19,017.0	2504.0	14,781.0	14,331.0	3157.0	834.0	89.0
332	0.7	8.3	140.0	174.0	19,017.0	753.0	19,017.0	1907.0	14,781.0	14,331.0	3157.0	834.0	85.0
333	0.4	0.1	114.0	404.0	19,017.0	753.0	19,017.0	8819.0	2868.0	4150.0	3157.0	834.0	55.0
334	0.1	0.0	20.0	1745.0	19,017.0	328.0	19,017.0	1907.0	14,781.0	14,331.0	3157.0	834.0	41.0
335	0.8	596.4	140.0	695.0	19,017.0	1386.0	19,017.0	2504.0	2868.0	4150.0	3157.0	746.0	63.0
336	0.9	0.0	1.0	23.0	19,017.0	670.0	19,017.0	8819.0	14,781.0	14,331.0	3157.0	248.0	91.0
337	0.8	596.4	35.0	25.0	19,017.0	753.0	19,017.0	1833.0	14,781.0	14,331.0	3157.0	834.0	79.0
338	0.5	1231.0	101.0	96.0	19,017.0	2963.0	19,017.0	8819.0	14,781.0	14,331.0	3157.0	834.0	40.0
339	0.2	0.0	21.0	174.0	19,017.0	344.0	19,017.0	1833.0	14,781.0	14,331.0	3157.0	248.0	47.0
340	0.1	1827.5	15.0	1745.0	19,017.0	2963.0	19,017.0	1907.0	14,781.0	14,331.0	3157.0	855.0	35.0
341	0.4	0.0	25.0	71.0	19,017.0	2963.0	19,017.0	8819.0	14,781.0	14,331.0	3157.0	834.0	61.0
342	0.1	0.0	59.0	114.0	19,017.0	753.0	19,017.0	8819.0	14,781.0	14,331.0	3157.0	834.0	76.0
343	0.1	485.2	241.0	404.0	19,017.0	753.0	19,017.0	8819.0	14,781.0	14,331.0	3157.0	834.0	38.0
344	0.0	111.5	241.0	326.0	19,017.0	473.0	19,017.0	8819.0	14,781.0	14,331.0	3157.0	834.0	38.0
345	3.4	0.0	53.0	326.0	19,017.0	473.0	19,017.0	8819.0	14,781.0	14,331.0	3157.0	834.0	76.0
346	0.6	0.0	45.0	25.0	19,017.0	753.0	19,017.0	1907.0	14,781.0	14,331.0	3157.0	834.0	55.0
347	0.0	0.0	3.0	23.0	19,017.0	670.0	19,017.0	8819.0	14,781.0	14,331.0	3157.0	834.0	31.0
348	1.1	752.3	160.0	1745.0	19,017.0	473.0	19,017.0	409.0	14,781.0	14,331.0	3157.0	834.0	71.0
349	0.0	0.0	160.0	82.0	19,017.0	473.0	19,017.0	409.0	14,781.0	14,331.0	3157.0	834.0	48.0
350	0.7	0.0	84.0	661.0	19,017.0	473.0	19,017.0	8819.0	14,781.0	14,331.0	3157.0	834.0	85.0
351	0.1	596.6	70.0	326.0	19,017.0	473.0	19,017.0	429.0	14,781.0	14,331.0	3157.0	42.0	27.0
352	0.8	0.0	140.0	1745.0	19,017.0	753.0	19,017.0	1907.0	14,781.0	14,331.0	3157.0	834.0	56.0

(continued)

(continued)

	Opened	Changed	Reporter	Assignee	Product	Component	Status	Resolution	Hardware	OS	Severity	Version	Summary
353	0.0	0.0	140.0	1745.0	19,017.0	753.0	19,017.0	1907.0	14,781.0	14,331.0	3157.0	834.0	57.0
354	0.0	0.0	140.0	114.0	19,017.0	753.0	19,017.0	1907.0	14,781.0	14,331.0	3157.0	834.0	74.0
355	0.2	0.0	140.0	1745.0	19,017.0	753.0	19,017.0	1907.0	14,781.0	14,331.0	3157.0	834.0	56.0
356	0.1	596.4	149.0	695.0	19,017.0	1386.0	19,017.0	1833.0	14,781.0	14,331.0	3157.0	834.0	73.0
357	3.1	0.0	149.0	695.0	19,017.0	1386.0	19,017.0	1633.0	14,781.0	14,331.0	3157.0	855.0	53.0
358	1.0	0.3	2.0	25.0	19,017.0	328.0	19,017.0	1833.0	14,781.0	14,331.0	3157.0	834.0	93.0
359	0.7	0.0	189.0	695.0	19,017.0	1386.0	19,017.0	8819.0	14,781.0	14,331.0	3157.0	248.0	53.0
360	0.0	0.0	8.0	7.0	19,017.0	473.0	19,017.0	8819.0	14,781.0	14,331.0	3157.0	248.0	69.0
361	3.8	0.0	228.0	404.0	19,017.0	753.0	19,017.0	2504.0	14,781.0	14,331.0	3157.0	834.0	76.0
362	1.2	596.5	228.0	114.0	19,017.0	753.0	19,017.0	1833.0	14,781.0	14,331.0	3157.0	834.0	42.0
363	0.3	0.0	37.0	326.0	19,017.0	473.0	19,017.0	1633.0	14,781.0	14,331.0	3157.0	834.0	26.0
364	1.6	592.5	85.0	188.0	19,017.0	2067.0	19,017.0	8819.0	14,781.0	14,331.0	3157.0	855.0	81.0
365	0.6	0.0	241.0	404.0	19,017.0	753.0	19,017.0	1907.0	14,781.0	14,331.0	3157.0	834.0	41.0
366	0.5	149.9	8.0	27.0	19,017.0	44.0	19,017.0	533.0	14,781.0	14,331.0	3157.0	1424.0	90.0
367	0.0	72.5	8.0	661.0	19,017.0	389.0	19,017.0	1833.0	14,781.0	14,331.0	3157.0	1424.0	95.0
368	0.0	1672.3	8.0	96.0	19,017.0	2963.0	19,017.0	1833.0	14,781.0	14,331.0	3157.0	1424.0	89.0
369	0.0	0.0	8.0	695.0	19,017.0	1386.0	19,017.0	8819.0	14,781.0	14,331.0	3157.0	1424.0	91.0
370	0.0	0.6	8.0	103.0	19,017.0	451.0	19,017.0	1907.0	14,781.0	14,331.0	3157.0	1424.0	93.0
371	0.0	0.0	8.0	326.0	19,017.0	473.0	19,017.0	1633.0	14,781.0	14,331.0	3157.0	1424.0	20.0
372	0.6	8.3	1013.0	203.0	19,017.0	2067.0	19,017.0	533.0	14,781.0	14,331.0	3157.0	1424.0	40.0
373	3.7	223.1	70.0	132.0	19,017.0	344.0	19,017.0	1633.0	14,781.0	14,331.0	3157.0	834.0	93.0
374	0.4	0.0	228.0	69.0	19,017.0	451.0	19,017.0	2504.0	14,781.0	14,331.0	3157.0	834.0	84.0

(continued)

(continued)

	Opened	Changed	Reporter	Assignee	Product	Component	Status	Resolution	Hardware	OS	Severity	Version	Summary
375	0.1	0.0	228.0	103.0	19,017.0	451.0	19,017.0	2504.0	14,781.0	14,331.0	3157.0	834.0	68.0
376	1.2	0.0	114.0	190.0	19,017.0	418.0	19,017.0	2504.0	2868.0	4150.0	3157.0	834.0	95.0
377	7.8	0.0	51.0	198.0	19,017.0	328.0	19,017.0	8819.0	14,781.0	14,331.0	3157.0	834.0	54.0
378	0.0	0.1	51.0	326.0	19,017.0	473.0	19,017.0	8819.0	14,781.0	14,331.0	3157.0	834.0	76.0
379	0.4	0.0	43.0	198.0	19,017.0	328.0	19,017.0	2504.0	14,781.0	14,331.0	3157.0	834.0	57.0
380	1.0	596.8	20.0	695.0	19,017.0	1386.0	19,017.0	2504.0	14,781.0	14,331.0	3157.0	834.0	53.0
381	1.5	0.0	76.0	132.0	19,017.0	344.0	19,017.0	2504.0	2868.0	4150.0	3157.0	834.0	66.0
382	1.4	1827.3	0.0	96.0	19,017.0	2963.0	19,017.0	8819.0	14,781.0	14,331.0	3157.0	834.0	56.0
383	1.0	0.0	189.0	661.0	19,017.0	389.0	19,017.0	1633.0	14,781.0	14,331.0	3157.0	834.0	47.0
384	1.5	0.0	34.0	2.0	19,017.0	328.0	19,017.0	1833.0	14,781.0	14,331.0	3157.0	834.0	51.0
385	0.8	399.9	7.0	404.0	19,017.0	753.0	19,017.0	8819.0	14,781.0	14,331.0	3157.0	746.0	31.0
386	4.6	1230.5	1.0	211.0	19,017.0	2963.0	19,017.0	1833.0	14,781.0	14,331.0	3157.0	855.0	102.0
387	2.0	0.0	52.0	404.0	19,017.0	753.0	19,017.0	429.0	14,781.0	14,331.0	3157.0	855.0	52.0
388	4.3	32.3	34.0	8.0	19,017.0	473.0	19,017.0	1907.0	14,781.0	14,331.0	3157.0	248.0	44.0
389	0.8	1466.1	0.0	114.0	19,017.0	753.0	19,017.0	2504.0	14,781.0	14,331.0	3157.0	834.0	56.0
390	0.1	0.0	16.0	695.0	19,017.0	1386.0	19,017.0	628.0	14,781.0	14,331.0	3157.0	834.0	56.0
391	1.1	0.6	241.0	130.0	19,017.0	205.0	19,017.0	2504.0	14,781.0	14,331.0	3157.0	834.0	34.0
392	0.0	0.0	241.0	54.0	19,017.0	451.0	19,017.0	8819.0	14,781.0	14,331.0	3157.0	834.0	39.0
393	0.0	1230.7	241.0	96.0	19,017.0	2963.0	19,017.0	8819.0	14,781.0	14,331.0	3157.0	834.0	33.0
394	0.0	0.0	241.0	695.0	19,017.0	1386.0	19,017.0	8819.0	14,781.0	14,331.0	3157.0	834.0	35.0
395	0.0	0.0	241.0	190.0	19,017.0	418.0	19,017.0	8819.0	14,781.0	14,331.0	3157.0	834.0	35.0
396	0.0	0.0	241.0	661.0	19,017.0	389.0	19,017.0	8819.0	14,781.0	14,331.0	3157.0	834.0	39.0

(continued)

(continued)

	Opened	Changed	Reporter	Assignee	Product	Component	Status	Resolution	Hardware	OS	Severity	Version	Summary
397	0.0	0.0	241.0	15.0	19,017.0	670.0	19,017.0	8819.0	14,781.0	14,331.0	3157.0	834.0	33.0
398	0.0	231.0	241.0	3.0	19,017.0	218.0	19,017.0	8819.0	14,781.0	14,331.0	3157.0	834.0	35.0
399	0.0	365.6	241.0	130.0	19,017.0	205.0	19,017.0	2504.0	14,781.0	14,331.0	3157.0	834.0	34.0
400	0.0	0.2	241.0	54.0	19,017.0	451.0	19,017.0	8819.0	14,781.0	14,331.0	3157.0	248.0	39.0
401	0.0	12.6	241.0	695.0	19,017.0	1386.0	19,017.0	8819.0	14,781.0	14,331.0	3157.0	834.0	35.0
402	0.0	0.0	241.0	190.0	19,017.0	418.0	19,017.0	8819.0	14,781.0	14,331.0	3157.0	834.0	35.0
403	0.0	0.0	241.0	661.0	19,017.0	389.0	19,017.0	8819.0	14,781.0	14,331.0	3157.0	248.0	39.0
404	0.0	596.4	241.0	188.0	19,017.0	2067.0	19,017.0	8819.0	14,781.0	14,331.0	3157.0	834.0	36.0

We discuss the linear discriminant analysis as the data preprocessing. We assume that the applied data have satisfied the following conditions:

1. The data set is based on the normal distribution.
2. Each class follows the same covariance matrix.
3. All variables are independent of each other.

We show numerical examples by using actual data set of OpenStack as the OSS [8]. A number of about 19,000 lines fault data are included in Figs. 1.1, 1.2 and 1.3, respectively. We discuss the potential of normal distribution for actual data sets. We show the estimated approximate curve based on normal distribution for actual fault big data of the edge OSS in Figs. 1.1, 1.2, and 1.3. From Figs. 1.1, 1.2, and 1.3, we found that the categories of "Product" and "Status" have one type only.

1.3 Linear Discriminant Analysis

The number of whole data is about 247,000 category data. The X-axes of Figs. 1.1, 1.2 and 1.3 is the number of occurrence for each factor. The Y-axes of Figs. 1.1, 1.2 and 1.3 show the occurrence rate for each factor. All data have been converted from actual data to the rate data by using the frequency encoding. Many software reliability assessment methods based on the stochastic models have been developed and proposed in the past. Almost all of these methods of software reliability assessment are based on the data in terms of the number of software faults and the time between software failures, e.g., the exponential software reliability growth model and the hazard rate model such as Jelinski-Moranda. Therefore, we focus on the data in terms of the number of faults. We found that almost all data factors are approximately based on the normal distribution from Figs. 1.1, 1.2 and 1.3 except for the categories of "Product" and "Status".

Moreover, several numerical illustrations based on linear discriminant analysis are shown in Figs. 1.4, 1.5, 1.6, 1.7, 1.8, 1.9, 1.10 and 1.11, respectively. We discuss the estimation results of Figs. 1.4, 1.5, 1.6, 1.7, 1.8, 1.9, 1.10 and 1.11 as follows:

Product: We cannot show the estimate based on linear discriminant analysis, because one factor is only shown.

Component: Four clusters with small groups are shown by estimation. In particular, the group with core component and main component is placed on the center part, the group of upper left side is the small size four groups.

Version: Two clusters are structured by the calculation. The left side cluster becomes large. In particular, the right side group means "unspecified" version. Therefore, the right side group is meaningless.

Reporter: Three clusters are grouped by the calculation. We can consider that three groups are the unbiased result. This software has been reported by three grouped reporters.

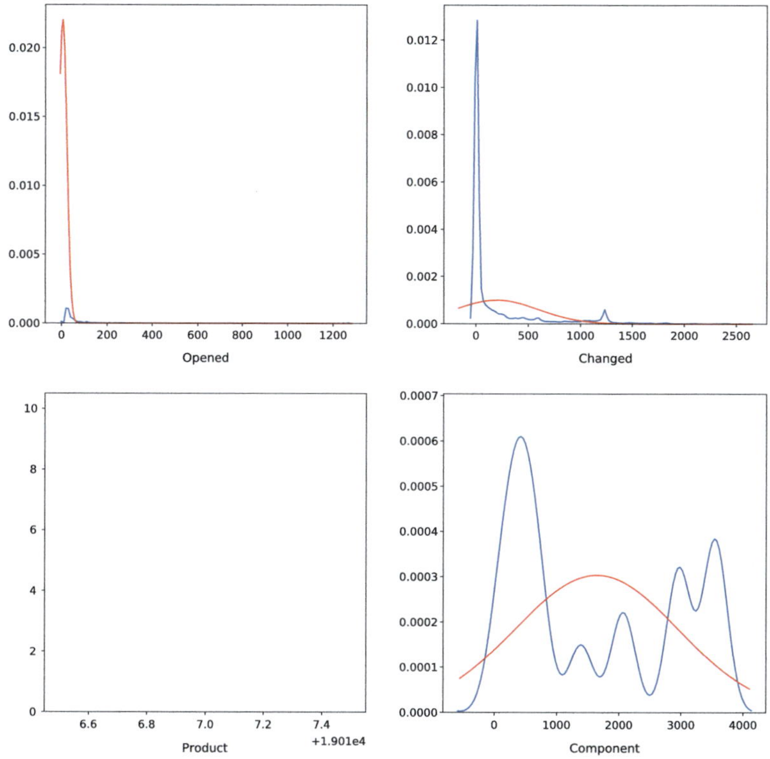

Fig. 1.1 The estimated approximate curve based on normal distribution for 1st group in actual fault big data

Severity: One cluster is estimated. In particular, the results of Severity are well-balanced.

Status: We cannot show the estimate based on linear discriminant analysis, because one factor is only shown.

Resolution: There are two types of clusters with small and large types.

Hardware: There are three types of clusters. In particular, we found that "x86_64", "Unspecified", and "All" are contained in the data set.

OS: The small three groups has contained in the data set.

Summary: The level of uniformity is very high.

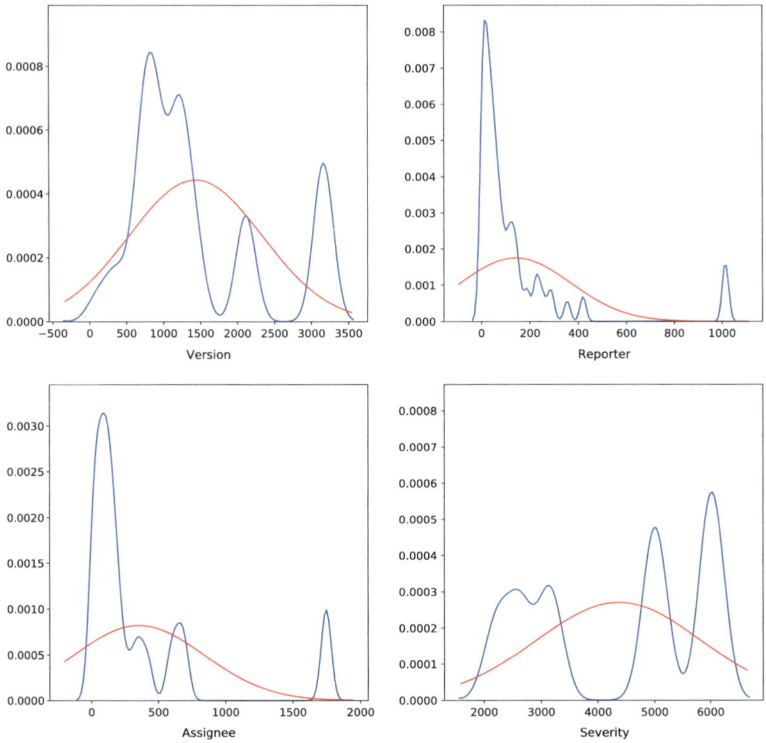

Fig. 1.2 The estimated approximate curve based on normal distribution for 2nd group in actual fault big data

Finally, we have discussed the trend of fault big data and the contribution rate as the data preprocessing. We have compared the data of OpenStack with Apache HTTP server.

1st Point: We have confirmed the trend of fault detection process by using the estimated approximate curve.
2nd Point: We analyze the trend of fault big data by using the discriminant analysis.
3rd Point: We discuss the importance of the encoding in terms of fault big data,

At first, we have confirmed that the fault detection process is based on the stochastic process. Then, we have found that the analysis of fault detection process should use the stochastic model. It is difficult to assess the OSS reliability, because OSS fault detection process is different for each OSS. From above results, we consider that the fault detection process is different from the general open data sets.

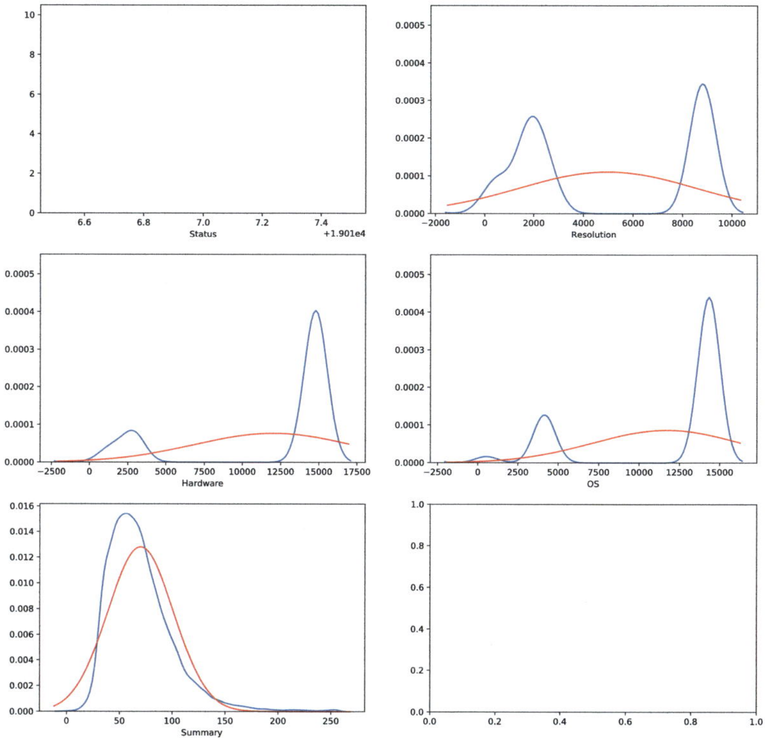

Fig. 1.3 The estimated approximate curve based on normal distribution for 3rd group in actual fault big data

1.4 Contribution Rate

In this paper, we estimate the contribution rate for all factors except for the categories of "Product" and "Status", because the contribution rate means the changing rate of each factor for changes in all data. The right side of Figs. 1.4, 1.5, 1.6, 1.7, 1.8, 1.9, 1.10 and 1.11 show the estimated contribution rate for each factor except for the categories of "Product" and "Status".

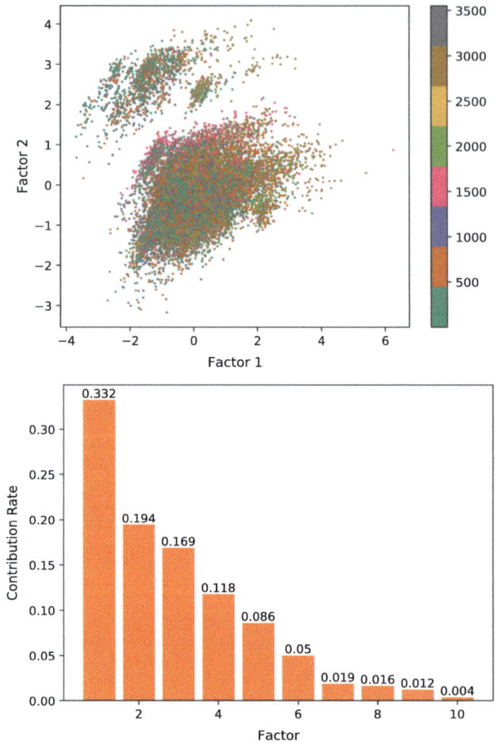

Fig. 1.4 The estimate based on linear discriminant analysis for Component of actual fault big data

We show the consideration in terms of the estimation results based on Figs. 1.4, 1.5, 1.6, 1.7, 1.8, 1.9, 1.10 and 1.11 as follows:

Component: The contribution rate is shown comparatively.
Version: The contribution rate is the third largest value.
Reporter: The contribution rate is shown comparatively. The result is similar to "Component".
Severity: The results of Severity are well-balanced. However, the value is not high.
Status: We cannot show the estimate based on linear discriminant analysis, because one factor is only shown.
Resolution: The contribution rate is shown comparatively.
Hardware: The contribution rate is the highest value in all data.
OS: The contribution rate is the highest value in all data.
Summary: The contribution rate is shown comparatively.

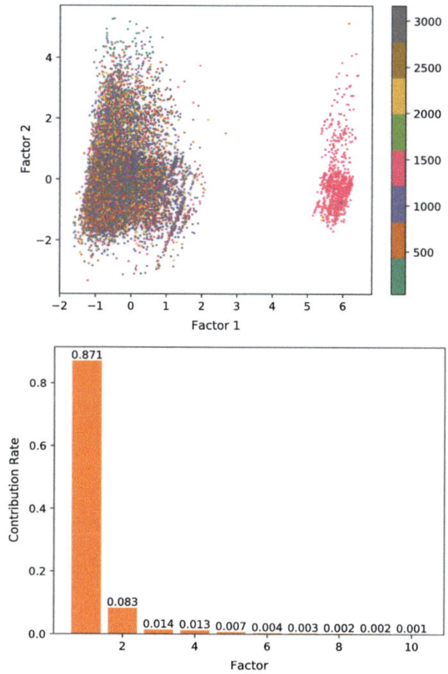

Fig. 1.5 The estimate based on linear discriminant analysis for Version of actual fault big data

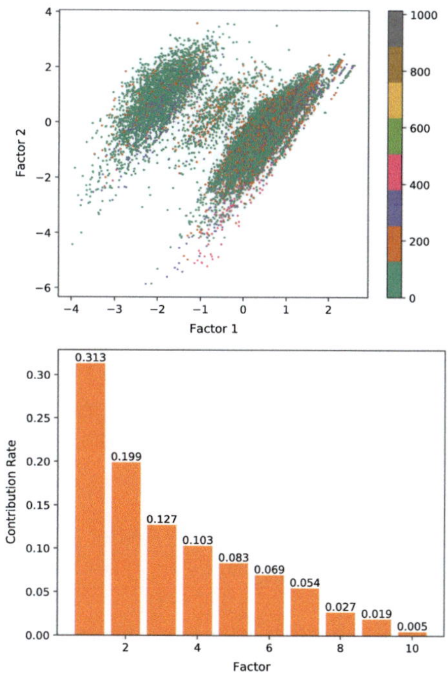

Fig. 1.6 The estimate based on linear discriminant analysis for Reporter of actual fault big data

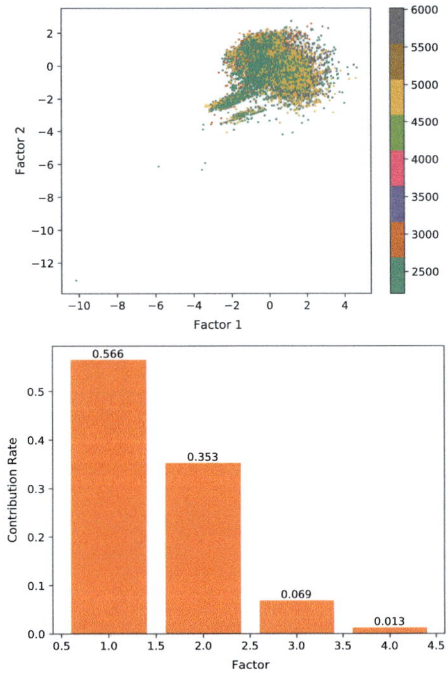

Fig. 1.7 The estimate based on linear discriminant analysis for Severity of actual fault big data

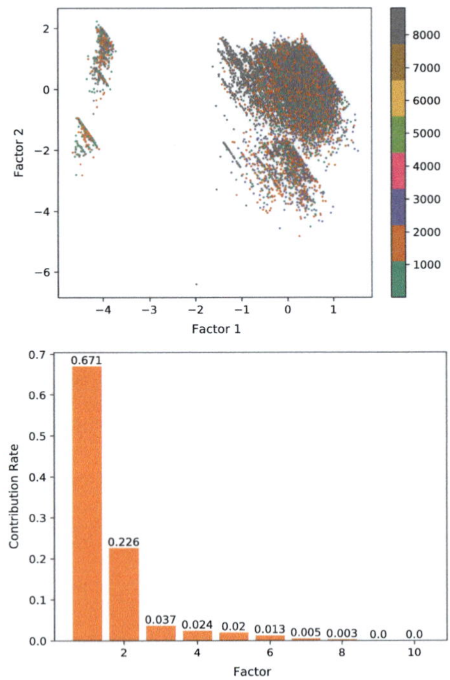

Fig. 1.8 The estimate based on linear discriminant analysis for Resolution of actual fault big data

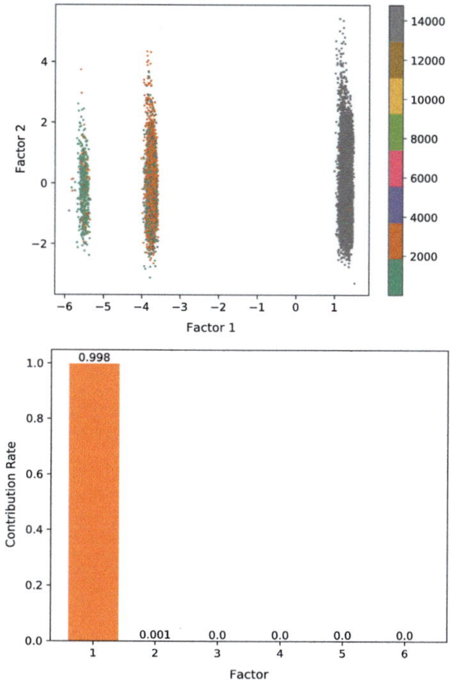

Fig. 1.9 The estimate based on linear discriminant analysis for Hardware of actual fault big data

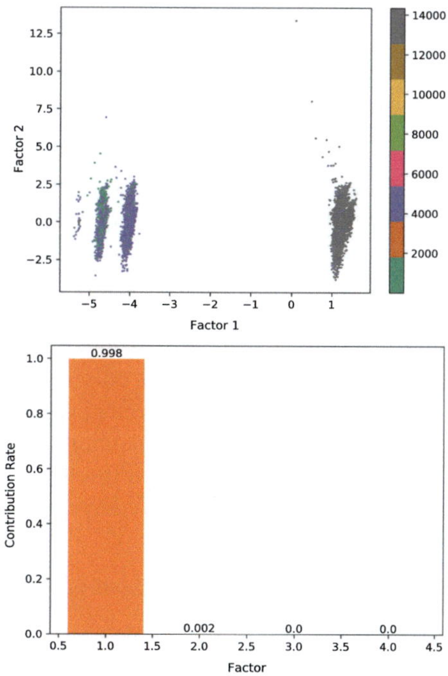

Fig. 1.10 The estimate based on linear discriminant analysis for OS of actual fault big data

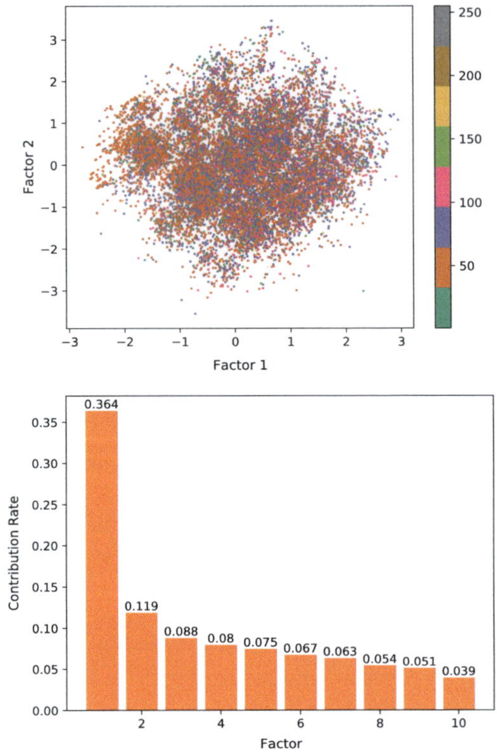

Fig. 1.11 The estimate based on linear discriminant analysis for Summary of actual fault big data

1.5 Conclusion

Our research group has proposed several reliability assessment methods based on deep learning. In particular, the estimation results based on deep learning depend on the data preprocessing of fault big data. Our research group has proposed the paper in terms of discriminant analysis in the past [9]. Then, we have analyzed the fault big data for the Apache HTTP Server [10] in the past paper. In this paper, we have obtained the different results in terms of discriminant analysis. In particular, the category of "Summary" is only similar to the estimation result of data in Apache HTTP Server. The trend of category in OpenStack is unlike Apache HTTP server because of the word count encoding.

Finally, we have discussed the trend of fault big data and the contribution rate as the data preprocessing. We have compared the data of OpenStack with Apache HTTP server.

1. We have confirmed the trend of fault detection process by using the estimated approximate curve.
2. We analyze the trend of fault big data by using the discriminant analysis.
3. We discuss the importance of the encoding in terms of fault big data,

At first, we have confirmed that the fault detection process is based on the stochastic process. Then, we have found that the analysis of fault detection process should use the stochastic model. It is difficult to assess the OSS reliability, because OSS fault detection process is different for each OSS. From above results, we consider that the fault detection process is different from the general open data sets.

Acknowledgements This work was supported in part by the JSPS KAKENHI Grant No. 23K11066 in Japan.

References

1. Yamada, S. (2014). *Software reliability modeling: Fundamentals and applications*. Springer-Verlag.
2. Kapur, P. K., Pham, H., Gupta, A., & Jha, P. C. (2011). *Software reliability assessment with OR applications*. Springer-Verlag.
3. Yamada, S., & Tamura, Y. (2016). *OSS reliability measurement and assessment*. Springer International Publishing.
4. Singh, V. B., Sharma, M., & Pham, H. (2017). Entropy based software reliability analysis of multi-version open source software. *IEEE Transactions on Software Engineering*. https://doi.org/10.1109/TSE.2017.2766070
5. Sone, H., Miyamoto, S., Kashihara, Y., & Tamura, Y., & Yamada, S. (2022). *Deep learning approach based on fault correction time for reliability assessment of cloud and edge open source software. Predictive Analytics in System Reliability*, Springer, pp. 1–18
6. Tamura, Y., & Yamada, S. (2022). *Prototype of 3D reliability assessment tool based on deep learning for edge OSS computing*. Mathematics, (Vol. 10, No. 9, pp. 1–12), Multidisciplinary Digital Publishing Institute.
7. Tamura, Y., Sone, H., Anand, A., & Yamada, S. (2021). A method of vulnerability assessment based on deep learning and OSS fault big data. In *Proceedings of the IEEE International Conference on Industrial Engineering and Engineering Management*, Singapore, December 13–16, 2021, CD-ROM (Reliability and Maintenance Engineering 2, Virtual mode).
8. The OpenStack project, OpenStack, http://www.openstack.org/
9. Tamura, Y., Watanabe, H., & Yamada, S. (2020). OSS project assessment based on discriminant analysis and jump diffusion process model for fault big data. *American Journal of Operations Research, 10*(6), 269–283.
10. The Apache Software Foundation, The Apache HTTP Server Project. http://httpd.apache.org/

Chapter 2
Towards a Practical Semantic Web of Things Framework for Underwater Sensor Network

Hatem A. Alharbi, Ahmad J. Showail, Khalid S. Aloufi, and Omar H. Alhazmi

Abstract The Internet of Things (IoT) is built over well-established technologies and protocols. Although various frameworks and protocols have been developed to enhance its portability and reliability, further advancements are required and developed by researchers and industrial communities. Moreover, efforts to expand and integrate it with the web in what is called the Web of Things (WoT) are being made. Transitioning to the WoT would result in obtaining and consequently benefiting from well-structured web technologies. One of the primary recent web standards is the semantic web. Hence, the IoT is receiving more attention; this has resulted in the creation of the Semantic Web of Things (SWoT). This study facilitates discussion of integrating semantic web to underwater sensors network and presents a framework for the integration of semantic technologies to provide a practical semantic WoT framework for underwater sensor network.

Keywords Internet of things · Web of things · Semantic web of things · Semantic web

H. A. Alharbi · A. J. Showail · K. S. Aloufi
Department of Computer Engineering, Taibah University, Madinah 12512, Saudi Arabia
e-mail: hmehmadi@taibahu.edu.sa

A. J. Showail
e-mail: ashowail@taibahu.edu.sa

K. S. Aloufi
e-mail: koufi@taibahu.edu.sa

O. H. Alhazmi (✉)
Department of Computer Science, Taibah University, Madinah 12512, Saudi Arabia
e-mail: ohhazmi@taibahu.edu.sa

P. K. Kapur et al. (eds.), *Data-centric Approaches to Industrial Decisions*, Asset Analytics, https://doi.org/10.1007/978-981-96-7556-2_2

2.1 Introduction

The semantic web (SW) adds annotations to resources, such as objects, locations, or events, to create new web and Internet services and applications. The Internet of Things (IoT) or physical computing enables interconnecting various computing devices over the internet using ubiquitous or pervasive computing. Ubiquitous or pervasive computing is defined as computing everywhere or by everything (device) in any format and poses challenges like heterogeneous constrained devices.

The Web of things (WoT) enables IoT devices access to web standards and protocols. The Semantic Web of Things (SWoT) integrates the SW, WoT, and IoT. It depends on knowledge-based systems (KBS), thereby posing a challenge to constrained devices because they will require a fog layer to enhance computational requirements.

In the IoT, resource discovery happens either by centralized brokers like the message queuing telemetry transport (MQTT), or decentralized protocols like the internet protocol (IPv6), low-power wireless personal area networks (6LoWPAN) and the constrained application protocol (CoAP) [1–4]. An annotation for the IoT was developed by ontologies from different research projects [5–9].

Ruta et al. presented a framework for the SWoT called ubiquitous knowledge base (u-KB) without explicit management of KB or centralized coordination of the devices in a mobile ad hoc network (MANET). Because u-KB does not require an internet connection, an ontology universally unique identifier (OUUID) [10] is used to identify ontologies. Uniform resource identifiers (URIs) and uniform resource locators (URLs) are adopted for ontology identification [11]. u-KB comprises a two-levelled infrastructure: the field and discovery layers. The latter involves several stages namely: the diffusion of resource parameters, resource discovery, and the extraction of selected resource annotations. In the proposed framework, each resource is characterized by a 96-bit global ID, a 64-bit OUUID, data-oriented attributes, and semantically compressed annotations in the resource description framework (RDF). Previous studies using the frameworks with Bluetooth, EPCglobal RFID, ZigBee and EIB/KNX on different application scenarios and case studies showed performance gains [12–18].

The IoT application layer integrates the web, the WoT, recent advanced developments in web technologies, the SW, and linked data (LD) with the help of data management, ontology engineering, and stream processing, resulting in the Semantic Web of Things (SWoT). The IoT is supported by cloud infrastructure middleware and sensor networks. The convergence of the IoT and WoT results in several benefits. The IoT is composed of things or devices that are connected with the internet to share their data or to obtain some commands, which in turn are executed, and some actions are performed depending on the context. The data published by the devices are shared over the web using the semantic web. Devices connected to the internet can provide information about their existence and services using different protocols, such as MQTT and CoAP.

The IoT data streams are generated by different devices with very limited information. For example, the data generated from a temperature sensor only provides the temperatures measured at a specific time. A machine receives the temperature data and adds the limited information, such as the time and the source ID, to the data. Additionally, the devices must also provide the information required by one of the ontologies; for example, extra information is needed at the application level. The data received from the IoT devices are either event-driven or time-driven. Each data type has different requirements and delivery methodologies. Most of the proposed frameworks consider event-driven systems; time-driven systems are not considered directly. Standard query language and protocol (SPARQL) is a recursive acronym for SPARQL Protocol and Resource Description Framework (RDF) Query Language proposed by the World Wide Web Consortium (W3C). SPARQL is a structured query language that retrieves and manipulates data stored in RDF format. SPARQL engines are integrated with the SWoT and enquire streams based on events. Some applications require time-driven systems, which must be configured with other parts of a framework for the SWoT.

Underwater Sensors Network can be integrated into the existing network to achieve various applications such as automatic non-pollution monitoring and providing an early warning of water pollution. The key advantage is that it does not require paved roads for implementation. The placement of this system does not affect the traffic flow and its functions are generally environmentally friendly. The system can be utilized in industries like agriculture and forestry, water quality management, defense, etc., and can be applied widely in projects such as mobile communications (including 5G mobile), environmental protection, and natural resources conservation. Challenges in an underwater environment can include a harsh environment and the latency. Therefore, selecting appropriate protocols is essential because different protocols might work differently in an underwater environment. For example, Durante et al. [19] have shown in an experiment of 40 nodes wireless network that MQTT-SN consumes 10% less energy than CoAP with 30% less latency.

This study presents a framework for the SWoT. Its primary contribution is to show which protocol is better in different situations. The IoT devices are discovered by their protocols based on their location and services. Services can be defined by semantics in the RDF format. An integration layer is required to make the semantics readable by the IoT protocols. A client can subscribe to any published service falling within the recommended services. The framework can be implemented in various IoT infrastructures. Real time processing is an essential part of such services. Redundant data are another challenge for IoT streams, where similar data can be gathered from different sources. The challenge is to filter such redundant data to save bandwidth and batteries. Scalability and low latency are other metrics for the IoT.

The remainder of this paper is organized as follows. Section II provides a background on the IoT, WoT, and SWoT. Sections III and IV present the various frameworks. The paper concludes with some recommendations.

2.2 Background

In this section, we provide a summary of IoT, WoT, and SWoT architectures.

2.2.1 *Internet of Things (IoT)*

The IoT is a rapidly developing technology that promises to transform the way people work and live. It refers to a network of physical objects (things) that contain embedded systems with connectivity and computing power for the purpose of exchanging data with other devices and systems over the Internet. By 2025, the number of IoT devices is projected to be 100 billion with an economic impact of more than 11 trillion dollars [14].

Cisco divided the IoT architecture into seven layers, as shown in Fig. 2.1 [17]. Control information moves from the higher (seven) to the lower (one) levels. Monitoring information moves from the lower to the higher levels. In most systems, the movement is in both directions. The IoT reference model starts with level one, which consists of physical devices and controllers that might control the devices. Communications, connectivity, and information transmission are performed in the second level. In the third and the fourth levels, data are converted into information (level 3) that is suitable for processing and storage (level 4). In the fifth level, data is abstracted to enable the development of simpler, performance-enhanced applications. This abstracted data is then represented and controlled in level six. The involvement of people and businesses is done within the seventh level.

The Internet of Underwater Things (IoUT) is a special category of IoT where most of the system is underwater. This introduces several challenges like communication latency, various energy optimization challenges, and the harsh underwater environment. IoUT is defined as a "world-wide network of smart interconnected underwater objects that enables to monitor vast unexplored water areas" [20].

2.2.2 *Web of Things (WoT)*

The growing number of IoT devices globally has presented challenges in creating platforms that can contain them. The solution is to provide virtual representation platforms through the WoT [21]. It describes a set of standards set by the W3C for solving data interchangeability and connectivity issues, and accessibility of things issues in various IoT applications. Several building blocks have been defined in the WoT to support and address these challenges [22]:

- WoT Thing Description: This is a key part of the WoT building blocks. It provides a readable data format (usually in JavaScript Object Notation (JSON [16]) format) for describing the metadata of things.

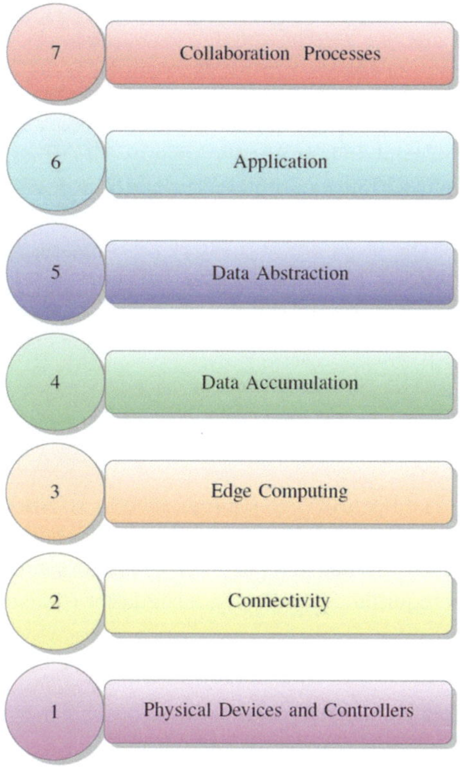

Fig. 2.1 Cisco IoT architecture

- WoT Binding Templates: These provide informational guidelines to handle multiple protocols and interaction mechanisms available in IoT ecosystems.
- WoT Scripting API: This is an optional building block that reduces the complexity of the implementation of the application, and is an application programming interface (API).
- Security and Privacy Guidelines: These provide guidelines for secure execution. Analysis issues must be considered in any WoT system.

2.2.3 *Semantic Web of Things (SWoT)*

The SWoT [18] adopts semantic web standards for the IoT data to become machine readable when transmitted over the web. The data are managed using data stores. A set of standards and protocols were developed to identify resources, encode data, and provide a meaning to each datapoint. Adopting semantics in IoT ecosystems has been presented by researchers to semantically describe their contents using ontology modeling like the semantic sensor network (SSN) ontologies. Examples of IoT data sources include sensors, observations, and actuators [23].

2.3 The Proposed Framework

Different frameworks have been designed for the emergence of the IoT with the SW; however, the proposed framework considers the performance of the stream from different IoT protocols, mainly MQTT and CoAP. The SWoT represents the IoT data within a complete semantic protocol from one or more of the recommended ontologies, such as M3 [24] (Fig. 2.2).

The proposed framework presents the dynamic connectivity between devices and services. It comprises four different layers: the data source, IoT, presentation layer, and application layer.

2.3.1 Data Source Layer

It is a layer of data sources such as sensors or events. It includes any source of data or intermediate source, such as a fog layer. Since constrained devices are not ready to publish data over the web or process it, an intermediate layer with high processing is required. Specifically, a fog layer is designed to perform this task [26].

2.3.2 IoT Layer

It is responsible for the connectivity and the delivery of the IoT data. Different IoT devices publish their data to the cloud, where only interesting information should be kept. Additionally, the IoT devices should send their information upon request. When the semantic web is added to the framework of the IoT, two concerns arise.

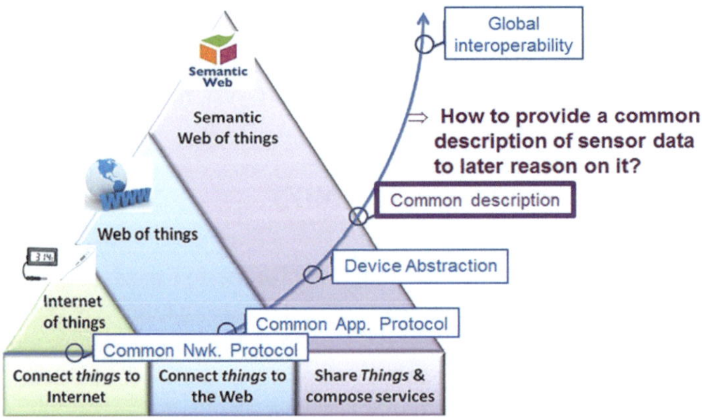

Fig. 2.2 General integration model of the IoT, WoT, and SWoT [25]

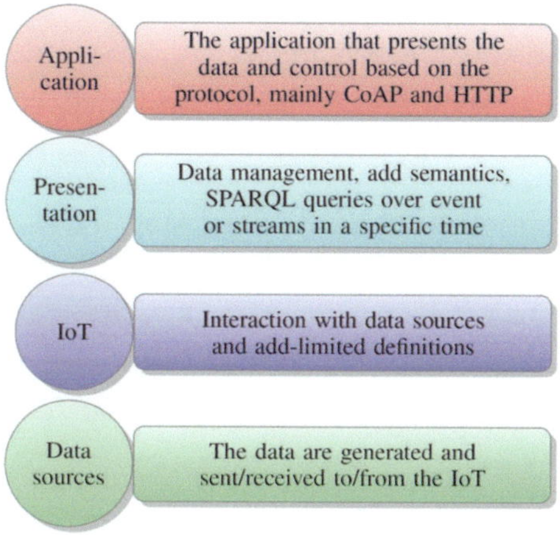

Fig. 2.3 The proposed framework

First, the representation of the IoT-generated data is denoted using one of the SW standards, mainly RDF. Second, the control and spread of information is done by the SPARQL endpoint that should be available to the subscriber of the end IoT device. The subscriber is either subscribed by their location or a similar functionality when the subscriber requests information as needed. Different examples can be listed for such a situation. Furthermore, the subscribers are notified for an event. For instance, when the ITC room temperature reaches a certain threshold, a notification is sent and automatic actions may be taken, such as shutting down some servers for security and maintenance to save data center (DC) equipment (Fig. 2.3).

2.3.3 Presentation Layer

It acts as a main layer for the representation of the IoT data. It adds semantics to the data to enable to query the streams, or to integrate and interpret the data with other data sources. The SWoT may not require SPARQL to obtain IoT live stream data in certain situations, as will be shown later; however, SPARQL is required to add meaning and extra recommendations. Adding the semantic web to the IoT will bring changes in the communication between the devices and the world. This layer also integrates the storage and management of data. The fog layer can be a part of the presentation layer. The fog layer contains edge nodes to perform the processing required for IoT constrained devices before forwarding the data to the internet, as shown in Fig. 2.4.

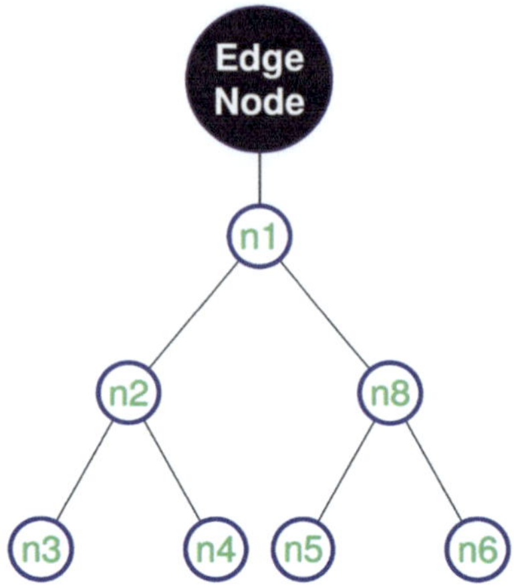

Fig. 2.4 WSN of IoT nodes connected to an edge node

2.3.4 *Application Layer*

It is related to the delivery and interactions of different events controlled by IoT protocols, such as CoAP. The HTTP protocol can be used by humans to manage and control IoT devices, and to query data streams. Additionally, web applications can be developed for developing decision-making systems. While CoAP is used for machine-to-machine connectivity, HTTP is used for human-to-machine connectivity.

This layer comprises different protocols such as CoAP, MQTT, and universal plug and play (UPnP). UpnP allows different network components to discover each other and share some network services. Because the internet is growing and due to the addition of the IoT, Ipv6 is the recommended protocol for IoT devices. However, it is too heavy for constrained devices. To overcome this limitation, 6LoWPAN is introduced to allow constrained devices to connect over IPv6 and low-power wireless personal area networks (LoWPANs). Therefore, constrained devices with low power and limited processing capabilities can share data in wireless ad hoc networks (WANETs) and, more importantly, use IP-enabled networks. Fig. 2.2 shows the general framework of integrating the IoT and SW, and Fig. 2.3 shows the proposed framework.

2.4 Other Frameworks

2.4.1 *Spitfire*

Pfister proposed a scalable framework for the SWoT called SPITFIRE [1]. It achieves scalability by avoiding the entity registries. Additionally, it does not require a certain data format when communicating with the Linked Open Data (LOD) cloud. One of the unique features of SPITFIRE is its ability to map detests to each other to define overly complicated processes. Moreover, the authors invented a new method that is slightly automatic to annotate sensors as they join the network. This mechanism is built on the level of correlation between the annotations of sensors in the same region. Clustering is used to classify data streams that are generated by various sensors. Sensors with similar annotations form a stand-alone cluster. New sensors will join the closest cluster. The distance between two sensors is simply the similarity score of the two fuzzy sets, i.e., the output time series of the sensors). Furthermore, semantic entities are introduced to link sensors to real-world states. Finally, prediction models are employed to estimate the correct high-level state for a specific semantic entity.

2.4.2 Semantic Publish/Subscribe (SPS)

The semantic public/subscribe (SPS) framework [27] provides an information level to exchange data in IoT applications. The framework is built on top of SPARQL where regular updates and queries are used. This framework is intended to be built on the edge of the network where the data are generated. The proposed framework provides a scalable solution for semantic event processing by limiting overhead networking through limiting the number of notifications (added or deleted events) on the side of the subscriber. The notifications are sent only if new updates have occurred since the previous one. The authors developed a novel event detection algorithm to detect continuous updates of heterogeneous events. The efficiency of the framework has been proven in a smart city scenario, where more than thousands of subscriptions can be triggered by a simple update and more than three thousand subscriptions can be triggered by complex updates.

2.5 Conclusions

The proposed framework integrates different technologies to provide one solution for the SWoT. The COAP protocol or MQTT can be integrated with any other protocol to deliver a single solution. In this paper, the problem of connectivity of the IoT is presented. IoT devices can use the well-established protocols of the web to send and receive data. Various vendors and research communities provide different frameworks. However, none has presented a framework for the IoT over the web using SW technology. A framework is thus presented for IoT device connectivity over web technologies. Various applications can be developed following this framework.

We recommend future studies to present systems based on this framework. Further evaluations and comparisons are required for such a system to find the best trade-offs to be used differently. One of the known issues with the IoT is the problem of connecting IoT devices from different vendors. Using such a framework can solve this issue to increase the number of possible applications. In the future, smart city projects will require such a framework to allow IoT devices from different vendors to run in complete open ecosystems.

References

1. Pfisterer, D., Romer, K., Bimschas, D., et al. (2011). SPITFIRE: Toward a semantic web of things. *IEEE Communications Magazine, 49*(11), 40–48. https://doi.org/10.1109/MCOM.2011.6069708
2. Beek, W., Rietveld, L., Bazoobandi, H. R., et al. (2014). LOD laundromat: A uniform way of publishing other people's dirty data. In *The Semantic Web ISWC at the 13th International Semantic Web Conference*, (pp. 213–228). Riva del Garda, Italy, Part I.
3. SAUDI National e-Government Portal. (2021). National Portal. saudi.gov.sa, 2015. [Online]. Available: http://www.saudi.gov.sa/wps/portal/. [Accessed: Jul. 30, 2021].
4. SAUDI National e-Government Portal. (2001). *Saudi Government Open Data portal*. data.gov.sa. [Online]. Available: http://data.gov.sa/. [Accessed: Jul. 30, 2021].
5. Berners-Lee, T. (2009). *Linked Data*. [Online]. Available: http://www.w3.org/ [Accessed: Jul. 30, 2015].
6. Linking Open Data W3C SWEO Community Project. [Online]. Available: http://www.w3.org/ [Accessed: Jul. 30, 2021].
7. RDF Working Group. (2014). "RDF". w3.org. [Online]. Available: http://www.w3.org/RDF/. [Accessed: Jul. 30, 2021].
8. Prud'hommeaux, E., & Seaborne, H (eds). (2013). *SPARQL Query Language for RDF*. W3C Recommendation. Available at: https://www.w3.org/TR/rdf-sparql-query/.
9. Maali, F., Erickson, G. H., Tetherless World Constellation (RPI), et al. (2014). *Data Catalog Vocabulary (DCAT)*. [Online]. Available: http://www.w3.org/TR/vocab-dcat/. [Accessed: Jul. 30, 2021].
10. DCMI Usage Board. (2012). *DCMI Metadata Terms*. [Online]. Available:http://dublincore.org/documents/2012/06/14/dcmi-terms/?v=terms. [Accessed: 30 Jul 2021].
11. Domingo, M. C. (2012). An overview of the internet of underwater things. *Journal of Network and Computer Applications, 35*(6), 1879–1890. https://doi.org/10.1016/j.jnca.2012.07.012
12. Nucivic. (2015). *DKAN: The all in one open data platform*. [Online]. Available: http://nucivic.com/dkan/. [Accessed: 30 Jul 2021].
13. W3C. (2015). *World Wide Web Consortium (W3C)*. [Online]. Available: http://www.w3.org/. [Accessed: 30 Jul 2021]
14. "GIV 2025 - Huawei. (2018). [Online]. [Accessed: 1 Apr 2021].
15. Friesen, J. (2019). *Introducing JSON*. Java XML JSON, (p. 187203).
16. IoT Refence Model, Cisco Systems. (2014). [Online]. Available: http://cdn.iotwf.com/resources/71/IoTReferenceModelWhitePaperJune42014.pdf
17. Scioscia, F., & Ruta, M. (2009). Building a semantic Web of Things: Issues and perspectives in information compression. In *IEEE Int. Conf. Semantic Comput. (ICSC)*, (p. 589594).
18. Francesco, A., & Viola, F. (2019). Building the semantic Web of things through a dynamic ontology. *IEEE Internet of Things Journal, 6*(6), 10560–10579.
19. Zou, L., Liu, L., Özsu, M. T. (eds). (2018). SPARQL. Encyclopedia of Database Systems. Springer. https://doi.org/10.1007/978-1-4614-8265-9_80803
20. Durante, G., Beccaro, W., & Peres, H. (2018). IoT protocols comparison for wireless sensors network applied to marine environment acoustic monitoring. *IEEE Latin America Transactions, 16*(11), 2673–2679. https://doi.org/10.1109/TLA.2018.8795107
21. Dave, R. (2015). The web of things: Challenges and opportunities. *Computer, 48*(5), 26–32.
22. Semantic Web. (2015). *Semantic Web - W3C - World Wide Web Consortium)*. w3.org, 2015. [Online]. Available: http://www.w3.org/standards/semanticweb/. [Accessed: Jul. 30, 2021].

23. Luca, R., et al. (2016). A semantic publish-subscribe architecture for the Internet of Things. *IEEE Internet of Things Journal, 3*(6), 1274–1296.
24. Aloufi, K. S., & Alhazmi, O. H. (2020). Secure IoT resources with access control over RESTful web services. *Jordan Journal of Electrical Engineering, 6*(1), 63–77.
25. Van der Schaaf, H., & Herzog, R. (2014). *Mapping the OGC Sensor Things API onto the OpenIoT Middleware, Interoperability and Open-Source Solutions for the Internet of Things: International Workshop, FP7 OpenIoT Project, SoftCOM*, (pp. 62–70). Springer International Publishing.
26. Ruta, M., Scioscia, F., & Sciascio, E. D. (2012). Enabling the semantic web of things: framework and architecture. In *IEEE sixth international conference on semantic computing*, (pp. 345–347).
27. Olivieri, A. J., Bocchi, A., Jung, Y., et al. (2014). Semantic Web of Things: An analysis of the application semantics for the IoT moving towards the IoT convergence. *International Journal of Web and Grid Services., 10*, 244–272. https://doi.org/10.1504/IJWGS.2014.060260

Chapter 3
From Solid Waste to Mathematics of OR: The Concept of Recycling

Santosh Kumar, Elias Munapo, Philimon Nyamugure, and Trust Tawanda

Abstract This chapter introduces protean systems, which have a typical characteristic, i.e., they change and thereby make decision making difficult and challenging, yet decisions must be made. Every system interacts with external systems and changes arise due to these external interactions. In this chapter we take information recycling approach from the solid waste management to mathematical analysis and illustrate recycle concept on a liner system and derive necessary information for the $(k + 1)^{th}$*interval* by using information available from the analysis of the $(k)^{th}$ interval system. Recycle concept has been used to find the solution to the changed problem not as a new problem but by making use of the information from the $(k)^{th}$ interval problem.

Keywords Protean systems · Information recycling in mathematical modeling · Recycling concept from waste management to mathematics

S. Kumar (✉)
Department of Mathematical and Geospatial Sciences, School of Sciences, RMIT University, Melbourne, Australia
e-mail: santosh.kumarau@gmail.com

E. Munapo
School of Economics and Decision Sciences, North West Ucniversity, Mafikeng Campus, Mafikeng, South Africa
e-mail: emunapo@gmail.com

P. Nyamugure · T. Tawanda
Department of Statistics and Operations Research, National University of Science and Technology, Box AC 939, Ascot, Bulawayo, Zimbabwe
e-mail: philimon.nyamugure@nust.ac.zw

T. Tawanda
e-mail: trust.tawanda@nust.ac.zw

P. K. Kapur et al. (eds.), *Data-centric Approaches to Industrial Decisions*, Asset Analytics, https://doi.org/10.1007/978-981-96-7556-2_3

3.1 Introduction

Recycling is a well-accepted concept in solid waste management, it has taken a long time for its acceptance. Recycling in waste management has been accepted as it can reduce the bulk of solid waste, decrease the landfill and, in some instances, reprocessed waste can be reutilized as a resource, Grover et al. [1]. This, 'waste-to-resource' journey in solid waste has taken time for its acceptance in waste management, and it has hardly been accepted by other disciplines such as science, management, and engineering. In fact, instead of converting waste into resource, in many situations, good resources are wasted in different forms. As a necessity, in developing countries, energy requirements are met by using wood and agricultural waste, straws, and cow dung, see Oladimeji et al. [2]. This is a simple illustration where waste has been used to satisfy energy requirements. Distribution of energy in urban areas is dependent on some infrastructure, such as waterpipe, electric and gas supply networks which are missing in rural areas. This lack of infrastructure forced people in rural areas to think alternatives, and they developed their own strategy to convert some of their waste into energy. However, when waste can be used as a resource in one situation, it opens avenues for further investigations in other areas of commerce and industry. Lack of infrastructure is one reason; another situation arises when infrastructure exists, but it is not able to provide the required service. For example, see Rasmi [3] where electric grid in the city of Beirut is not performing well, industry faces long periods of time out when power is not available. Renewal energy by the solar panel is capturing use of renewal energy, which is growing in this city. Rasmi [3] has stated that in 2010, Beirut generated zero solar power, but had 90 MW of solar energy in 2020, and added 100 MW in 2021 and 500 MW in 2022. Khoury, CEO, Lebanese Center for Energy Conservation says that 30% of its energy will be from renewable sources by 2030.

In this chapter, we intend to take the recycling concept from waste management and examine if the information can be reused for analysis of the problem after it changes. This recycling concept from the solid waste management has a potential for its use in other areas, and particularly our investigations are to explore the recycling concept in the mathematics of OR? In context of mathematical modeling of a changing system, we have a choice either to analyze the changed problem as a new problem or use the recycle concept, to find the new solution to the changed problem. This chapter explores the following:

(a). In Sect. 3.2, we identify a few real-life systems, which change, and never remain in a study state. These changing systems have been called as 'protean' systems. Analysis of changing systems, and decision making in changing environment, is a challenge frequently faced by commerce and industry.

(b). In Sect. 3.3, we illustrate the complexity of a changing system where use of analytical tools may not be easy, hence heuristics may be an answer. We illustrate use of heuristic approach for analysis of a changing system and point out limitations.

(c). In Sect. 3.4, we deal with the situation of software support which is a protean system and point out some adverse consequences when software support is delayed for associated theoretical developments.

(d). Finally, in Sect. 3.5, we discuss use of the recycling concept from the solid waste management and take it to the mathematics of OR and apply it to a linear model. The linear programming model has been successfully applied to many industrial situations, which change. Here we derive the optimal solution to the model at the $(k + 1)^{th}$ interval using the information available from the optimal solution of the model for the kth interval.

The chapter is concluded by making a few concluding remarks in Sect. 3.6.

3.2 Real-Life Systems that Change or Protean Systems

Most real-life situations operate in an environment that interacts with some external systems, and consequently are not fully under the control management. For example, a peace-loving country may not plan to attack any other country, but is forced to enter a war when it faces an attack from a neighboring country. Similarly, management attempts to avoid all reasons that can bring a change, yet system changes and management must safeguard themselves by appropriate preparedness, take necessary action, when required. However, they need to develop their own strategies to deal with the challenges created by these external interactions. As an additional example, consider that a production process is planned on the raw material supply that arrives every interval before the start of the production process. This system works satisfactorily, however, on a particular day, due to some road work, supply of some resources maybe delayed. These changing systems have been described as protean systems, which are of immense importance in describing many day-to-day situations faced by commerce and industry. The word protean originated from the word Proteus—the legendary God of sea, who is believed to have the power of assuming different shapes. The name 'Protean' for the 'changing systems' was suggested to the author (SK) by G. Woolsey, when he was at the Colorado School of Mines and was visiting Melbourne, see Kumar [4]. A close examination of many real-life systems reveals that not many systems can exist that are not protean, i.e., these systems display inbuilt characteristics of a change, which is beyond control of the management. We need more powerful tools to analyze these changing systems. Here are a few simple illustrations of protean systems.

3.2.1 *The Job Allocation Problem*

Consider a system where the demand on its services changes frequently, however, demand is kept under control by the management by means like holding people waiting on the phone or in a queue or by shifting demand to some future date by making some alternative bookings. In some cases, none of the above is possible, and one is left with no choice but to turn down the opportunity. In all such alternatives, there is always a risk of losing that business for ever. When a demand pattern establishes a definite increasing pattern, the management may take actions such as increasing manpower required and upgrading service facility to meet the increased demand. However, in a short term, extra resources may not be justified and may not even be feasible. In a commercial venture, when demand exceeds the facility available, management can either postpone that demand to some future date or give incentives to existing staff to handle the excess demand on limited facility available. Increasing staff and/or facility takes time and has cost implications, see, Kalyan and Kumar [5]. If 'n' denoted the number of jobs and 'm' denotes the number of persons, they discussed a short-term situation by looking at the value (n/m). If this value is equal to an integer k, every person gets k number of jobs until necessary facilities are increased. However, if (n/m) is not equal to an integer, we find the value of $\left(\frac{n}{m}\right) = k,\ \textit{next higher integer}$. In this case all workers get 'k' or (k − 1) number of jobs, until facilities and manpower is adequate for handling all jobs to be attended adequately.

3.2.2 Capacity Expansion

Consider the road network in a metropolitan city, where due to gradual demographic changes, demand on various routes changes gradually. Assuming that improvement in road capacity is possible by adding extra lanes but construction of all additional lanes to satisfy the traffic requirement at the same time is neither possible nor desirable due to financial considerations and traffic hazards caused by construction work. The problem is to distribute limited resources and prioritize associated improvements. A simple mathematical model for a single road network can be seen as:

$$Maximize x_o = f_{1N}$$

$$Subject\ to:$$

$$\sum_i x_{ij} - \sum_k x_{jk} = \begin{cases} -f_{1N}, for & i = 1 \\ 0, \quad for & i, j = 2, 3, \ldots (N-1) \\ +f_{1N}, for & k = N. \end{cases}$$

$$0 < x_{ij} < b_{ij} + \sum_k y_{ijk} b_k \forall (i, j)$$

$$\sum_k y_{ijk} = 1 \forall (ij)$$

$$\sum_{ij} y_{ijk} = 1 \forall (k)$$

$$x_{ij} \geq 0 \textit{ and integer}; \; y_{ijk} \textit{ is } 0 \textit{ or } 1. \tag{3.1}$$

Here node 1 represents the source and node N denotes the destination. f_{1N} denotes the number reaching from the origin node 1 to the destination node N. $x_{ij}denotes$ the flow in section joining the nodes $iandj$. Mathematical model for a complex network will be more complicated and accordingly its analysis will be not easy. Road network improvement have been attempted by Vijay, Konda and Vrat [6, 7].

Traffic congestion also arises by variations in government policies, for example, in Melbourne suburbs, the local government is moving towards higher density of population. For example, when an existing house is demolished, more than one dwelling is constructed on the same land. All new constructions end up with a larger number of dwellings on the same block where only a single house existed. This action is slowly increasing density per unit area, which is increasing with respect to all requirements needed at a household and is heading towards congestion that will lead to disruption in services eventually. This gradual change is seen throughout the world, single story is converted to double story, double to multiple stories and cluster developments, so much so that in one block hundreds of families live in one multi-story complex developments. They all are using the same road network, same grid of electricity, same pipe for gas and water flow, etc., leading to unbearable congestions.

3.2.3 Protean Reliability Requirements

Consider a network and assume that maximum reliability of a route between two specific nodes is a known probability. However, due to change in technology, the reliability of a particular route may not be considered adequate, and in that case, reliability can be improved to meet the new demand by some parallel duplication or some other similar action to increase the reliability. Similarly, changes in technology force network updates, which is observed in communication networks, various service centers such as airports and other service networks. This type of situations has been analyzed by Kumar [8–10], Talukdar and Kumar [11, 12], Talukdar et al. [13] and Kumar et al. [14].

Here, we also intend to point out that systems also undergo an aging process and require major redevelopments and infrastructures when existing infrastructure becomes inadequate to handle the changed system. Mobile phone network is an example. This discussion is not intended to be exhaustive; it is just intended to point out causes that bring a change to an existing system, making analysis and decision difficult.

3.3 Improving the Terminal Pair Reliability

The terminal pair reliability (TPR) of a communication network is an important measure of its performance. TPR is the probability of successful communication between two specified nodes through any path. Kim, Case, and Ghare [15] suggested a method to evaluate the TPR by using the formula (3.2) and (3.3)

$$P(TPR) = \sum_{i} P_i - \sum_{i \neq j} [P_i \cdot P_j]^* + \cdots + (-1) \wedge (n+1)[P_1 P_2 \ldots P_n]^* \quad (3.2)$$

Note

$$P_i = \prod_{j} p_j \quad (3.3)$$

where P_i represents the reliability of the path i, which is the product of the reliabilities that constitute this path. The approach by Kim, Case, and Ghare [15] considers all possible paths between the two points. Here the * indicates that if a link is common to more than one path, it is considered only once. For details, see Kalyan and Kumar [16]. They obtained all possible paths and conducted a simulation experiment under three strategies, which were:

Strategy 1: Duplicate a component by parallel redundancy which is common to many paths.

Strategy 2: Parallel duplication is provided to the minimum reliability component in the maximum reliability path.

Strategy 3: Duplicate the minimum reliable non-duplicated component in the system, until the components available for duplication have been exhausted.

Several experiments were carried out under 'k' number of redundancy support for $1 \leq k \leq 5$. Analytical optimal solution is hard for large complex reliability network. Therefore, analysis was carried out by using heuristic strategies to make necessary decisions.

3.4 Software Support is a Protean System

Software development is a never-ending protean system, as capability of theoretical results keep advancing. Software provides computational support for analysis of complex models, and when theory advances, software also needs updates. Software reliability is a well-accepted research area, see Kapur et al. [17]. Researchers are dealing with software reliability growth models, mathematical models have been developed to find optimal release policies, reliability growth is attained by detecting and debugging software errors. Software is subject to testing phase before it is released. Problems that have been tackled are dealing with optimal release time, optimal cost, and optimal reliability of the software. Testing a software during its development phase improves its reliability, however, the software may not have all desired features, yet it is released to its users. New features are continuously added, and software capability needs updating. This cycle never ends. Ideally, software must keep changing for the better.

From the above, it may be noted that software development is never likely to reach in its steady state, as further improvements are always in the process and this makes software development a never-ending changing system, a typical protean system. However, software companies generally fail to satisfy this requirement.

Another aspect of software is to support some existing accepted methodology; however, theoretical development and advancement never come to rest, hence the existing software is always a bit behind. It is always waiting for new updates to be added to the existing software. Theoretical advances come from different researchers and get published in different journals around the world. However, an existing software belongs to some software company, which is a commercial entity. The only way these new theoretical developments can reach practitioners by timely software support by these companies. Many researchers may argue that why depend on development of the software of these companies, why not supply the supporting software along with the theoretical development? Kumar and Munapo [18] argue that it does not happen for the following reasons:

1. A new coding will not stand compared to an existing coding developed by coding professionals, and that has been tested, modified, and eventually it has become a bug-free coding.
2. Scientific community does not consider coding of existing methods as a significant scientific contribution.
3. Adding a new feature to an existing method, requires previous software, which is not made available whereas theoretical development is well documented.

In other words, mathematical development is like a building block, but the associated software in the present environment is not. Consequences are many mathematical developments do not reach their potential users. Theoretically, advancement will reach the practitioners when software support is available and this way when appropriate software support is missing, the advantage of new research and development, and new methodology is missed out by many of its users.

One can understand a defined time-leg between the theory and software, but many good ideas have never drawn attention of the software companies. Many such instances have been pointed out by Kumar and Munapo [18].

3.5 Recycling Concept in OR Methodology: A Linear Program

Here, we discuss a well-known LP model in OR, which has been used in commerce and industry for the insight before decision making in commerce and industry related situations. We illustrate use of the information recycling in a changing environment. This concept is appropriate for situations where input to a mathematical model may change (as it may be beyond control of the management) form one period to the next period, but the structure of a mathematical model remains the same. Protean term reflects changes in the input parameters which can fluctuate, but mathematical structure of the model is governed by the situation, which remains unchanged. For example, we consider a linear program for a particular production system, where the resource input may be influenced by external interactions, the production process is stable. The management can only do their best to control the input but deal with it when it changes. We illustrate this using the LP model in Sect. 3.5.1. In a LP model, the central inverse of the basic matrix is central, which is available from the model before the change. Details of the recycling approach are discussed in Sect. 3.5.1.

3.5.1 Recycle Concept Applied to a LP Model

3.5.1.1 Numerical Illustration 1

Consider a production system represented by a LP model (3.4).

$$\begin{aligned} &\text{Maximize } x_0 = \sum_j c_j x_j \\ &\text{Subject to } \sum_{j=1}^{n} a_{ij} x_j \le b_i,\ i = 1, 2, \ldots, m \textit{ and } x_j \ge 0 \textit{ for all } j. \end{aligned} \tag{3.4}$$

Here b_i, a_{ij}, $and c_j$ are assumed known quantities and one is required to find the value of the unknown decision variables x_j. We are assuming that the input parameters b_i, a_{ij}, $and c_j$ may fluctuate due to external interactions. Therefore, it is reasonable to state that on a typical k^{th} interval the above problem (3.4) can be stated as LP model (3.5) given below:

$$\begin{aligned} &Maximize x_0^k = \sum_j c_j^k x_j^k \\ &\text{Subject to} \sum_{j=1}^{n} a_{ij}^k x_j^k \le b_i^k,\ i = 1, 2, \ldots, m \textit{ and } x_j^k \ge 0,\ j = 1, 2, \ldots, n. \end{aligned} \tag{3.5}$$

Due to fluctuations and uncontrollable situations, the same model in the next period becomes as shown by (3.6)

$$Maximize x_0^{k+1} = \sum_j c_j^{k+1} x_j^{k+1}$$

$$\text{Subject to} \sum_{j=1}^{n} a_{ij}^{k+1} x_j^{k+1} \le b_i^{k+1},\ i = 1, 2, \ldots, m \ and \ x_j^{k+1} \ge 0, j = 1, 2, \ldots, n. \tag{3.6}$$

It maybe noted that the optimal solution for the model (3.5) in the kth interval has been determined by using an appropriate LP method or software, see Dantzig [19]. The model (3.6) requires a new solution, if some or all values $a_{ij}^{k+1}, b_i^{k+1}, andc_j^{k+1}$ changed compared to $a_{ij}^k, b_i^k, andc_j^k$. These two sets of values have a relation among them as shown in (3.7).

$$c_j^{k+1} = c_j^k + \overline{c_j^k}, \forall j, a_{ij}^{k+1} = a_{ij}^k + \overline{a_{ij}^k}, b_i^{k+1} = b_i^k + \overline{b_i^k}, \forall i, j, \ and\,(ij) \tag{3.7}$$

Note that the changed amount is represented by $\overline{c_j^k}, \forall j, \overline{a_{ij}^k}, forall j and\,(ij), and \overline{b_i^k}, \forall i$. These quantities can be positive, negative or zero. They all depend on the change that occurred when the system moved from the period k to (k + 1).

This model has been analyzed by Kumar [11]. The algorithm developed by Kumar [11] is as follows:

Step 0

Record for the kth interval the values of various parameters

$$n_k, m_k$$

$$c_j^k \forall j, a_{ij}^k \forall (ij), b_i^k \forall i$$

$$x_{Bl}^k, l = 1, 2, \ldots, m \text{ (Basic variables)}$$

And the final simplex tableau giving the optimal solution.

Record the problem for the $(k + 1)^{th}$ interval problem, i.e., value of

$$n_{k+1}, m_{k+1}$$

$$c_j^{k+1} \forall j, a_{ij}^{k+1} \forall (ij), b_i^{k+1} \forall i$$

$$x_{Bl}^k, l = 1, 2, \ldots, m$$

Step 1

If a variable present in the kth model and is not required in the $(k+1)^{th}$ interval model, create a duplicate column with negative values for the sparse column of the model (3.6) and use the substitution (3.8):

$$\begin{aligned}\overline{c_j^k} &= -c_j^k \\ \overline{a_{ij}^k} &= -a_{ij}^k\end{aligned} \tag{3.8}$$

If a new variable x_{n+1}^{k+1} is required in the $(k+1)^{th}$ interval model compared to the model in the kth interval, it is simply added to the sparse problem (3.6) and use the substitution (3.9).

$$\begin{aligned}\overline{c_{n+1}^k} &= c_{n+1}^{k+1} \\ \overline{a_{i,n+1}^k} &= a_{i,n+1}^{k+1}\end{aligned} \tag{3.9}$$

Step 2

Compare constraints in the kth interval model and the model for the $(k+1)^{th}$ interval. If there is no change in the number of constraints, go to Step 3. If the number of constraints are different, one of the following will apply:

Case 1: Suppose the lth constraint is no longer required in the $(k+1)^{th}$ interval, substitute

$$\overline{b_l} = M \tag{3.10}$$

where M represents the usual large quantity, making that constraint in active.

If a new $(m+1)^{th}$ constraint must be added to the problem in the $(k+1)^{th}$ interval, this can be achieved by the following substitutions:

$$\begin{aligned}b_{m+1}^k &= 0 \\ \overline{b_{m+1}^k} &= b_{m+1}^{k+1} \\ \overline{a_{j,m+1}^k} &= a_{j,m+1}^{k+1} for\ all\ j\end{aligned} \tag{3.11}$$

Step 3

Rearrange the problem in the $(k+1)^{th}$ interval into two parts as shown in (7). If the sparse part of the problem does not exist, the optimal solution of the problem in the kth interval remain as the optimal solution for the $(k+1)^{th}$ interval. Go to Step 6; otherwise go to step 4.

Step 4

Using the B_k^{-1} from the kth interval, carry out for each sparse column $\overline{P_j} findB_k^{-1} \cdot \overline{P_j}$. Add the column P_j from the kth interval problem and $B_k^{-1} \cdot \overline{P_j}$ for each j. Re-establish the pivot operation with respect to the kth interval. If the resulting solution does not satisfy feasibility of the solution, use the dual-simplex iterations to restore feasibility of the solution and go to Step 5.

Step 5

Once the solution is feasible, now by using the simplex method, one can move towards the optimality of the solution. Go to Step 6.

Step 6

The current solution is optimal to the LP model for the $(k + 1)^{th}$ interval. Print this solution, and make the $(k + 1)^{th}$ as the kth, return to Step 0 for analysis of the next interval model.

3.5.1.2 Numerical Illustration 1

For numerical illustration, consider a typical LP model for the kth and the $(k + 1)^{th}$ interval as given by (3.12) and (3.13). We assume that the optimal solution to the problem (3.12) has been obtained by using an appropriate LP software and objective is to find the optimal solution to the problem (3.13) by using available information from the solution of the model (3.12). We will be using the algorithm discussed above.

$$\begin{aligned} &Maximize x_0^k = 3x_1^k + 2x_2^k + 2x_3^k \\ &\text{Subject to } x_1^k + 3x_2^k + x_3^k \le 16 \\ &2x_1^k + x_2^k + 2x_3^k \le 18 \\ &x_1^k, x_2^k, x_3^k \ge 0 \end{aligned} \tag{3.12}$$

The problem for the $(k + 1)^{th}$ interval is given by (3.13).

$$\begin{aligned} &Maximize x_0^{k+1} = 3x_1^{k+1} + 1x_2^{k+1} + 3x_3^{k+1} \\ &\text{Subject to } 2x_1^{k+1} + 3x_2^{k+1} + 2x_3^{k+1} \le 16 \\ &2x_1^{k+1} + x_2^{k+1} + 2x_3^{k+1} \le 14 \\ &x_1^{k+1}, x_2^{k+1}, x_3^{k+1} \ge 0 \end{aligned} \tag{3.13}$$

Consider, optimal solution to (3.12) is known by the application of the simplex method, given below in Table 3.1.

Table 3.1 LP optimal solution to the problem in k^{th} interval

Basic	x_1^k	x_2^k	x_3^k	s_1^k	s_2^k	RHS
Obj	0.00	0.00	1.00	0.20	1.40	28.4
x_2^k	0.00	1.00	0.00	0.40	−0.20	2.80
x_1^k	1.00	0.00	1.00	−0.20	0.60	7.60

i.e., the optimal solution is $x_1^k = 7.60$ and $x_2^k = 2.80$ and $x_0^k = 28.40$.

Here $B_k^{-1} = \begin{bmatrix} 1 & .20 & 1.40 \\ 0 & .40 & -0.20 \\ 0 & -.20 & 0.60 \end{bmatrix}, \overline{P_1} = \begin{bmatrix} 0 \\ 1 \\ 0 \end{bmatrix}, \overline{P_2} = \begin{bmatrix} -1 \\ 0 \\ 0 \end{bmatrix}, \overline{P_3} = \begin{bmatrix} 1 \\ 1 \\ 0 \end{bmatrix}$.

This gives:

$$B_k^{-1}.\overline{P_1} = \begin{bmatrix} .20 \\ .40 \\ -.20 \end{bmatrix}, B_k^{-1} \cdot \overline{P_2} = \begin{bmatrix} -1.0 \\ 0.0 \\ 0.0 \end{bmatrix}, B_k^{-1} \cdot \overline{P_3} = \begin{bmatrix} 1.2 \\ 0.4 \\ -0.2 \end{bmatrix}.$$

This information for the $(k+1)^{th}$ interval can now be summarized in Table 3.2. Adding the columns x_j^k and $\overline{x_j} for j = 1, 2, 3$ we get Table 3.3

Note that in Table 3.3, we have no longer have unit columns with respect to the basic variables. We restore unit columns with respect to the two basic variables in Table 3.4.

The solution in Table 3.4 is not optimal, hence one simplex iteration must be performed, resulting in Table 3.5.

Thus, using the recycle concept, we have established the LP solution to the problem in the $\boldsymbol{(k+1)^{th}}$ interval by recycling the information already available from the optimal solution to the problem at the $\boldsymbol{k^{th}}$ interval.

Similarly, the information in Table 3.5 can be used for deriving a solution to the model in the next interval.

Table 3.2 The data for the $(k+1)^{th}$ interval

Basis	x_1^k	x_2^k	x_3^k	$\overline{x_1}$	$\overline{x_2}$	$\overline{x_3}$	s_1^k	s_2^k	RHS
Obj	0.0	0.0	1.0	0.2	−1.0	1.2	0.2	1.4	28.4–5.6
x_2^k	0.0	1.0	0.0	0.4	0.0	0.4	0.4	−0.2	2.8 + .8
x_1^k	1.0	0.0	1.0	−0.2	0.0	−0.2	−0.2	0.6	7.6–2.4

Table 3.3 Simplex tableau for the $(k+1)^{th}$ interval

Basis	x_1^{k+1}	x_2^{k+1}	x_3^{k+1}	S_1^{k+1}	S_2^{k+1}	*RHS*
	0.2	−1.0	2.2	0.2	1.4	22.8
x_2^{k+1}	*0.4*	1.0	0.4	0.4	−0.2	3.60
x_1^{k+1}	*0.8*	0.0	0.8	−0.2	0.60	5.20

Table 3.4 Unit columns with respect to the basic variable

Basis	x_1^{k+1}	x_2^{k+1}	x_3^{k+1}	S_1^{k+1}	S_2^{k+1}	*RHS*
	0.0	0.0	2.0	−0.25	1.75	20.50
x_2^{k+1}	*0.0*	1.0	0.0	0.5	−0.5	1.00
x_1^{k+1}	*1.0*	0.0	1.0	−0.25	0.75	6.50

Table 3.5 Optimal solution for the model in the $(k + 1)^{th}$ interval

Basis	x_1^{k+1}	x_2^{k+1}	x_3^{k+1}	S_1^{k+1}	S_2^{k+1}	*RHS*
	0.0	0.0	2.0	0.00	1.50	21.0
S_1^{k+1}	*0.0*	1.0	0.0	1.0	−1.0	2.00
x_1^{k+1}	*1.0*	0.0	1.0	0.00	0.50	7.00

3.5.1.3 Numerical Illustration 2

Consider the problem in the $\boldsymbol{k^{th}}$ interval problem as given by relation (3.14) and Table 3.6. It is given below:

$$\begin{aligned} &Maximize x_0^k = x_1^k + 2x_2^k \\ &3x_1^k + 2x_2^k \le 24 \\ &2x_1^k + 4x_2^k \le 32, x_1^k, x_2^k \ge 0 \end{aligned} \tag{3.14}$$

The optimal solution is given in Table 3.7.

Table 3.7 indicates that the problem has more than one optimal solution. The other solution can be easily seen as: $x_1^k = 4, and x_2^k = 6$, shown in Table 3.8.

Table 3.6 Illustration 2

Basic	x_1^k	x_2^k	s_1^k	s_2^k	RHS
s_1^k	3	2	1	0	24
x_2^k	2	4	0	1	32
$z_j^k - c_j^k$	−1	−2	0	0	0

Table 3.7 Optimal solution to Illustration 2

Basic	x_1^k	x_2^k	s_1^k	s_2^k	RHS
s_1^k	2	0	1	−1/2	8
x_2^k	0.5	1	0	1/4	8
$z_j^k - c_j^k$	0	0	0	0.5	16

Table 3.8 Alternative optimal solution

Basic	x_1^k	x_2^k	s_1^k	s_2^k	RHS
x_1^k	1	0	0.5	−0.25	4
x_2^k	0	1	−0.25	0.25	6
$z_j^k - c_j^k$	0	0	0	0.5	16

Suppose the problem in the next period becomes as given by (3.15):

$$\begin{aligned} &Maximize x_0^{k+1} = x_1^{k+1} + 3x_2^{k+1} \\ &3x_1^{k+1} + 3x_2^{k+1} \le 25 \\ &2x_1^{k+1} + 5x_2^{k+1} \le 33, x_1^{k+1}, x_2^{k+1} \ge 0 \end{aligned} \tag{3.15}$$

The optimal solution to (3.15) is given in Table 3.9.

Here, our interest is to establish the solution in Table 3.9 from the information available from solution in Table 3.7. A close look at problems (3.14) and (3.15) gives us:

$$B_k^{-1} = \begin{bmatrix} 1.0 & -0.50 \\ 0.0 & 0.25 \end{bmatrix}, \overline{P_2^k} = \overline{b^k} = \begin{bmatrix} 1 \\ 1 \end{bmatrix}, B_k^{-1} \cdot \overline{P_2^k} = B_k^{-1} \cdot \overline{b^k} = \begin{bmatrix} 0.50 \\ 0.25 \end{bmatrix} \tag{3.16}$$

From Table 3.7 and the result (3.16), we get the following for Table 3.10.

Re-establishing basis with respect to the variable x_2^{k+1}, we obtain the next Table as was obtained earlier by the direct solution, i.e., Table 3.9. Hence, the optimal solution by the recycle approach is given by:

$$x_1^{k+1} = 0,\ x_2^{k+1} = 33/5,\ s_1^{k+1} = 26/5,\ s_2^{k+1} = 0,\ \text{and } x_0^{k+1} = 99/5. \tag{3.17}$$

Table 3.9 Optimal solution to problem (3.15)

Basic	x_1^{k+1}	x_2^{k+1}	s_1^{k+1}	s_2^{k+1}	RHS
s_1^{k+1}	9/5	0	1	−3/5	26/5
x_2^{k+1}	2/5	1	0	1/5	33/5
$z_j^k - c_j^k$	1/5	0	0	3/5	99/5

Table 3.10 Updated values to the problem (3.15)

Basic	x_1^{k+1}	x_2^{k+1}	s_1^{k+1}	s_2^{k+1}	RHS
s_1^{k+1}	2	$0 + 0.50$	1.0	−0.50	$8 + 0.50$
x_2^{k+1}	0.5	$1 + 0.25$	0.0	0.25	$8 + 0.25$
$z_j^k - c_j^k$	0.5	0.75	0	0.75	16

3.6 Concluding Remarks

In this chapter, the recycling concept has been applied to a linear programming model. This idea has been extended to a quadratic programming model, and second order differential equation, see Kumar [11] and Geometric programming model, see Kumar [20]. It is desirable to investigate recycling concept application to other mathematical models in OR.

It is equally desirable to develop supporting software for each methods using the recycling concept for industrial applications of various OR methods.

Determination of a moving average is also a recycling concept, Hamilton [21]. Recycling concept is also noted in dynamic programming approach, when a n-variable problem is converted to a one variable 'n' problems, see Bellman [22], Bellman and Dreyfus [23], and Dreyfus and Law [24]. For example, a simple illustration is a classical problem of dividing a constant 'C' into n > 1 parts such that the product is maximum, and sum is equal to C. Mathematically this problem can be stated as a 'n' dimensional problem (3.18):

$$Max \prod_{i=1}^{n} x_i, \text{ such that } \sum_{i=1}^{n} x_i = C, x_i > 0 \forall i. \tag{3.18}$$

Instead of dealing with the above as a 'n' dimensional problem, we consider n = 2, i.e., divide C into two parts such that the product is maximum. Let x be the value of one part, then (C − x) will the value of the second part. The product {x(C − x)} is maximum when x = (C/2). This policy can be recycled, and when we consider n = 3, we divide into three parts, however, let one such part be denoted by x, a total of the remaining two parts will be (C − x). However, this remaining quantity will be maximum when they are equal and hence product of the three will be given by: $x\left(\frac{C-x}{2}\right)^2$, which is maximum when x = (C/3) and max product will be (C/3)^3. This is how the recycle concept has been used and a 'n' dimensional problem has been converted to 'n' problems, each being a single dimension problem [25–27].

References

1. Grover, V. I., Grover, V. K., & Hogland, W. (2002). *Recovering energy from waste: Various aspects*. Science Publishers.
2. Oladimeji, A. A., Kumar, S., Parekh, C. T., & Bhala, M. J. (2002). Energy-from-waste: A perspective from developing countries. In Grover & Hogland (Eds.), *Chapter 3 in recovering energy from waste: Various aspects*, (pp. 21–29). Science Publishers, Ebfield.
3. Rasmi, A. (2023). *Asolar revolution driven by power cuts*. Time Magazine March 27 and April 2(double issue), (pp. 18–19).
4. Kumar, S. (1995). Optimizations of protean systems: A Review. In M. Fushimi & K. Tone (Eds.), *APORS Proc 1994* (pp. 139–146). World Scientific Publishers.
5. Kalyan, R., & Kumar, S. (1987). System relaxation and an assignment problem. *Indian Journal of Management and Systems, 3*(1), 11–17.

6. Vijay, R., Konda, A., & Vrat, P. (1992). An algorithm for capacity expansion of urban road network. *Opsearch, 29*(3), 181–202.
7. Vijay, R., Konda, A., & Vrat, P. (1993). Multi-period capacity expansion od Road network. *Opsearch, 30*(2), 117–140.
8. Kumar, S. (1979). Maximum reliability route subject to m improvements in anon-directed reliability network. *IEEE Transaction on Reliability, R28*, 127–128.
9. Kumar, S. (1982). Sequential evaluation of terminal pair reliability in a reliability network. *Microelectronics and Reliability, 22*, 151–162.
10. Kumar, S. (1985). Maximum flows subject to improvements. *Indian Journal of Management and Systems, 1*(1), 26–38.
11. Kumar, S. (2005). Information recycling mathematical methods for protean systems. *South African Journal of Industrial Engineering, 16*(2), 81–1001.
12. Talukdar, H. M., & Kumar, S. (2005). *Circuits free of specified links in a protean distribution network, discovery and innovation*, (Vol. 17, pp. 37–44). Academy Science Publishers.
13. Talukdar, H. M., Kumar, S., & Jones, B. C. (2006). Information recycling mathematical methods: An analysis of link weighted protean communication networks. *Indian Journal of Mathematics, 48*(2), 221–238.
14. Kumar, S., Munapo, E., Lesaoana, M., and Nyamugure, P. (2016). Identification and application of virtual directions in a non-directed network: A labelling method for determination of a maximum reliability route. *Communications in Dependability and Qualityy Management, CDQM 19*(1), 85–95.
15. Kim, Y. H., Case, K. E., & Ghare, P. M. (1972). A method of computing complex system reliability. *IEEE Transactions on Reliability, 21*, 215–219.
16. Kalyan, R., & Kumar, S. (1991). A study of protean systems—Some heuristics strategies for redundancy optimization. In: S. Kumar (Ed.), *Chapter 12 in Recent Developments in Mathematical Programming*, Gordan and Breach Science Publishers, ISBN 2-88124-800-4.
17. Kapur, P. K., Garg, R. B., & Kumar, S. (1999). *Contributions to Hardware and software reliability*. World Scientific Publishing.
18. Kumar, S., & Munapo, E. (2021). Interdependence of software and progress of mathematics of OR: Some illustrative cases and challenges. *American Journal of Operations Research, 11*, 110–119.
19. Dantzig, G. B. (1963). Linear programming and extensions, Princeton Universaity Press, Princeton.
20. Kumar, S. (2006). Information recycling mathematical methods: A path-way approach to geometric program, South African. *Journal of Industrial Engineering, 17*(2), 127–43. (ISSN1012-277X).
21. Hamilton, J. D. (1994). *Time series analysis*. Princeton University Press.
22. Bellman, R. (1957). *Dynamic programming*. Princeton University Press.
23. Bellman, R., & Dreyfus, S. E. (1962). *Applied dynamic programming*. Princeton University Press.
24. Dreyfus, S. E., & Law, A. (1977). *The art and theory of dynamic programming*. Academic Press.
25. Kalyan, R., & Kumar, S. (1989). Comparison of a simulation and an exact method for reliability evaluation of large network using a personal computer. *Micro-Electron and Reliability, 29*(2), 133–136.
26. Kumar, S., Arora, H., Chiera, C., & Raviganesh, G. (1999a). Path for self-healing protean network. *Bulletin of Allahabad Mathematical Society, 14*.
27. Kumar, S., Raviganesh, G., McPhee, J., & Nyati, T. M. (1999b). Terminal pair reliability in a protean network. *International Journal of Operations and Quantitative Management, 5*(2).

Chapter 4
Human Emotion Detection using Speech Recognition in Real Time

Priya Singh, Asif Khan, and Aviral Tiwari

Abstract Human Speech Processing is one of the most important areas in Digital Signal Processing and is used in Human Computer Interface, Telecommunications, Assistive Technology, Audio Production, Security, etc. to perform many operations. Human Speech Emotions Recognition (HSER) is a technique that can be utilized to detect emotions reflected in various speech patterns. The task is similar to text sentiment analysis, differing only in the data modalities. From the Machine learning perspective, it can be treated as a classification problem where input speech patterns must be grouped into some already defined emotions. Multiple factors are to be considered while working with speech data, such as speaker, language, region, and emotions, to name a few, that make the problem of HSER complex. The complexity increases multifold when the input speech patterns are ambiguous, making it challenging to apply HSER in real-world. The current paper proposes a novel Convolutional Neural Network (CNN) based HSER framework to detect emotions in human speech effectively. It is evaluated on the popular RAVDESS dataset. The results received are further compared with the MLP classifier on accuracy.

Keywords Human speech · Emotion recognition · Speech recognition · Human emotion detection · Sentiment analysis

4.1 Introduction

Understanding and accurately detecting human emotions have long been challenging tasks in the fields of psychology, cognitive science, and artificial intelligence. Emotions are complex and multi-dimensional phenomena that are expressed through various channels, including facial expressions, physiological responses, vocal intonations, and textual cues. Analysing and interpreting these cues provides valuable insights into individuals' affective states, enabling personalised and adaptive human–computer interaction systems.

P. Singh (✉) · A. Khan · A. Tiwari
Department of Software Engineering, Delhi Technological University, Delhi, India
e-mail: priya.singh.academia@gmail.com

P. K. Kapur et al. (eds.), *Data-centric Approaches to Industrial Decisions*, Asset Analytics, https://doi.org/10.1007/978-981-96-7556-2_4

Advances in computer vision, signal processing, and machine learning [10–17] have revolutionised the field of emotion detection, empowering researchers to develop sophisticated models that can automatically recognise and classify human emotions. By leveraging multimodal data sources, researchers have achieved remarkable progress in detecting emotions with high accuracy, robustness, and generalizability.

The use of speech recognition in real-time for human emotion detection can provide several benefits. Firstly, it allows for real-time analysis and interpretation of speech, which enables immediate response and feedback. Secondly, it is non-invasive and can be used remotely, making it a convenient tool for various applications. Lastly, it may enhance human–computer connection by enabling more individualised and sympathetic replies. Overall, the use of speech recognition in real-time for human emotion detection is an exciting and rapidly developing field with numerous potential applications. With continued research and development, it has the capacity to revolutionise the way we engage with generation and apprehend human feeling.

In this paper, the two models based on CNN and MLP have been studied to draw comparison between them using the benchmark dataset 'RAVDESS' by calculating the accuracy for the human emotion detection. In the next section, we have discussed the literature review in the field of emotion detection. Section 4.3 describes the methodology for the mentioned approaches. Section 4.4 discusses the results of the different models. Section 4.5 portrays limitations for the current research followed by conclusions and future scope of development in the next section.

4.2 Literature Review

In the fields of machine learning and artificial intelligence, emotion detection is a hot topic for research. Several studies and initiatives are now underway with the goal of enhancing the precision and dependability of emotion detection systems [16–21].

Some of the current research areas fall into mainly 2 categories-

1. **Machine learning algorithms for emotion detection**: To increase the precision of emotion recognition, researchers are utilising machine learning algorithms including deep neural networks, decision trees, and support vector machines.
2. **Multimodal emotion detection**: To increase the accuracy of emotion identification, researchers are looking into the use of several data sources, including physiological signals, voice cues, and facial expressions.

Following are the literatures studied regarding the human emotion detection -

Dung Nguyen [1], motivated by the deep learning techniques, uses deep learning techniques to address the challenges of understanding human emotional behaviour and enhance multimodal and unimodal emotion identification systems' performance.

Mingyong Li et al. [2], suggested a multi-objective optimisation algorithm-based approach for multimodal emotion recognition. The video imaging modality is based on an Enhanced Deep Separation Convolutional Network (DSCNN), while the voting module is based on a modified deep convolutional neural network (DCNN).

Chen Jie. [3], in order to boom the feature variations among emotional speech, carried out the mel-frequency cepstral coefficients function statistics matrix acquired from speech sign preprocessing and converted to enhance the speech emotion popularity charge.

Rajdeep Chatterjee et al. [4] provide a comprehensive approach to a total of emotional assessments based on human speech. One-dimensional CNNs have been used to analyse and classify emotions associated with human speech.

Sadil Chamishka et al. [5] used a state-of-the-art recurrent neural network (RNN)-based emotion detection model that captures the states of individual parties and context of the conversation to make categorical predictions in real-time.

Li Zheng et al. [6] proposed a new network model (CNN-RF) based on CNN combined with Random Forest. First, utilise a CNN as a feature extractor to pull spoken thought from the normalised spectrogram, and then classify thought features using a random forest classification technique.

Kun-Yi Huang et al. [7] focuses specifically on the importance of emotional awareness in real-life communication, where nonverbal sounds in speech play an important role in emotional awareness. This study suggests that current emotional intelligence may ignore nonverbal sounds such as laughter, crying, or other emotional disturbances that occur in everyday conversations. To overcome this limitation, researchers developed an emotional recognition method that takes into account verbal and nonverbal communication in real life.

Jonathan Rintala [8] focuses on addressing the limitations of traditional Speech Emotion Recognition (SER) models that heavily rely on manually engineered features and intermediate representations. These manual feature engineering approaches often require expert domain knowledge and significant resources. In contrast, in order to learn directly from the unprocessed voice input, the study investigates the possibility of end-to-end deep learning models that can automatically extract features.

Sitong Zhou, Homayoon Beigi [9] introduces a Time-Delay Neural Network architecture-based transfer learning method for speech emotion identification. The scarcity of labelled data is the main obstacle in speech-based emotion recognition. The suggested method uses transfer learning techniques and makes use of a significant amount of data from the automated speech recognition (ASR) task to overcome this problem.

Dias Issa, M. Fatih Demirci et al. [11] highlights a research work focused on speech emotion recognition, which is considered a challenging task in the field of data science. The study introduces a novel architecture that extracts various acoustic features, including mfcc (mel-frequency cepstral coefficients), chromagram, mel-scale spectrogram, Tonnetz representation, and spectral contrast features, from sound files. These features are utilised as inputs for a one-dimensional CNN for the identification of emotions.

4.3 Proposed Work

The sound input will be taken from the microphone and it will be recorded as a sound file using python's py-audio library. After saving the audio file, it is passed in the model as a numpy array and then the model predicts the emotion reflected by the user's voice.

4.3.1 Dataset Description

The work uses the RAVDESS dataset [22] for human speech emotion detection and recognition. It is published under the Creative Commons Attribution Licence.It contains Speech files which contain only speech from twenty-four different persons(16bit, 48 kHz.wav).RAVDESS dataset contains: 60 trials for each actor of the 24 actors which means it has total: $24 * 60 = 1440$ files. The 24 professional actors consist of 12 female and 12 male vocalising two syntactically identical utterances with a neutral North American tone [22]. It has the following speech emotions: joy, anger, calmness, disgust, fear, surprise. It has 2 emotional densities: strong and normal.

4.3.2 Multilayer Perceptron (MLP) Classifier

An artificial neural network for classification tasks is called an MLPClassifier. It is made up of numerous layers of perceptrons, which are synthetic neurons. Each neuron in a layer is connected to every neuron in the layer below it in a fully connected neural network. The input layer is the top layer, the output layer is the bottom layer, and the layers in between are referred to as hidden layers. The weighting of each neuron's inputs by an activation function causes the network to enter nonlinear states (Fig. 4.1).

Before extracting features, the cleaning of data is done, to remove the noises present in the speeches. This is done by using a mask and setting the threshold to 0.0005. We have used the MLP Classifier algorithm (Fig. 4.2) to train the speech emotion detection and recognition model. To do so load data, and extract features from it. Divide the dataset into a test set(30%) and training set(70%). Next, initialise the MLP classifier and train the speech emotion model. Calculate the accuracy of the model to check the performance using test data.

Stepwise procedure is discussed below -

1. Import necessary libraries such as: • Librosa • sound files • MLPClassifier neural network from scleral • os, pickle, glob • numpy • sklearn train_test_split from.model_selection • precision_score from sklearn.metrics.

Fig. 4.1 Model listening to input voice

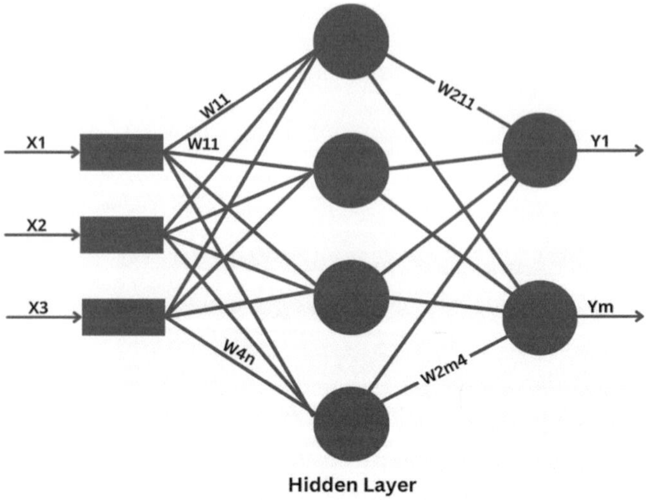

Fig.4.2 MLP classifier

2. Define the extract_feature function to fetch mfcc, chroma, and mel functionalities from speech files. The function consists of four arguments - first one is the name of the file and three boolean parameters for the three functions: **mfcc** representing the brief power spectrum of the sound, **mel** representing Mel spectrogram frequency and **chroma,** associated with 12 pitch levels.
3. Initialise a dictionary to map the integers to emotions in the RAVDESS dataset, and a list to store the desired ones (calm, happy, fearful, disgusted).
4. Use a function to load the records that take the test set's relative length as an argument. Lists x and y are empty.
5. It is time to divide the dataset into a training set and a test set. Keep the test set to 25% of everything. Then observe the shape of the training and test datasets and get the number of extracted features.
6. Initialise the MLP classifier. It is a multilayer perceptron classifier. Optimise a log loss function and fit/train a model using stochastic gradient descent or LBFGS (Figs. 4.2, 4.3).

4.3.3 Convolutional Neural Network (CNN)

The use of CNNs (Fig. 4.3) for human emotion detection in speech involves extracting features from speech signals and using these features as input to the CNN. In this study, MFCCs were used to represent audio signals, and the CNN was trained to classify emotions in speech, such as happy, sad, angry, and neutral.

The use of CNNs in voice recognition for human emotion detection has a number of benefits, including the capacity to handle enormous volumes of data and the

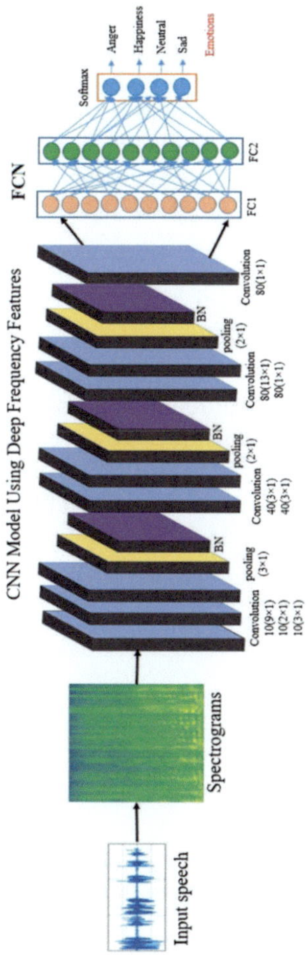

Fig. 4.3 CNN model

Table 4.1 Accuracy for different versions of MLPClassifier

S. No	alpha	batch_size	hidden_layer_size	max_iterations	warn_start	Accuracy
1	0.01	256	300	500	True	71.25
2	0.01	256	400	1500	True	73.75
3	0.01	512	300	2000	True	71.66
4	0.01	512	400	2000	True	68.33

ability to automatically learn features from the data. The variation in emotional displays between people and the difficulty in gathering substantial, diverse datasets of emotional speech are two obstacles that must be overcome.

For this paper, the dataset was divided into 70%(training) and 30%(testing) ratio. It was further tested by variating numerous parameters including-Epsilon, learning_ rate, batch_size, max_iterations, early_stopping, hidden_layer_sizes, warm_start.

4.4 Results and Discussions

This section describes the accuracy of different models of MLPClassifier and CNN, further comparing the two and discussing the possible reasons for the results. For the calculation of accuracy, various parameters are used as discussed above. The results show MLPClassifier received a higher accuracy of 73.5% in comparison to that of CNN model which has an accuracy of 68.75%.

4.4.1 MLP Classifier

Table 4.1 represents accuracy for different versions of MLPClassifier with different parameters values. Several tuning of the parameters, given in the table was performed to get the best-fit model for the best accuracy, a highest accuracy of 73.75% was achieved in the model no.2.

As seen from Fig. 4.4, the precision for the emotion 'happy' is lowest and "calm" is highest, which can be a matter of study to train the model in a better way.

4.4.2 Convolutional Neural Network (CNN)

In the case of CNN, we noticed the highest accuracy reaching 68.75 as seen in Table 4.2.

MLP provided the better results in comparison to CNN in terms of accuracy. It reached a maximum of 73.75%. We can add more hidden layers to our neural network

	precision	recall	f1-score	support
angry	0.74	0.71	0.72	48
calm	0.91	0.83	0.87	52
disgust	0.66	0.82	0.73	45
fearful	0.76	0.62	0.68	50
happy	0.64	0.71	0.67	45
accuracy			0.74	240
macro avg	0.74	0.74	0.74	240
weighted avg	0.75	0.74	0.74	240

Fig. 4.4 Confusion matrix for MLPClassifier

Table 4.2 Accuracy for different versions of CNN model

S. No	Number of network layers	Performed data cleaning	Epochs	batch_size	Accuracy
1	1	No	70	64	53.47
2	2	No	70	64	59.03
3	2	No	70	32	60.87
4	2	No	70	8	65.04
5	2	Yes	200	8	68.75

for the betterment of the accuracy. For a higher accuracy, we require a more cleaned data which contains minimum distortions and noise. Network connectivity is a must. The voice for recognition should be loud enough for accurate results. Model cannot train itself but needs voluntary help.

The comparative analysis of the MLPClassifier and CNN models revealed interesting insights into their respective performances. Based on the experimental results, it is evident that the MLPClassifier achieved a higher accuracy rate of 73.5% as compared to the CNN model, which achieved an accuracy of 68.75%. These findings suggest that for the specific task and dataset used in this study, the MLPClassifier outperformed the CNN model in terms of predictive accuracy. When deciding between these two models, it is crucial to take into account other aspects including model complexity, interpretability, and computational efficiency. To get a better grasp of these models' respective strengths and shortcomings, additional study might examine the applicability of these models on various datasets and tasks.

4.5 Limitations

It is a challenging task, and several studies and advancements have been made in the field of human speech recognition over the years. Despite significant industry improvements, reliable and efficient speech recognition systems still encounter a number of challenges and limitations. Some of these challenges and limitations include:

1. Variability in speech: The variation in speech among individuals and languages is one of the main issues with voice recognition, which might have an impact on the systems' accuracy.
2. Background noise: Speech recognition systems can also be affected by background noise, which can obstruct speech recognition.
3. Ambiguity in speech: Another challenge is the ambiguity in speech, where words can have multiple meanings and be used in different contexts.
4. Notwithstanding these demanding situations, there was sizable development within the area of speech recognition, and researchers and developers continue to paint on developing new methods and strategies to triumph over these boundaries and achieve reliable and strong speech recognition structures.

4.6 Conclusions and Scope of Future Development

In conclusion, this paper presents an investigation into human voice emotion detection to comprehend and decipher the emotional information communicated through speech signals. We have made progress towards reliably classifying emotions in speech by analysing a variety of acoustic variables and applying machine learning algorithms. The human speech emotion detection in real-time has a lot of industrial applications. In this study, the CNN and MLP classification techniques have been compared. Our study demonstrates that accuracy will always depend upon the amount of noise in the speech. Noise cancellation techniques can be used in future to avoid recording of noises in the voice. The proposed study shows that MLP classifier performs better in real-time speech emotion detection having accuracy of 73.5% which can be further increased by reducing noise.

Our study does have certain drawbacks, such as the tiny dataset we employed, which resulted in the restricted variety of speakers' emotions or voices, as well as the noise that was present in the different samples. By overcoming these obstacles, voice emotion recognition algorithms may become even more accurate and reliable.

In the future, the development of more sophisticated and effective deep learning algorithms is expected to continue to drive progress in the field of human speech emotion detection. Additionally, the integration of speech emotion detection with other modalities such as facial expression analysis and body gesture analysis promises to provide a fuller understanding of human emotions. Furthermore, the development of speech emotion detection in multiple languages and across different cultures is expected to expand the impact and applicability of this technology.

References

1. Dung, N. (2020). Multimodal emotion recognition using deep learning. *Journal of Applied Science And Technology Trends.*
2. Mingyong, L., Xue, Q., Shuang, P. (2021). *Multimodal emotion recognition model based on a deep neural network with multiobjective optimization.*
3. Chen, J. et al. (2021). *Speech emotion recognition based on convolutional neural network.* https://doi.org/10.1109/NetCIT54147.2021.00028
4. Rajdeep, C. et al. (2021). *Real time speech emotion analysis for smart home assistants.* https://doi.org/10.1109/TCE.2021.3056421
5. Sadil, C. et al. (2022). *A voice-based real-time emotion detection technique using recurrent neural network empowered feature modelling.* https://doi.org/10.1007/s11042-022-13363-4
6. Li, Z., Qiao, L. (2018). Speech emotion recognition based on convolution neural network combined with random forest. In *2018 Chinese Control And Decision Conference (CCDC)* https://doi.org/10.1109/CCDC.2018.8407844
7. Kun-Yi, H. et al. (2019). Speech emotion recognition using deep neural network considering verbal and nonverbal speech sounds. In *ICASSP 2019—2019 IEEE International Conference on Acoustics, Speech and Signal Processing (ICASSP)* https://doi.org/10.1109/ICASSP.2019.8682283
8. Jonathan, R. (2020). *Speech emotion recognition from raw audio using deep learning.*
9. Sitong, Z., Homayoon, B. (2020). *A transfer learning method for speech emotion recognition from automatic speech recognition.*
10. Malhotra, R., & Singh, P. (2023). Recent advances in deep learning models: a systematic literature review. *Multimedia Tools and Application.* https://doi.org/10.1007/s11042-023-15295-z
11. Dias Issa, M., Fatih, D. et al. (2020). *Speech emotion recognition with deep convolutional neural networks.* https://doi.org/10.1016/j.bspc.2020.101894
12. Fadlilah, A. F., Djamal, E. C. (2019). Speaker and speech recognition using hierarchy support vector machine and backpropagation. In *6th International Conference on Electrical Engineering, Computer Science and Informatics (EECSI)*, (pp. 404–409). IEEE.
13. Al Moubayed, S. et al. (2022) A survey of affective computing for stress detection from speech. *Journal of Ambient Intelligence and Humanized Computing*
14. Al Jumaily, A. M. et al. (2020) Emotion recognition using voice analysis and machine learning: A systematic literature review. *Journal of Healthcare Engineering*
15. Sharma, P. et al. (2021) Analysis of human emotion recognition from speech using deep neural networks. *Journal of Ambient Intelligence and Humanized Computing*
16. Jerry, J., Aparna, K., Shreya, R., Rama, S. (2020). *Speech emotion recognition using neural network and MLP classifier.* https://doi.org/10.1155/2021/6971100.
17. Ittichai Charoen, C. (2012). *Speech recognition using MFCC.* https://doi.org/10.13140/RG.2.1.2598.3208
18. Al-Talabani, A., Sellahewa, H., & Jassim, S. A. (2015). Emotion recognition from speech: Tools and challenges. *Mobile Multimedia/Image Processing, Security, and Applications, 9497*, 94970N. https://doi.org/10.1117/12.21916232021
19. Sezgin, M. C., Gunsel, B., & Kurt, G. K. (2012). Perceptual audio features for emotion detection. *EURASIP Journal on Audio, Speech, and Music Processing, 2012*(1), 16. https://doi.org/10.1186/1687-4722-2012-16
20. Zadeh, S. S., et al. (2018) Affective computing in the era of deep learning: A review and future directions. *IEEE Journal of Biomedical and Health Informatics*
21. Zhong, H. et al. (2020) *Voice emotion recognition: A survey of recent advances.*
22. Livingstone, S. R. (2019). RAVDESS Emotional speech audio. Kaggle.com.https://www.kaggle.com/datasets/uwrfkaggler/ravdess-emotional-speech-audio

Chapter 5
Automatic Sorting of Apparels for Efficient Inventory Management: A Deep Learning Approach

Anchal, Tripti Mahara, V. L. Helen Josephine, and Vandana Srivastava

Abstract Sorting of garments is a time consuming and monotonous task. Manual sorting method have limitations in terms of low throughput rate, human error and less productivity. To overcome this, Convolutional Neural Network can be used to automate sorting with better efficiency. Thus, the study aims to develop a sorting mechanism for a garment warehouse. In this study, the performance of three popular CNN architectures: InceptionV3, MobileNet, and ResNet50 are compared for efficient sorter. The required data for the study is sourced from various online clothing websites like "Myntra", "Flipkart", "Ajio", etc. Various experiments were performed to evaluate the performance of the three models. As per the results, ResNet50 model has better accuracy as compared to Inception V3 and MobileNet. Hence, it should be used to develop the automatic sorter that will reduce the manual work and will result in better efficiency.

Keywords Deep learning · Automatic sorting · Convolutional neural network (CNN) · Warehouse · ResNet50 · MobileNet · Inception V3

5.1 Introduction

Object sorting has a wide range of uses in a variety of scenarios in the fashion industry like for visual inspection of cloth for quality approval [1], fashion image retrieval to support visual search [2], customer wish list image based fashion product

Anchal · V. L. Helen Josephine · V. Srivastava
Christ (Deemed to Be University), Bangalore, Karnataka, India
e-mail: anchal.aneja305@gmail.com

V. L. Helen Josephine
e-mail: helen.josephine@christuniversity.in

T. Mahara (✉)
Prin L.N.Welingkar Institute of Management Development and Research, Bengaluru, Karnataka, India
e-mail: tripti.mahara@welingkar.org

P. K. Kapur et al. (eds.), *Data-centric Approaches to Industrial Decisions*, Asset Analytics, https://doi.org/10.1007/978-981-96-7556-2_5

recommendation system [3], categories clothes based on presence and position of logos in the garment production house [4], etc. Traditionally in the fashion industry, garments are the objects that are needed to be sorted. Traditional approach involves manual sorting of garments/ apparel. In manual sorting of garments, warehouse relies on human resources to visually identify and the garments. These can involve sorting it after they are ready or on the conveyor belt for quality inspection. It becomes difficult for an individual to categorize the apparels efficiently, as attributes for classification needs to be addressed every time. This manual approach is adopted as it is cost effective but for high-throughput procedures, manual sorting is not an option. It has also been noted that the constant turnover of sorting line workers has affected the consistency of the separation process. Also, this being a tedious and mundane task, the efficiency reduces with the number of hours of working. Thus, it is important that automatic sorting is deployed for reliable and efficient sorting of apparels.

An automated sorting system is a method of automatically separating mixed items using a sorter system based on image recognition technology. This can be a solution for the garment houses to sort the clothes automatically before packing or for quality check purpose. Deep Learning, one of the subclasses of machine learning is one of the popular and fast-growing technology that uses Artificial Neural Network to aid in reliable detection and classification of clothing type by classifying the images into relevant categories. There are various deep learning approaches that are best suited to the challenge at hand, particularly Convolutional Neural Networks (CNN).

Apparel for women dominate the market in terms of variety. For instance, there are at least fifteen different types of tops that a woman can wear. This includes tunic tops, ruffled tops, off-shoulder tops, tank tops, crop tops, cross-over tops, blouses, belted tops, kaftan tops, shirt, backless tops, peasant tops, and so on. Sorting of these tops is a real tedious work as there are lots of categories in which sorting can be done. Along with this sorting can also involve considering attributes like colors, patterns, sizes, etc. Thus, it is difficult to manually sort these many categories. Hence there is a need to have an automatics sorter that is far efficient and less time consuming.

Thus, this research focuses on identifying the women-top-wear under four categories, namely—"crop top", "off-shoulder top", "shirts" and "backless tops" based on the input images provided. To achieve this Deep Learning Networks, specifically CNN network is utilized. Performance of three deep networks namely: InceptionV3, MobileNet and ResNet50 are compared to build the classifier that can be used for efficient sorting of women top.

5.2 Literature Survey

Deep Learning has been used significantly for image classification in various domains [5–8]. CNN is the most widely used algorithm for image classification. It is a multi-layer neural network with convolutional layers. 'Convolution' is the mathematical function that is used to extract the features. Convolutional layer, Pooling layer and the Fully connected layers are the building blocks of a CNN.

Convolutional layer is the first layer of the network and is used to extract features from the image using the image matrix and a kernel/ filter. The image matrix is combined with the filter matrix, resulting in the "Feature Map." The pooling layer uses either Max Pooling or Average Pooling for dimensionality reduction. This is achieved by reducing the number of parameters. The matrix is flattened and converted into a vector that is the input to the fully connected layer. The fully connected layers are usually the end layers in a CNN that aids in classification.

With respect to apparel classification, limited work has been reported in the literature. For instance, Convolutional Neural Network was used to classify images for apparels into 5 classes namely: t-shirts, trousers, sarees, women's kurtas, and footwear [9]. A successful model has been built for traditional clothing of Bangladesh that retrieves the relevant photos similar to the query image using DenseNet121, ResNet50, SEResNet50, and SEResNeXt50 models [10]. Similarly, a recommender system was built on the basis of body and the style attributes [11] to guide the user to select the right apparel. A VGG19 architecture was used to build a model that automatically classified clothing features, which is important for organizing marketing campaigns, avoiding duplicates, and categorizing apparel products for e-commerce purposes, among other things [12]. In [5], authors classified the style on the apparel into 5 categories: plaid, block, vertical, plain and horizontal.

5.3 Methodology

5.3.1 Data Collection

The required data is sourced from online shopping websites like Flipkart, Amazon, Myntra, and Ajio. The objective is to identify the women-top-wear under four categories, namely—"crop top", "off-shoulder top", "shirts", and "backless tops". Backless top are the tops that has design on the back. Off-shoulder top are the ones in which shoulder area is not covered and neckline is usually elasticized and gathered and may be decorated with frills/ruffles. Crop top is a short upper-body garment that exposes the waist, navel, or abdomen whereas Shirt style top is a garment the upper body with a collar and sleeves, and with buttons down the front.

Few sample images of Off-shoulder tops, Backless tops, Shirt Style, and Crop tops are presented in Figs. 5.1, 5.2, 5.3, and 5.4 respectively. For each clothing category, 300 images have been collected. 4 -5 angles and viewpoints of the same image is collected. Also, almost all the colors under each category have been selected.

Fig. 5.1 Off-shoulder tops

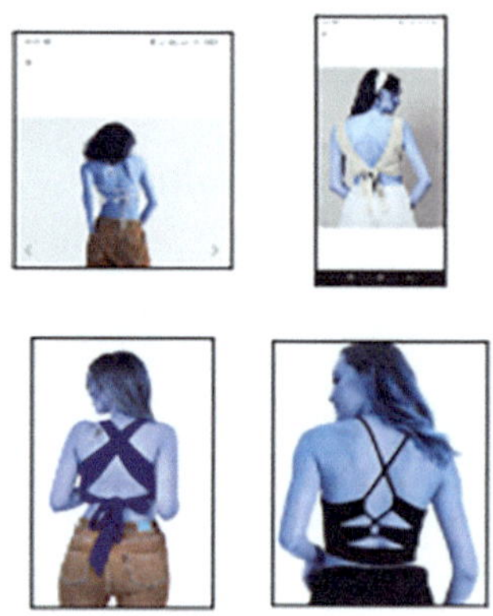

Fig. 5.2 Backless tops

5.3.2 Data Preprocessing

The data gathered is undergone pre-processed before applying the CNN model architecture. The preprocessing involves image resizing, data augmentation and normalization.

Image resizing involves cropping of an image. To "crop" an image implies removing or adjusting the outside edges of an image (typically a photo) so that it improves framing or composition. This is achieved by changing the size or aspect ratio. In this research, the images have been standardized to 224X244 to ensure that each input parameter has the same data distribution. This helps in faster convergence of the network during the training process.

Fig. 5.3 Shirt style

Fig. 5.4 Crop tops

Data augmentation is a technique that alters original picture by applying various transformations to it. It results in different copies of the same image. This can be achieved by applying rotation, flipping, and shifting to the images.

A high number of labelled samples are typically utilized to train the networks, reducing the danger of over-fitting and enhancing generalization. The time investment required to get a lot of labelled data is too expensive. Hence, the data augmentation algorithm produces latest samples by direct interpolation. In this research, Affine based data augmentation technique is applied to the images to enhance the volume of the dataset [13]. After augmentation, 500 images in each category are generated resulting in total 2000 images.

By dividing the picture pixel values by 255, the image is normalized. The main goal of normalization is to speed up computation by limiting the values in the range between 0 and 1. Normalization is done for better convergence. In case the data is not normalized, the training process becomes very slow and the weights might never converge.

5.3.3 Model Building

5.3.3.1 Convolution Neural Network

Convolutional Neural Networks (CNNs) are fascinating and powerful. They belong to a group of deep learning models Primarily, they work with grid data, like images. CNNs have become super important in computer vision because they can learn features from images all on their own. This means they do a great job with tasks like image classification, object detection, and segmentation. At the core of a Convolutional Neural Network (CNN) are its convolutional layers. These layers use filters, sometimes called kernels. Each filter glides over the input image. While doing this, it performs a dot product at every spot. The result is a feature map that shows where certain features appear in the image. These filters learn during training and detect things like edges and textures.

CNNs also have pooling layers. These layers help shrink the feature maps. This process often uses something called max pooling. It makes the whole thing a bit lighter to handle and helps stop overfitting by making sure small changes in the input don't throw the model off too much. Next, it contains fully connected layers. These dense layers typically sit towards the end of the network. After all that feature extraction from convolutional & pooling layers, these fully connected layers gather those high-level features together to classify them in the end. Activation functions are essential and must not be overlooked. ReLU, which stands for Rectified Linear Unit, is used in various places in the network. This function adds non-linearity, helping the model learn more complex patterns. There are also techniques like dropout that randomly switch off some neurons during training to prevent overfitting.

CNNs come in various architectures, from simpler ones like LeNet to more advanced models such as VGG, ResNet, and Inception. Each architecture is designed for specific tasks and shows outstanding performance across numerous computer vision applications. These networks have fundamentally changed how machines process and understand visual information, making them crucial for modern AI applications.

5.3.3.2 Inception V3

InceptionV3 is a really advanced convolutional neural network (CNN) model. It has made big strides in computer vision. Researchers at Google created it. This model builds on what InceptionV1 and InceptionV2 achieved, providing better accuracy & efficiency when recognizing images. One major breakthrough in InceptionV3 is the use of factorized convolutions. Traditional convolutions can be costly in terms of computation, especially when the filter sizes are large. The clever design of InceptionV3 tackles this by breaking down larger convolutions into smaller ones. For instance, a 5×5 convolution may be split into two 3×3 convolutions. This method cuts down on computational expenses & helps deepen the network without significantly adding to its parameters.

Another interesting aspect of InceptionV3 is the inclusion of an "auxiliary classifier". This extra classifier comes into play during training and sits in between layers. It helps with the vanishing gradient problem and offers regularization benefits. By having a second objective while training, the network becomes sturdier against overfitting. InceptionV3 uses techniques like batch normalization & RMSprop optimization. Batch normalization works to stabilize and speed up training by standardizing each layer's input. RMSprop, a type of gradient descent, adjusts learning rates for different parameters which enhances performance and convergence.

The structure of InceptionV3 is modular. It consists of several "Inception modules" that enable parallel processing of data through various filter sizes. Because of this design, the model captures a range of spatial features and boosts its ability to generalize across various image types. InceptionV3 has found a wide range of uses—everything from image classification to object detection, including transfer learning. It stands out for delivering high accuracy with relatively low computational needs, making it a preferred option for researchers and practical applications alike in deep learning. InceptionV3 marks a notable advancement in CNN architecture by blending innovative techniques to enhance performance, efficiency, & scalability in image processing tasks.

5.3.3.3 ResNet50

ResNet50 stands out as a crucial convolutional neural network (CNN architecture, crafted by talented researchers at Microsoft Research. This is part of the ResNet family which brought forward a fresh way to train very deep neural networks. As hinted by its name, ResNet50 is a 50-layer deep network. It shines in image recognition tasks, famously achieving impressive accuracy while addressing some typical challenges faced when training deep networks.

The main breakthrough of ResNet50 is its application of residual learning. Traditional deep networks frequently encounter the vanishing gradient problem. This happens when gradients become extremely small, making training less effective as the network grows deeper. The ResNet design tackles this concern by integrating residual blocks. These blocks feature shortcut connections that enable inputs to skip one or more layers before they are added directly to the output. Because of this setup, gradients can move through the network more freely, thus aiding in the training of much deeper models.

Within a typical residual block of ResNet50, you will find three layers of convolutions. Each block executes a series of convolutions, employs batch normalization, and applies ReLU activations—then it adds back in the original input through that shortcut connection. This arrangement allows the network to learn these residual functions better. The focus shifts towards honing the mappings from inputs to outputs rather than learning them directly. The structure of ResNet50 contains many of these residual blocks stacked together to create a 50-layer network. It kicks off with an initial convolutional layer and a max-pooling layer before proceeding with several residual blocks. At the end is the global average pooling layer. Fully connected layer follows it along with softmax activation to handle classification tasks.

One big benefit of ResNet50 is its strong ability to generalize across various datasets. Its versatility comes into play for multiple computer vision tasks that includes object detection, image classification, and segmentation. Because of its resilience and excellent performance, it has become a popular choice for transfer learning too. Pre-trained models based on ResNet50 can be fine-tuned for specific applications, offering added flexibility. ResNet50 has significantly influenced deep learning. It proves that very deep networks can indeed be trained effectively while achieving remarkable performance—thanks largely to its clever use of residual connections.

5.3.3.4 Mobile Net

MobileNet is a series of convolutional neural network (CNN) architectures crafted specifically for efficient performance on mobile & edge devices. These devices often have limited computational resources & memory. Created by Google, MobileNet models focus on cutting down the size of the image and the complexity of computations while still achieving high accuracy in important tasks like image classification and object detection. The primary difference in MobileNet is because of the use of depth wise separable convolutions. This technique divides a standard convolution into two parts: depth wise convolutions & pointwise convolutions. In depth wise convolutions, a single filter is applied to each input channel. On the other hand, pointwise convolutions use a 1×1 convolution to combine the outputs from depth wise convolutions. By separating these processes, MobileNet greatly reduces the number of parameters and the amount of computation needed. This makes it much lighter and faster than traditional convolutional networks.

MobileNet has a streamlined design that features a series of depth wise separable convolutional layers, along with batch normalization & ReLU activation functions. It comes in various versions like MobileNetV1, V2, and V3. Each new version brings improvements over the last. For instance, MobileNetV2 adds linear bottleneck layers and introduces a new activation function called ReLU6, which boosts both performance & efficiency. Meanwhile, MobileNetV3 takes it further by blending neural architecture search with hardware-aware design to improve speed and accuracy.

These MobileNet models are especially ideal for use in settings where resources are limited—like mobile devices—allowing them to handle tasks such as image classification, object detection, and segmentation with minimal delays. Their effectiveness has made them quite popular for applications needing real-time processing on mobile and edge devices.

5.3.3.5 Model Building

Three CNN architectures (InceptionV3, ResNet50, and MobileNet) [14–17] are used to develop the model for categorizing the images into their respective categories: Crop Top, Off-shoulder Top, Shirt Style Top, or Backless Top based on the input image provided. Convolutional Neural Network, is a type of deep feedforward artificial neural network. Deep learning models are pre-trained on large-scale image data, such as the ImageNet dataset from the ImageNet Large Scale Visual Recognition Challenge (ILSVRC), in order to incorporate deeper CNN architectures with better outcomes. After the pre-training, the fine-tuning of the last fully connected layer with our own fashion image dataset is done. For the training network, we use InceptionV3, ResNet50, and MobileNet module that includes Convolutional layers, pooling layers and fully connected layers. Convolutional layer is used to extract different features from the input image and the output of this layer is termed as Feature Map. A feature map provides information about the image in terms of corners and edges and acts as the input to other layers. Pooling layer follows the convolutional layer. It is used to reduce the size of the convolved feature map. Techniques like average pooling, max pooling are used to achieve this. A fully connected layer is used to predict the class of the image on basis of previous features.

After passing through the pre-trained network, the training image is sorted into 4 fine-grained categories. Output activations for each level predictions depend on the problem at hands, i.e., as the category are multi-class problems, softmax activation function is used. The network is trained by minimizing a weighted cross-entropy loss for each level in order to estimate the parameters that originate the most correct predictions for the category. The loss functions are optimized via backpropagation and batched-based Adam optimizer using different learning rates.

After data augmentation, total 500 images were generated in each category, resulting in 2000 images. For the purposes of creating and validating models, 70–30 split of data set was done with 70% training data and 30% test data.

The Inception V3 model consists of 42 layers. It consists 1×1, 3×3, and 5×5 convolutional layers. Feature extraction in the convolutional layers is done using 5X5 filters. It also has 3×3 max-pooling layers that are used to reduce spatial sizes and prevent overfitting Rectified Linear Unit (ReLU) activation function is used after each convolutional layer to increase non-linearity.

MobileNet has 13 convolutional layers, one average pooling layer, one fully connected layer and a softmax layer. For the convolutional layers, there are thirteen, 3 × 3 depth wise convolutional layers, one, 3 × 3 layer and thirteen 1 × 1 convolutional layer. Except for the first layer, MobileNet is constructed using depth wise separable convolutions. A fully convolutional layer makes up the first layer. The classifier uses Fully Connection layer FC-128 (new layer) and a Softmax classifier.

ResNet50 is a ResNet model version that incorporates 50 layers. There are 48 Convolution layers, one MaxPool layer, and one Average Pool layer. There is one 7 × 7 kernel convolutional layer with 64 kernels and a stride of size 2 followed by Max-pooling layer with a stride size of two. The following convolutions are as follows.

9, 3 × 3 64 kernel convolutional.

1, 1 × 1, 64 kernel convolutional.

1, 1 × 1, 256 kernel convolutional.

These three layers are repeated 3 times resulting in 9 layers. The other convolutions include.

1, 1 × 1, 128 kernels.

1, 3 × 3, 128 kernels.

1, 1 × 1, 512 kernels.

These three layers are repeated 4 times resulting in 12 additional layers.

The next 18 layers comprise of following 3 layers repeated 6 times.

1, 1 × 1, 256 cores.

1, 3 × 3. 256 kernels.

1, 1 × 1, 1024 kernels.

The final 9 layers are created using the following 3 layers repeated 3 times.

1, 1 × 1, 512 kernels.

1, 3 × 3, 512 kernels.

1, 1 × 1, 2048 kernel.

The average pooling layer consists of 1000 fully connected nodes. One output layer is the final outcome of the softmax function.

To estimate the parameters that lead to the best accurate predictions for the category, the networks are trained by minimizing a weighted cross-entropy loss for each level. Using various learning rates, the loss functions are optimized using backpropagation and the batched-based Adam optimizer. Hyper parameter tuning is performed for searching an ideal model architecture via changing number of hidden layers, learning rate, number of epochs and addition or removal of dropout layer.

5.4 Results and Discussions

Learning rate is the first hyperparameter that is evaluated for optimal performance. All the three models were trained using the three learning rates: 1E-03, 1E-05, and 1E-06. Most optimum results were obtained for 1E-05, and it was selected for further experimentation. Figures 5.5, 5.6, 5.7, 5.8, 5.9 and 5.10 represents and compares the results of all the three models with and without including the drop out layer.

The Table 5.1 summarizes these results and as observed the performance of all the three models increase after introducing the dropout layer. This table presents a comparative analysis of three Convolutional Neural Network (CNN) architectures: Inception V3, MobileNet, and ResNet50, evaluated both with and without dropout layers. Initially, without dropout, Inception V3 demonstrated a higher training loss of 1.6744 and validation loss of 2.6346, with training accuracy at 0.7226 and validation accuracy at 0.59, indicating potential overfitting. MobileNet performed better, with a training loss of 1.5044 and a validation loss of 2.4646, achieving slightly improved accuracies of 0.7326 and 0.64 for training and validation, respectively. ResNet50 displayed the lowest losses in this setup, with training and validation losses of 1.4744 and 2.4346, and equal train and validation accuracies of 0.7426 and 0.64. The inclusion of dropout layers significantly enhanced model performance across the board. Inception V3's training and validation losses dropped to 0.605 and 1.098, respectively, with corresponding accuracy improvements to 0.79 for training and 0.77 for validation, showcasing reduced overfitting. MobileNet, with dropout, achieved training and validation losses of 0.8045 and 0.769, and accuracies increased to 0.8 for training and 0.81 for validation, indicating strong generalization capabilities. ResNet50 also benefited from dropout, showing a training loss of 0.825 and validation loss of 0.789, with both accuracies at 0.79 and 0.80. Overall, the dropout layer proved effective in enhancing the generalization of these models, with MobileNet achieving the highest validation accuracy, suggesting its robustness in mitigating overfitting compared to the other architectures.

The next set of experimentation includes introducing a greater number of hidden layers along with keeping the dropout layer. Table 5.2 summarizes the results. This table provides a detailed comparison of three CNN architectures—Inception V3, MobileNet, and ResNet50—evaluated with varying numbers of hidden layers: one, two, and three. When using one hidden layer, Inception V3 shows a training loss of 0.915 and a validation loss of 1.098, with corresponding training and validation accuracies of 0.79 and 0.77, respectively. MobileNet showed better performance in this configuration, with lower training and validation losses of 0.805 and 1.058, and higher accuracies of 0.82 for training and 0.79 for validation. ResNet50 outperformed both architectures, achieving the lowest losses of 0.625 for training and 0.889 for validation, and the highest accuracies, 0.85 for training and 0.81 for validation.

As the number of hidden layers increased to two, there was a noticeable improvement in performance across all architectures. Inception V3's training and validation losses decreased significantly to 0.3654 and 0.498, respectively, with accuracies rising to 0.9238 for training and 0.9011 for validation. MobileNet continued to show improvement, with its training and validation losses dropping to 0.2254 and 0.368, and its accuracies increasing to 0.944 for training and 0.891 for validation. ResNet50 maintained its lead, achieving training and validation losses of 0.2254 and 0.228, with accuracies of 0.9518 for training and 0.91 for validation.

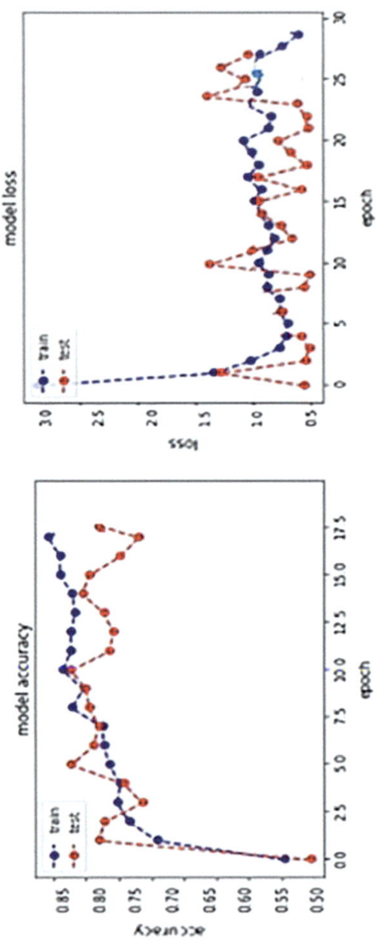

Fig. 5.5 Inception V3 model accuracy and loss without dropout

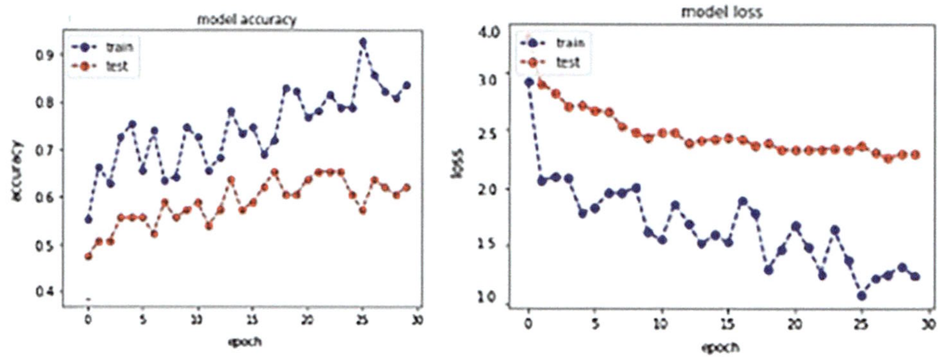

Fig. 5.6 Inception V3 model accuracy and loss with dropout

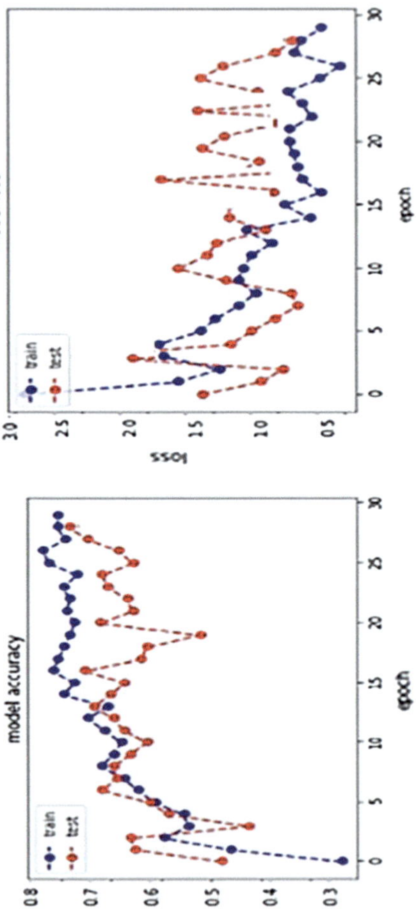

Fig. 5.7 MobileNet model accuracy and loss without dropout

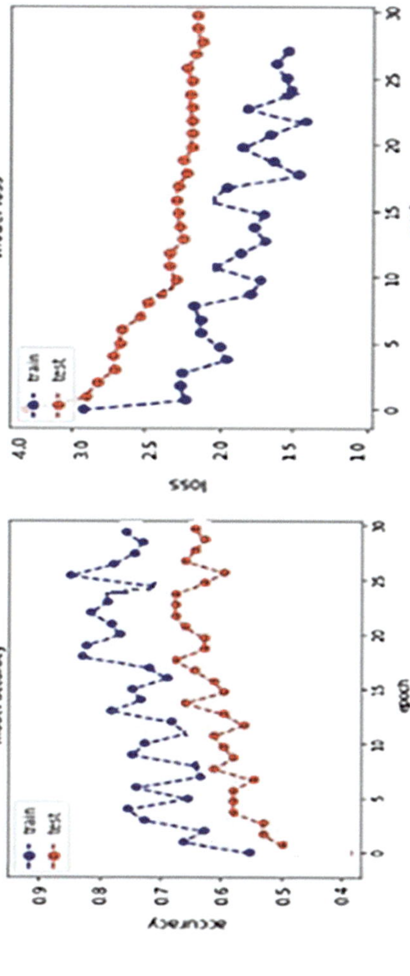

Fig. 5.8 MobileNet model accuracy and loss with dropout

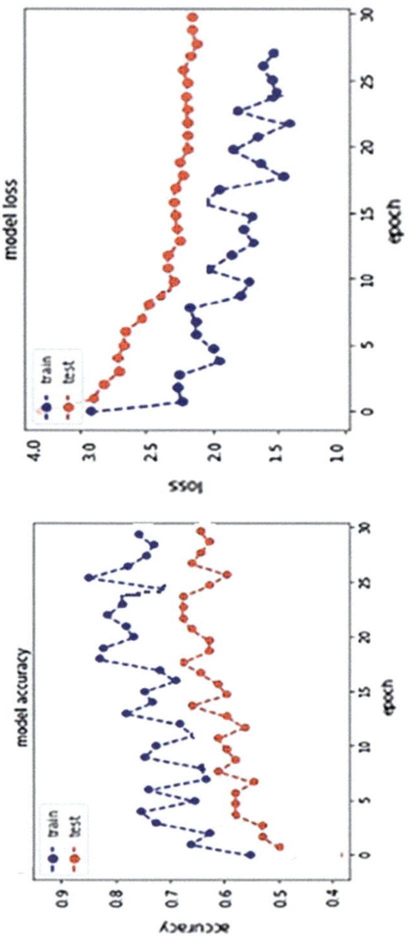

Fig. 5.9 ResNet 50 model accuracy and loss without dropout

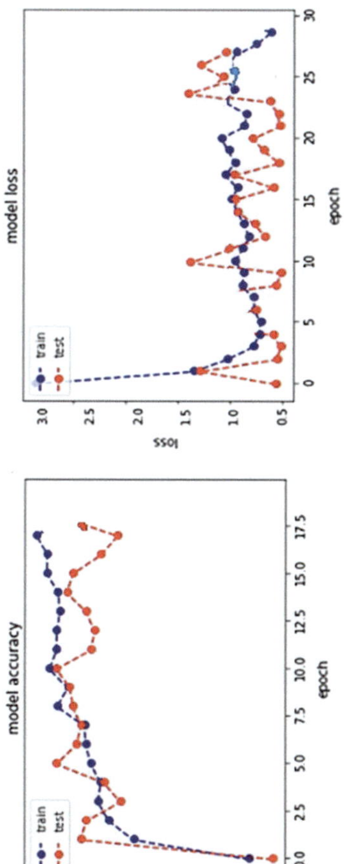

Fig. 5.10 ResNet 50 model accuracy and loss with dropout

Table 5.1 Comparing models with and without dropout layers

CNN architecture	Without dropout layer				With dropout layer			
	Train loss	Val loss	Train accuracy	Val accuracy	Train loss	Val loss	Train accuracy	Val accuracy
Inception V3	1.6744	2.6346	0.7226	0.59	0.605	1.098	0.79	0.77
MobileNet	1.5044	2.4646	0.7326	0.64	0.8045	0.769	0.8	0.81
ResNet50	1.4744	2.4346	0.7426	0.64	0.825	0.789	0.79	0.80

With three hidden layers, the trend of improved performance continued. Inception V3 achieved training and validation losses of 0.7831 and 0.9348, with training accuracy reaching 0.875 but validation accuracy decreasing slightly to 0.7067, suggesting potential overfitting. MobileNet showed a slight increase in losses to 0.465 for training and 0.335 for validation, but maintained strong accuracies of 0.905 for training and 0.86 for validation. ResNet50 exhibited the best performance with three hidden layers, with minimal training and validation losses of 0.3531 and 0.3348, and high accuracies of 0.905 for training and 0.87 for validation, indicating robust generalization. Overall, adding hidden layers improved the performance of all models, with ResNet50 consistently delivering superior results.

As observed, when the number of hidden layers increases from one to two, there is a drastic improvement observed in the performance of all the models. But, when more hidden layers are added, again the model performance decreases. Along with it, the gap in the training accuracy and the test accuracy also starts increasing which again leads to the issue of overfitting of the model, since adding more layers will increase the number of neurons, which increases the complexity of the model. So finally, the model having 2 hidden layers with dropout and learning rate of 1E-05 for each type is further evaluated to compare the performance of each type of the model.

Figures 5.11, 5.12, and 5.13 depicts the model accuracy and loss for Inception V3, MobileNet, and RestNet 50 respectively. The Table 5.3 summarizes the results.

The table presents a comparative analysis of three Convolutional Neural Network (CNN) architectures: Inception V3, MobileNet, and ResNet50. Their performance is evaluated on both training and validation datasets using loss and accuracy metrics. ResNet50 demonstrates superior performance with the lowest training loss of 0.2254 and validation loss of 0.228. Additionally, it achieves the highest training accuracy of 0.9518 and validation accuracy of 0.91, indicating strong generalization capabilities. MobileNet, while exhibiting lower training loss of 0.2254 compared to Inception V3, shows slightly lower training accuracy of 0.944 and validation accuracy of 0.891. This trade-off is likely due to its architecture prioritizing computational efficiency. Inception V3 offers a balance between the two, with training loss of 0.3654, validation loss of 0.498, training accuracy of 0.9238, and validation accuracy of 0.9011. The table shows that ResNet50 outperforms the other models in both loss and accuracy, making it the most effective for image classification among the three architectures tested. Specifically, ResNet50 achieves the highest accuracy for sorting apparel

Table 5.2 Comparison of number of hidden layers for model performance

CNN architecture	1 hidden layers				2 hidden layers				3 hidden layers			
	Train loss	Val loss	Train accuracy	Val accuracy	Train loss	Val loss	Train accuracy	Val accuracy	Train loss	Val loss	Train accuracy	Val accuracy
Inception V3	0.915	1.098	0.79	0.77	0.6354	0.498	0.9238	0.9011	0.7831	0.9348	0.875	0.7067
MobileNet	0.805	1.058	0.82	0.79	0.2254	0.368	0.944	0.891	0.465	0.335	0.905	0.86
ResNet50	0.635	0.889	0.85	0.81	0.2254	0.228	0.9518	0.91	0.3531	0.3348	0.905	0.81

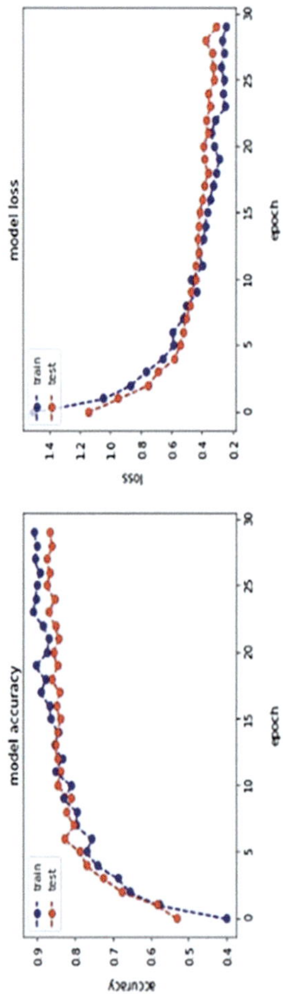

Fig. 5.11 Inception V3 model accuracy and loss

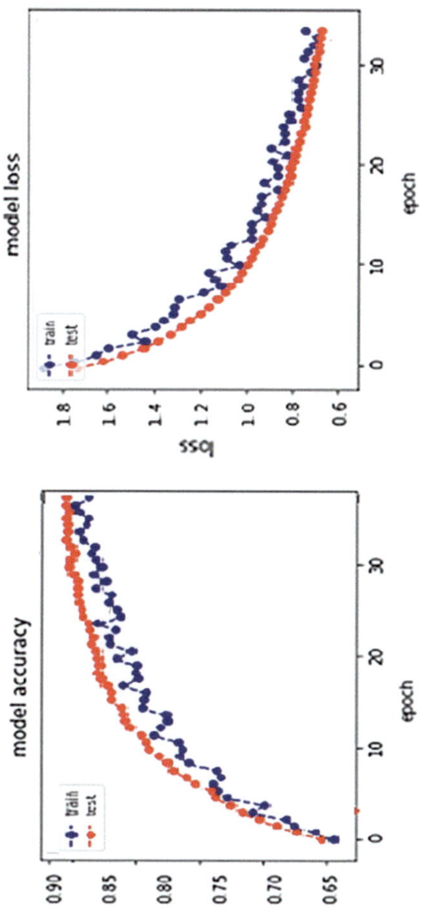

Fig. 5.12 MobileNet model accuracy and loss

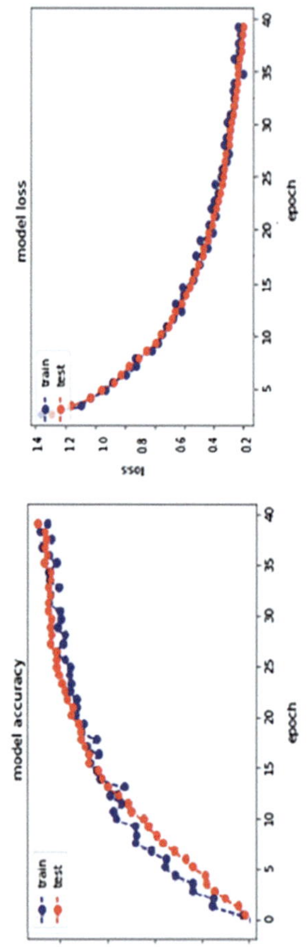

Fig. 5.13 RestNet 50 model accuracy and loss

Table 5.3 Evaluation of models

CNN architecture	Train loss	Val loss	Train acc	Val acc
Inception V3	0.3654	0.498	0.9238	0.9011
MobileNet	0.2254	0.368	0.944	0.891
ResNet50	0.2254	0.228	0.9518	0.91

into four categories, surpassing Inception V3 and MobileNet. Therefore, ResNet50 is the recommended choice for applications requiring high accuracy and efficient sorting of clothing items, depending on the balance needed between accuracy and computational efficiency.

5.5 Conclusion

In this research, the classification of women's top-wear fashion images using Convolutional Neural Network (CNN) architectures Inception V3, MobileNet, and ResNet50 was analytically evaluated. The results show that ResNet 50 outperforms the other models and can be used to develop an automatic sorter system to sort women-top-wear before packing in garment house to overcome the limitations of manual sorting.

Future work will focus on analyzing more models and also consider other categories possibly that are available in women-top-wears and other apparel types. This will result in a sorter system that will be able to sort any type of the garment in the relevant category.

References

1. Liu, L., Zhang, J., Fu, X., Liu, L., & Huang, Q. (2019). Unsupervised segmentation and elm for fabric defect image classification. *Multimedia Tools and Applications*. https://doi.org/10.1007/s11042-018-6786-7
2. Park, S., Shin, M., Ham, S., Choe, S., & Kang, Y. (2019). *Study on fashion image retrieval methods for efficient fashion visual search*.
3. 2019 IEEE/CVF Conference on Computer Vision and Pattern Recognition Workshops (CVPRW).
4. Tuinhof, H., Pirker, C., & Haltmeier, M. (2018). Image *based fashion product recommendation with deep learning*. Retrieved from http://arxiv.org/abs/1805.08694
5. Donati, L., Iotti, E., Mordonini, G., & Prati, A. (2019). Fashion product classification through deep learning and computer vision. *Applied sciences (Basel, Switzerland), 9*(7), 1385. https://doi.org/10.3390/app9071385
6. Patle, N., Dhake, S., & Ausekar, D. (2018). Automatic object sorting using deep learning. *International Research Journal of Engineering and Technology (IRJET), 05*(08), 289–292. Retrieved from https://www.irjet.net/

7. Xu, H., Wang, D., Zhao, R., & Zhang, Q. (2019). AdaRF: Adaptive RFID-based indoor localization using deep learning enhanced holography. *Proceedings of the ACM on Interactive, Mobile, Wearable and Ubiquitous Technologies.* https://doi.org/10.1145/3351271
8. Joshua, J.; Hendryli, J.; Herwindiati, D. E. (2020) Automatic license plate recognition for parking system using convolutional neural networks. In *Proceedings of the 2020 International Conference on Information Management and Technology (ICIMTech)*, (pp. 71–74). Bandung, Indonesia, 13–14 August 2020.
9. Du, W., Z, J., Li, W., Liu, Z., Wang, H., Han, X. (2022). Efficient recognition and automatic sorting technology of waste textiles based on online near infrared spectroscopy and convolutional neural network. *Resources, Conservation and Recycling, 180.* https://doi.org/10.1016/j.resconrec.2022.106157.
10. Eshwar, S. G., Rishikesh, A. V., Charan, N. A., & Umadevi, V. (2016). Apparel classification using convolutional neural networks. In *Proceedings of IEEE International Conference on ICT in Business Industry & Government*, (pp. 1–5).
11. Park, S., Shin, M., Ham, S., Choe, S., & Kang, Y. (2019). Study on fashion image retrieval methods for efficient fashion visual search. In *2019 IEEE/CVF Conference on Computer Vision and Pattern Recognition Workshops (CVPRW).*
12. Guan, C., Qin, S., & Long, Y. (2019). Apparel-based deep learning system design for apparel style recommendation. *International Journal of Clothing Science and Technology, 31*(3), 376–389.
13. Chang, Y.-H., & Zhang, Y.-Y. (2022). Deep learning for clothing style recognition using YOLOv5. *Micromachines, 13*, 1678. https://doi.org/10.3390/mi13101678
14. Bali, M., & Mahara, T. (2023). Comparison of affine and DCGAN-based data augmentation techniques for chest X-ray classification. *Procedia Computer Science, 218*, 283–290.
15. Szegedy, C., Vanhoucke, V., Ioffe, S., Shlens, J. & Wojna, Z. (2016). Rethinking the Inception Architecture for Computer Vision. In *2016 IEEE Conference on Computer Vision and Pattern Recognition (CVPR)*, (pp. 2818–2826).
16. He, K., Zhang, X., Ren, S. & Sun, J. (2016). Deep residual learning for image recognition. In: *Proceedings of 2016 IEEE Conference on Computer Vision and Pattern Recognition* (pp. 770–778). IEEE.
17. Howard, A. G., Zhu, M., Chen, B., Kalenichenko, D., Wang, W., Weyand, T., Andreetto, M. & Adam, H. (2017). *MobileNets: Efficient Convolutional Neural Networks for Mobile Vision.*

Chapter 6
A Time-Scaled ETAS Model for Earthquake Forecasting

K. Muralidharan and A. Das

Abstract The Himalayan region, including Nepal, is prone to frequent and large earthquakes. Accurate forecasting of these earthquakes is crucial for minimizing loss of life and damage to infrastructure. In this study, we propose various time-scaled Epidemic Type Aftershock Sequence (ETAS) models to forecast earthquakes in Nepal. The ETAS model is a statistical model that describes the temporal and spatial patterns of aftershocks following a main shock. A dataset of earthquake occurrences in Nepal from 2000 to 2020 was collected, and this data was used to fit the models showcased in this article. Our results show that the time-scaled ETAS model is able to accurately forecast earthquake occurrences in Nepal, and could be a useful tool for earthquake early warning systems in the region.

Keywords Earthquake · Forecasting · Epidemic type aftershock sequence models · Alternative time scales

6.1 Introduction

The Himalayan region, including Nepal, is prone to frequent and large earthquakes. Accurate forecasting of these earthquakes is crucial for minimizing loss of life and damage to infrastructure. Earthquake forecasting is a complex and interdisciplinary field, involving the integration of data from various sources, including seismology, geology, and statistics. One important aspect of earthquake forecasting is the study of aftershocks, which are smaller earthquakes that occur after a main shock. Aftershocks can provide valuable information on the state of the fault and the likelihood of future earthquakes. In this study, we propose a time-scaled Epidemic Type Aftershock Sequence (ETAS) model to forecast earthquakes in Nepal. The ETAS model

K. Muralidharan (✉) · A. Das
Department of Statistics, Faculty of Science, The Maharaja Sayajirao University of Baroda, Vadodara, Gujarat 390002, India
e-mail: muralikustat@gmail.com

A. Das
e-mail: agniva.d-statphd@msubaroda.ac.in

P. K. Kapur et al. (eds.), *Data-centric Approaches to Industrial Decisions*, Asset Analytics, https://doi.org/10.1007/978-981-96-7556-2_6

is a statistical model that describes the temporal and spatial patterns of aftershocks following a main shock. The study of aftershocks following a main shock is crucial for understanding the state of the fault and the likelihood of future earthquakes [1, 2]. One important aspect of earthquake forecasting is the study of aftershocks, which are smaller earthquakes that occur after a main shock. Aftershocks can provide valuable information on the state of the fault and the likelihood of future earthquakes.

In recent years, various statistical models have been proposed for forecasting earthquakes, including the Epidemic Type Aftershock Sequence (ETAS) model [3]. The ETAS model is a statistical model that describes the temporal and spatial patterns of aftershocks following a main shock. The model is based on the assumption that the occurrence of aftershocks is influenced by both the main shock and the previous aftershocks.

The ETAS model [3] has been widely used in various regions around the world, including Japan [4], California [1], and Italy [2] wherein, the ETAS model was found to be able to accurately forecast earthquake occurrences. However, the ETAS model has not been widely applied in the Himalayan region, including Nepal. The Himalayan region is characterized by a high level of seismic activity and a complex tectonic setting. Therefore, it is important to develop a model that is specifically tailored to the characteristics of this region.

In this study, we propose a time-scaled ETAS model to forecast earthquakes in Nepal. The time-scaled ETAS model takes into account the time-dependent behaviour of aftershocks, which is important for accurately forecasting earthquakes in regions with high seismic activity such as Nepal.

The dataset of earthquake occurrences in Nepal from 2000 to 2020 was collected and used to fit the time-scaled ETAS model. The model was then validated using the earthquake data from Nepal. The results of this study will provide valuable insights into the characteristics of earthquakes in Nepal and will be useful for earthquake early warning systems in the region.

6.2 Research Methodology

6.2.1 Dataset

The dataset used in this study was collected from ANSS global comprehensive catalogue (available at http://earthquake.usgs.gov/earthquakes/search/). It includes information on all earthquakes that occurred in Nepal between Jan 1, 1990 and May 31, 2022. The data includes the date, time, magnitude, and location (Nepal and surrounding areas including the rectangle formed by coordinates 27–30° N and 79–88° E) of each earthquake, with magnitude restriction $m_0 = 5.0$. We have used the depth variable for the proportional hazards time scaling.

The dataset was cleaned and pre-processed to ensure that it met the requirements of the time-scaled ETAS model. This included removing any duplicate or missing data, and ensuring that the magnitude and location of each earthquake were accurate.

The dataset used in this study is the most comprehensive and up-to-date dataset of earthquakes in Nepal, and provides a valuable resource for studying the characteristics of earthquakes in the region.

6.2.2 Epidemic Type Aftershock Sequence Models

The ETAS model is a statistical model that is used to describe the temporal and spatial patterns of earthquakes, specifically the aftershocks that follow a main shock event. The model is based on the assumption that the rate of aftershocks is influenced by both the magnitude and frequency of the main shock, as well as the spatial distribution of previous aftershocks [5]. It has been widely used in seismology to analyse seismic data and makes predictions about future earthquake activity.

Consider a temporal marked process $\boldsymbol{X}$ consisting of N events. Let (t_i, x_i, y_i, m_i) denote the vector containing the time, longitude, latitude and magnitude of the ith earthquake $i = 1, 2, \ldots, N$, whose distribution parametrized by constants β and θ, is marked by the conditional intensity function n $\lambda(t, x, y, m|\mathcal{H}_t)$, where $\mathcal{H}_t = \{\ (t_i, x_i, y_i, m_i) \in \boldsymbol{X}|t_i < t\}$ (occurrence history of the earthquakes up to time t), satisfying

$$\lambda(t, x, y, m|\mathcal{H}_t) = \nu_\beta(m).\lambda_\theta(t, x, y|\mathcal{H}_t)$$

where,

$$\nu_\beta(m) = \beta \exp[-\beta(m - m_0)],\ \beta > 0,\ m > m_0$$

is the probability density function of the magnitude of an event and

$$\lambda_\theta(t, x, y|\mathcal{H}_t) = \tilde{u}(x, y) + \sum_{i:t_i<t} \kappa_{A,\alpha}(m_i) g_{c,p}(t - t_i) f_{D,\gamma,q}(x - x_i; y - y_i; m_i)$$

where, $\tilde{u}(x, y)$ is the stationary background seismicity rate, $\kappa_{A,\alpha}(m_i)$ is the expected number of triggered events (aftershocks) generated from an eventof magnitude m_i and it can be expressed as

$$\kappa_{A,\alpha}(m_i) = A \exp[\alpha(m - m_0)],\ m > m_0$$

where $A > 0$ and $\alpha > 0$ are unknown (fixed) parameters, $g_{c,p}(t - t_i)$ is the probability density function of the occurrence time of a triggered event generated from an event of magnitude m_i occurring at time t_i, assumed to be a function of the time lag $(t - t_i)$ and independent of m_i, expressed as

$$g_{c,p}(t-t_i) = \left(\frac{p-1}{c}\right)\left(1+\frac{t-t_i}{c}\right)^{-p}, t > t_i$$

where $c > 0$ and $p > 1$ are unknown (fixed) parameters and $f_{D,\gamma,q}(x-x_i; y-y_i; m_i)$ is the probability density function of the occurrence location of a triggered event generated from an event of magnitude m_i occurring at the location (x_i, y_i) assumed to be dependent on each other, such that the radially symmetric joint density function is given by

$$f_{D,\gamma,q}(x-x_i; y-y_i; m_i) = \frac{q-1}{\pi D \exp[\gamma(m_i-m_0)]}\left(1+\frac{(x-x_i)^2+(y-y_i)^2}{D\exp[\gamma(m_i-m_0)]}\right)^{-q}$$

where, $D > 0$, $\gamma > 0$ and $q > 1$ are unknown (fixed) parameters.

Note: This assumption is overlooked in the case where we have used the time-scaled ETAS in Sect. 6.2.4. Instead, there we have assumed exponential and gamma distributions for event magnitudes.

It is assumed that each event is either a background (spontaneous) event or an event triggered by a previous event. The background events are generated by a Poisson process with intensity $\tilde{u}(x, y)$, which is stationary in time. The relaxing coefficient μ in $\tilde{u}(x, y) = \mu u(x, y)$ is introduced in order to fasten the convergence of the modelfitting algorithm. Previous events, whether they are background or triggered events, generate further events according to a non-stationary Poisson process with intensity function

$$\sum_{i:t_i<t} \kappa_{A,\alpha}(m_i) g_{c,p}(t-t_i) f_{D,\gamma,q}(x-x_i; y-y_i; m_i).$$

The expected number of triggered events generated from a typical event, regardless of magnitude is

$$\int_{m_0}^{\infty} \kappa_{A,\alpha}(m) v_\beta(m) dm = \frac{A\beta}{\beta-\alpha}.$$

Now, $\frac{A\beta}{\beta-\alpha} < 1$ implies the existence of a stationary model, which in turn results in the total spatial intensity function being of the form

$$\begin{aligned}\Lambda(x, y) &= \lim_{T\to\infty} \frac{1}{T}\int_0^T \lambda_\theta(t, x, y|\mathcal{H}_t) dt \\ &= \mu u(x, y) + \lim_{T\to\infty}\left(\frac{1}{T}\sum_{i:t_i<T} \kappa_{A,\alpha}(m_i) f_{D,\gamma,q}(x-x_i; y-y_i; m_i)\int_{t_i}^{T} g_{c,p}(t-t_i) dt\right) \\ &\approx \mu u(x, y) + \frac{1}{T}\sum_{i:t_i<T} \kappa_{A,\alpha}(m_i) f_{D,\gamma,q}(x-x_i; y-y_i; m_i)\end{aligned}$$

and the clustering (triggering) coefficient which quantifies the clustering effect relative to the total spatial intensity at any location (x, y) is given by

$$\omega(x, y) = 1 - \frac{u(x, y)}{\Lambda(x, y)}.$$

Zhang et al. [6] suggested the probability of event j being a background event as

$$1 - p_j = \frac{\mu u\left(x_j, y_j\right)}{\lambda_\theta\left(t_j, x_j, y_j | \mathcal{H}_{t_j}\right)}$$

where $p_j = \sum_{i:t_i<t_j} p_{ij}$ is interpreted as the probability that event j is triggered by a previous event, where

$$p_{ij} = \begin{cases} \frac{\kappa_{A,\alpha}(m_i) g_{c,p}(t-t_i) f_{D,\gamma,q}(x-x_i; y-y_i; m_i)}{\lambda_\theta\left(t_j, x_j, y_j \middle| \mathcal{H}_{t_j}\right)} & t_j > t_i \\ 0, & t_j \le t_i \end{cases}$$

where p_{ij} is interpreted as the probability that event j is triggered by event i.

Since, the log-likelihood function can be expressed as

$$l(\beta, \theta | \mathcal{H}_T) = \sum_{i=1}^{N} \delta_i \log(\lambda_{\beta,\theta}(t_i, x_i, y_i, m_i | \mathcal{H}_{t_i}))$$
$$- \int_{m_0}^{\infty} \int_{t_{start}}^{t_{start}+T} \iint_S \lambda_{\beta,\theta}(t, x, y, m | \mathcal{H}_t) dx dy dt dm$$

where $\delta_i = 1$ if event i is a target event and $\delta_i = 0$ otherwise, rendering it separable into two parts as $l(\beta, \theta | \mathcal{H}_T) = l_1(\beta | \mathcal{H}_T) + l_2(\theta | \mathcal{H}_T)$ the MLE of β is given by $\hat{\beta} = \frac{N'}{\sum_{i=1}^{N} \delta_i (m_i - m_0)}$ and $\hat{\theta}$ (MLE of θ) is obtained by minimizing $\xi(\theta) = -l_2(\theta | \mathcal{H}_T)$ iteratively using a suitable optimization algorithm.

We have followed the example of [4–6] by using the Davidon–Fletcher–Powell algorithm. An alternative approach [7] that has been used to fit some of the time-scaled models is described in Sect. 6.2.4.

6.2.3 Alternative Time Scales

Duchesne [8, 9] and Lawless [10] have both proposed methods for time scaling that are alternatives to the ideal time scale. One alternative time scaling method proposed by Duchesne [11] is called the "calibration time scale". Calibration time scale is a time scale that is used to make a non-identifiable model identifiable by adjusting the scaling of the time axis. The method uses a transformation of the time axis, which is chosen to make the model identifiable, without changing the underlying hazard function. Another alternative is the "Proportional Hazard Time scale" [12], which is based on the proportional hazards model. In this method, the hazard function is assumed to be proportional to a baseline hazard function and an explanatory variable. The time scale is chosen such that the coefficient of the explanatory variable is equal to one, which simplifies the interpretation of the model.

Here are a few examples of different time scale options [13–16] and their corresponding assumptions:

1. **Ideal Time Scale:** The hazard function is constant over time. This assumption is often made when there is no clear understanding of the underlying process generating the data, or when the data is too sparse to accurately estimate a more complex hazard function. This can be represented as $t^Z = \phi_1(t) = t$, say.
2. **Calibration Time Scale:** The hazard function is non-constant over time, but a transformation of the time axis is used to make the model identifiable without changing the underlying hazard function. The method assumes that the model is non-identifiable without the calibration time scale, but identifiable with it. Here,$t^Z = \phi_2(t, \omega) = \frac{t}{\omega}$, say.
3. **Proportional Hazards Time Scale:** The hazard function is proportional to a baseline hazard function and an explanatory variable. The time scale is chosen such that the coefficient of the explanatory variable is equal to one, which simplifies the interpretation of the model. This method assumes that the hazard function is proportional to a baseline hazard function and an explanatory variable. Mathematically, $t^Z = \phi_3(t, Z(t)) = \frac{t}{Z(t)}$, say.
4. **Log-Linear Time Scale:** The hazard function is a log-linear function of time. This method assumes that the hazard function is a log-linear function of time, which can be useful when the data exhibit a clear trend over time. Here, $t^Z = \phi_4(t) = \log t$, say.
5. **Power Time Scale:** The hazard function is a power function of time. This method assumes that the hazard function is a power function of time, which can be useful when the data exhibit a clear trend over time. Hence, we have $t^Z = \phi_5(t, \omega) = t^{\omega}$, say.

It is important to note that the choice of time scale can have a significant impact on the results and interpretation of the model and it is essential to validate the assumptions and check the results with other models or different time scales.

6.2.4 The Time-Scaled ETAS Model

The time-scaled ETAS model is a statistical model that describes the temporal and spatial patterns of aftershocks following a main shock, altering the time scale such that the time between events are such that it is easier to distinguish between whether the event is a background event or triggered event. The model is based on the Epidemic

Type Aftershock Sequence (ETAS) model [3] and it takes into account the time-dependent behaviour of aftershocks. The model is defined by a set of parameters, including the background rate of earthquakes, the intensity of the main shock, and the rate at which aftershocks decay with time. These parameters are estimated from the data using a maximum likelihood estimation technique. The time-scaled ETAS model also includes a spatiotemporal component, which accounts for the spatial distribution of earthquakes and the temporal clustering of aftershocks, and is defined by a set of spatial and temporal kernels, which are estimated via maximum likelihood estimation.

We have obtained results for model performance of the ETAS model via two approaches. The first approach leads to assuming the ETAS model as a Marked Point Process and maximizing the resulting log-likelihood (read: minimizing negative log-likelihood) iteratively using the optimization procedure proposed by Nelder–Mead (1965) [7]. By this approach, we have, for each type of time-scaled data, considered the ETAS model as a Ground Intensity Function for two models where the underlying probability distribution of event magnitudes is either exponential or gamma, and estimates of the parameters have been thus compared. The second approach [5] entails a similar path, except that for likelihood maximization (or negative log-likelihood minimization), we opt for the Davidon–Fletcher–Powell (DFP) algorithm for parameter optimization, similar to the model proposed by Ogata [3] Ogata and Zhuang [4], coined Iterative Stochastic De-clustering Method (ISDM) by A. Jalilian [5].

6.3 Results

All the calculations and model development discussed in this paper have been programmed in R language, and is available in the GitHub repository link: https://github6.com/dasagniva/time-scaled-etas6. Table 6.1 compares the outputs of the ETAS model (unscaled time) as fitted using the Iterative Stochastic De-clustering Method (assuming that event magnitudes follow a radially symmetric joint density function) and the alternative models where the ETAS model is used as a Ground Intensity Function, assuming exponential and gamma distributions of event magnitudes.

Clearly, we can see that both the models perform equally poorly in this regard, although the Gamma model fares marginally better than the Exponential model in this case. For both the models that we see above,μ or the background seismicity rate of the process represents the average number of events occurring per unit time in the absence of any triggering effect, which is very low, indicating that more often, the earthquakes occurring in this region have been due to some triggering event.α, the triggering rate parameter is seen to be almost identical, and is slightly higher in the exponential model than the gamma model indicating that using the gamma model may result in minute underestimation of the triggering rate. The gamma model does have a more pronounced clustering parameter c, which represents the probability of an event being able to trigger one or more consequent events. The parameter p, or

Table 6.1 Parameter estimates using ETAS model as ground intensity function $\{t^Z = \phi_1(t) = t\}$

Parameter	Exponential model	Gamma model
μ	0.001450572	0.001397770
A	0.00005300226	0.00006914710
α	1.713889	1.647376
c	0.5407705	0.5955604
p	1.133662	1.126354
D	0.2280044	0.2656319
β	–	0.3676430
Log-likelihood	−7837.827	−7795.654

the probability distribution parameter of the model which represents the distribution of the time between an event and its triggered events, is also slightly higher in the exponential model than the gamma model. This may be attributed to the parameter β, which is responsible for the variation in the magnitude of earthquakes in the region.D denotes the spatial decay parameter of the model and it represents the rate at which the triggering effect decreases with distance. As can be inferred from the table itself, we can see that they are almost identical in case of either exponential or gamma distribution of magnitudes. A may be interpreted as the minimum number of aftershocks expected in case of an earthquake at minimum threshold $m = m_0$. Again, as the estimate of A is very close to zero, we can expect that aftershocks are rare, and for number of aftershocks to be high, the magnitude of the triggering earthquake needs to be very high.

However, comparing the same model with the ETAS model fitted using ISDM and likelihood optimized using the DFP algorithm, we have the results in Table 6.2.

From Table 6.2, we can safely say that the ISDM approach yielded much better results from the point of view of fit. From the interpretation viewpoint as well, we

Table 6.2 Parameter estimates using ETAS model using ISDM optimized using DFP $\{t^Z = \phi_1(t) = t\}$

Parameter	Estimate	Standard error
μ	1.1153	0.0383
A	0.2102	0.0906
α	1.5979	0.0263
c	0.0071	0.1552
p	1.1395	0.0139
β	3.5912	0.1240
D	0.0033	0.1910
q	1.8398	0.0457
γ	1.2236	0.0730
Log-likelihood	−497.8743	(AIC = 1011.749)

can see that this set of parameter estimates indicate significantly higher background intensity, higher value of number of aftershocks at minimum magnitude threshold (approximately 1 in 5, in this case), almost identical α indicating that the triggering rate estimate may be accurate. We see a significantly lower value of c in this case, which may indicate low clustering rate, but we have applied an iteratively de-clustered algorithm here, so the lower value of c may be justified. The p parameter may be identical for both the models, indicating that the distribution of time between earthquakes may be similar, although the same cannot be said for D, which is lower than that of the previous model, indicating that the rate of decrease in triggering effect per unit distance is lower. Additionally, a significantly higher value of β indicates higher variation in the magnitude over the entire area.γ and q are the shape and the scale parameters of the radially symmetric PDF $f_{D,\gamma,q}(x - x_i; y - y_i; m_i)$, respectively, and may be treated as hyperparameters, to be used for tuning the model for better accuracy. Tables 6.3, 6.4, 6.5 and 6.6 are models fitted using the ETAS Ground Intensity Function with Exponential and Gamma event magnitude approach after using various time scales discussed earlier.

Here, we can see that increasing the time scale to near 1000 provides better log-likelihood. However, higher values of ω have not yielded better results. One may use a search algorithm to find a scale parameter to maximize the log-likelihood, but such algorithms may require costlier processing hardware. The parameter estimates

Table 6.3 Parameter estimates using ETAS model as ground intensity function

a. $\left\{t^Z = \phi_2(t, \omega) = \frac{t}{\omega}, \omega = 5\right\}$

Parameter	Exponential model	Gamma model
μ	0.01450549	0.01397782
A	0.0005300868	0.0006914417
α	1.713866	1.647391
c	0.05407687	0.05955264
p	1.133658	1.126352
D	0.2280044	0.2656316
β	–	0.1162570
Log-likelihood	−5433.928	−5391.755
b. At $\omega = 1000$		
μ	1.450546	1.397766
A	0.05301076	0.06915157
α	1.713874	1.647388e+00
c	0.0005407048	0.0005954518
p	1.133651	1.126342
D	0.2280044	0.2656312
β	–	0.01162555
Log-likelihood	−626.1306	−583.9576

Table 6.4 Parameter estimates using ETAS model as ground intensity function $\left\{t^Z = \phi_3(t, Z(t)) = \frac{t}{Z(t)}\right\}$

Parameter	Exponential model	Gamma model
μ	0.0002956871	0.0002937223
A	0.2932234	0.3480650
α	0.5707210	0.4834184
c	0.03981759	0.05134718
p	1.113115	1.117977
D	0.2280044	0.3000280
β	–	0.3114590
Log-likelihood	−5148.079	−5075.610

Table 6.5 Parameter estimates using ETAS model as ground intensity function $\{t^Z = \phi_4(t) = \log t\}$

Parameter	Exponential model	Gamma model
μ	0.6656586	0.2100070
A	0.1282819	2.197027
α	1.368082	0.9307006
c	5.471171	0.5841404
p	0.9164903	1.321349
D	0.2280044	1.576231
β	–	1.576231
Log-likelihood	−4461.112	−3637.195

Table 6.6 Parameter estimates using ETAS model as ground intensity function $\{t^Z = \phi_5(t, \omega) = t^\omega, \omega = 0.5\}$

Parameter	Exponential model	Gamma model
μ	2.996577e−01	2.603446e−01
A	2.401597e−01	3.450314e−01
α	1.483756	1.398046
c	0.0002384452	0.0002525749
p	0.9817204	0.9741814
D	0.2280044	0.2703550
β	–	0.01578156
Log-likelihood	−974.8525	−928.1033

in Table 3b is closer in interpretation to the ISDM model than the unscaled ETAS model with exponentially or gamma distributed magnitudes and also induces better fit in that regard. For the results in Table 6.4, we have used the average depth of all minor earthquakes, between two major quakes (magnitude > 5), as a usage measure.

Just like the rest of the cases, the Gamma model fares only slightly better than the Exponential model. However, the choice of the Depth variable as a usage measure is not found to be suitable for scaling purposes. The parameter estimates are closer to the unscaled ETAS model than the iteratively de-clustered ETAS model. Hence using depth as a covariate for time scaling seems unsuitable. One may opt for other usage measures like increase in height of a particular Himalayan range per unit time, which may be an interesting variable to use in this regard. Another variable that may be potentially used as a usage measure can be the rate of depletion of ice caps in the Himalayas with time. Some interesting outtakes may potentially arise from such analyses. However, collection of such data may be arduous at the ground level, as government records may not even be available in certain cases.

Although the log scale does not induce a good enough log-likelihood to compete with the calibration or proportional hazards time scale, it does fare significantly better than the unscaled model. This goes to show that a combination of other scales with the log scale may potentially provide a better model.

The Power Scale taken at $\omega = \frac{1}{2}$ give the best result at a log-likelihood in the neighbourhood of negative one thousand, which is still better than the log-linear scale or even the proportional hazards scale. Here again, the estimates of model parameters indicate a situation closer that estimated by the iteratively de-clustered model. One may note that usage of calibration time scale in this regard, translates to easier likelihood optimization as opposed to other scales. One may use, as in our case, iterative algorithms in search of calibration parameter values that maximize log-likelihood. Using the power-scale implies that a stronger algorithm may be required to find that optimum.

However, the calibration time scale performs best when compared with the other time scales in terms of model fit. Having said that, the best performance has been that of the model generated using ISDM optimized using FDP algorithm. The figures provided below may provide some insight onto the earthquake patterns observed in Nepal since January 11, 1990.

Figure 6.1 indicates an outlier at 25th April, 2015, where multiple great earthquakes occurred in quick succession in the regions as indicated by the circles in the above map. Figure 6.2 marks the spatial neighbourhoods where background intensity is higher and neighbourhoods that are clustering prone. From Fig. 6.2, it can be said that background seismicity rate is higher in the north-western part of Nepal (Karnali region), whereas the total spatial seismicity rate and the clustering coefficient is significantly higher in central Nepal (Kathmandu and Narayani region). This indicates a higher potential of powerful aftershocks in the central Nepal region. However, earthquakes are more frequent in the North-Western region. Figure 6.3 (top-left and top-right) shows the results of model diagnostics performed on the ETAS model. It has been observed that earthquakes closer to the Kathmandu region have low residuals, meaning that the background intensity and the clustering coefficient of that region is well represented by the model. However, as we move eastwards or westwards, we see a sharp increase in the residuals. This may be the result of varying levels of terrain elevation that have not been accounted for in our models. There may also be other underlying factors that need to be investigated in greater detail. In Fig. 6.3 (bottom-left, and bottom-right), it can be seen that points on the plots of transformed times τ_j and the Q-Q plot of U_j lie approximatelyon the $y = x$ line. To confirm, we performed a Kolmogorov–Smirnov test for goodness-of-fit. The p-value of 0.4386 (> 0.05) indicates that U_j follows a $U(0, 1)$ distribution, at a 5% level of significance.

Figure 6.4 indicates that most of the quakes near the Kathmandu–Janakpur region are triggered whereas quakes occurring in the North-West region (Karnali region) mostly consist of background seismic activity.

6.4 Conclusions

From the results obtained above, the following outtakes can be derived:

- The Davidon–Powell–Fletcher algorithm, fares better than the Nelder–Mead Optimization algorithm for maximum likelihood estimation in this case. Additionally, the probability distribution of the magnitude, if taken to be either exponential or gamma, fits very poorly on the data. The radially symmetric density assumption on the other hand yields a superior fit.
- Judging by interpretability, we should take into account, the fact that, in case of the exponential or gamma assumption of magnitude distribution, the inherent interpretability of the model is increased, which loosely translates to better foundations for decision-making. The parameters of the radially symmetric density are complex to interpret and to base decisions on. However, they may be used as hyperparameters in "deep"-er models, for obtaining better forecasts.
- Although by a very small margin, among the theoretical distributions that have been postulated for magnitude probability distribution, the gamma model fares better in all types of time scales than the exponential model. That may be due to an extra parameter which makes the difference. However, having said that, the cost of including an extra parameter may prove costlier in terms of likelihood per additional covariate, if too many instances are to be trained to the model.
- The residual map indicates lower values of residuals in the central part of the country whereas as we go from the centre outwards, residuals become higher. This may be a direct connection with the extent of urbanization in those areas to the magnitude and frequency. The areas having low variance in raw spatial residuals, are the more urbanized areas of the country, whereas relatively less inhabited areas have very high variation of magnitude. This may be solved by scaling the magnitudes along with time.

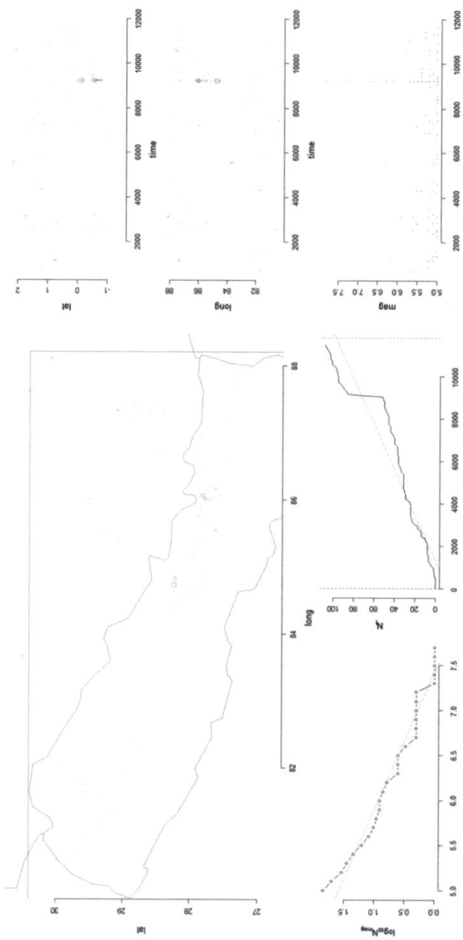

Fig. 6.1 Location of epicentres (top-left panel), logarithm of frequency by magnitude (bottom-left panel), cumulative frequency v/s time (bottom-middle panel) and latitude, longitude and magnitude v/s time (right panels) of earthquakes with magnitude ≥ 5 occurring between 1990-01-11 and 2022-05-20 in Nepal and its proximity (27– 30° N and 79–89° E), extracted from the ANSS global comprehensive catalogue

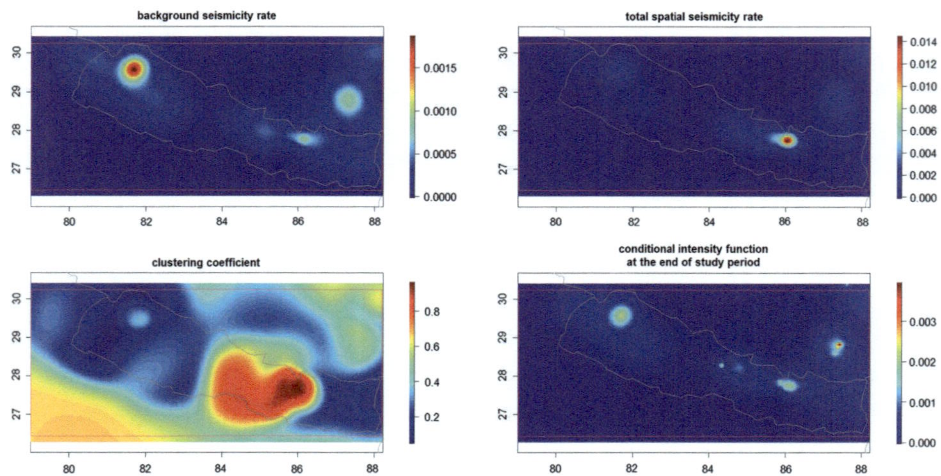

Fig. 6.2 Plots of estimates of the background seismicity rate, total spatial intensity, clustering (triggering) coefficient and conditional intensity at $t = t_{start} + T$ for the fitted ETAS model to the Nepal catalogue

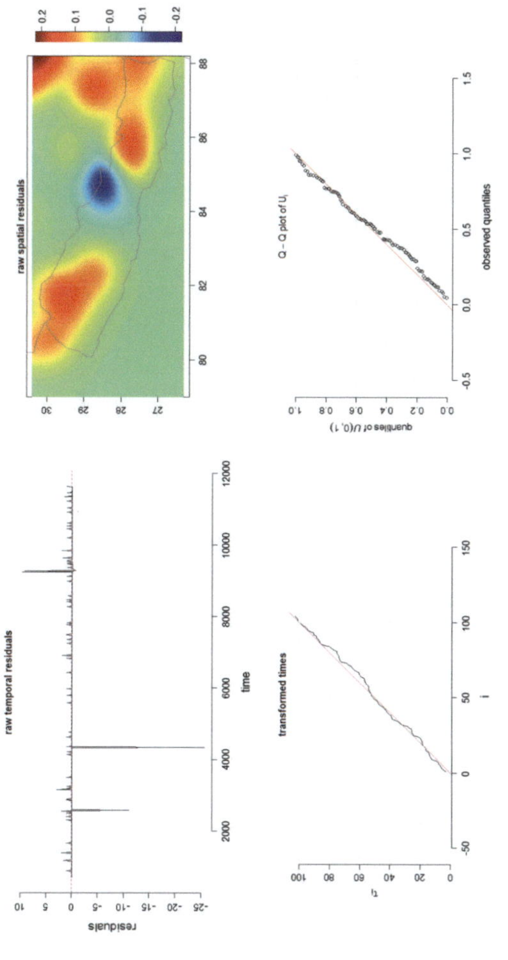

Fig. 6.3 Diagnostic plots for the fitted ETAS model to the Nepal catalogue: the temporal residuals (top-left), smoothed spatial residuals (top-right), transformed times τ_i against i (bottom-left) and the Q-Q plot of U_i (bottom-right)

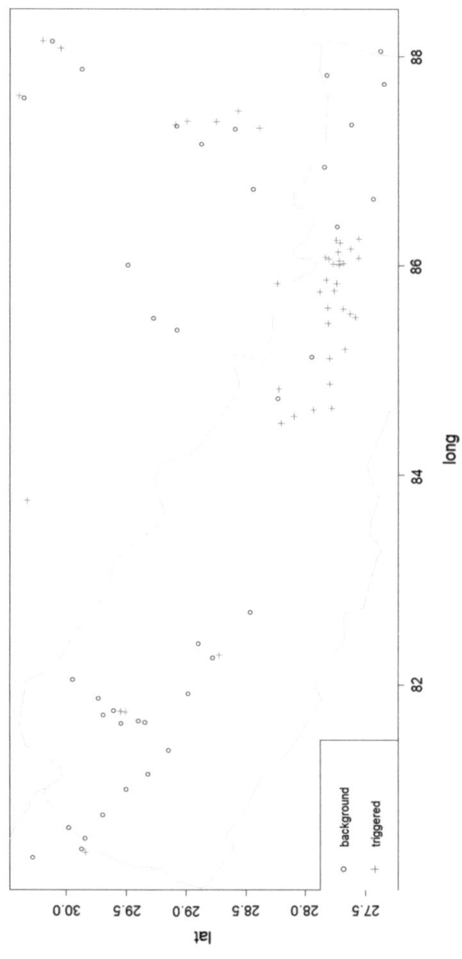

Fig. 6.4 Most likely (Probability > 0.95) background and triggered events detected by estimated de-clustering probabilities p_j

- The conditional density and the clustering coefficient for central locations of the country have a higher probability of aftershocks than the locations, and as we go eastwards or westwards from the centre. One may take the usage measure as some variable highly correlated with the magnitude for this purpose.

Although a lot of research in this sphere is centred around exploiting the flexibility offered by the ETAS model to accommodate additional constraints or to bypass assumptions so as to introduce robustness, as evident by the recent works [17–22], and the recent advances in the domain of time scaling for lifetime data [23–25], collaboration between researchers in these two spheres seldom occurs, resulting in dearth of opportunities and unexplored possibilities. Ours seems to be one of the earlier attempts at bridging these two domains in comprehensive manner. However, the future seems hopeful, as there lies tremendous prospect in cross-collaborations with the computational domain as well, given that research in deep modelling is at the peak, with deep models for both, seismicity as well as for time-between-event responses, is now commonplace.

References

1. Wiemer, S., & Katsumata, K. (1999). Spatial variability of seismicity parameters in aftershock zones. *Journal of Geophysics Research Solid Earth, 104*(B6), 13135–13151. https://doi.org/10.1029/1999JB900032
2. Gulia, L., & Wiemer, S. (2010). The influence of tectonic regimes on the earthquake size distribution: A case study for Italy. *Geophysics Research Letters, 37*, 1–6. https://doi.org/10.1029/2010GL043066
3. Ogata, Y. (1988). Statistical models for earthquake occurrences and residual analysis for point processes. *Journal of the American Statistical Association, 83*(401), 9–27. https://doi.org/10.1080/01621459.1988.10478560
4. Ogata, Y., & Zhuang, J. (2006). Space–time ETAS models and an improved extension. *Tectonophysics, 413*(1–2), 13–23. https://doi.org/10.1016/J.TECTO.2005.10.016
5. Jalilian, A. (2019). ETAS: An R package for fitting the space-time ETAS model to earthquake data. *Journal of Statistical Software.* https://doi.org/10.18637/jss.v088.c01
6. Zhuang, J., Ogata, Y., & Vere-Jones, D. (2004). Analyzing earthquake clustering features by using stochastic reconstruction. *Journal of Geophysical Research: Solid Earth, 109*(B5), 5301. https://doi.org/10.1029/2003JB002879
7. Shah, R., Muralidharan, K., & Parajuli, A. (2020). Temporal point process models for Nepal earthquake aftershocks. *International Journal of Statistics and Reliability Engineering, 7*(2), 275–285.
8. Duchesne, T., & Lawless, J. (2002). Semiparametric inference methods for general time scale models. *Lifetime Data Analysis, 8*(3), 263–276. https://doi.org/10.1023/A:1015853905091/METRICS
9. Duchesne, T., & Lawless, J. (2000). Alternative time scales and failure time models.
10. Lawless, J. F. (2011). *Statistical models and methods for lifetime data.* John Wiley & Sons.
11. Duchesne, T. (1999). *Multiple time scales in survival analysis.* University of Waterloo.
12. Kordonsky, K. B., & Gertsbakh, I. (1997). Multiple time scales and the lifetime coefficient of variation: Engineering applications. *Lifetime Data Analysis, 3*(2), 139–156. https://doi.org/10.1023/A:1009657101784/METRICS

13. Kordonsky, K. B., & Gertsbakh, I. B. (1993). Choice of the best time scale for system reliability analysis. *European Journal of Operational Research, 65*(2), 235–246. https://doi.org/10.1016/0377-2217(93)90336-L
14. Kordonsky, K. B., & Gertsbakh, I. B. (1995). System state monitoring and lifetime scales—I. *ReliabEngSystSaf, 47*(1), 1–14. https://doi.org/10.1016/0951-8320(94)00026-K
15. Oakes, D. (1995). Multiple time scales in survival analysis. *Lifetime Data Analysis, 1*(1), 7–18. https://doi.org/10.1007/BF00985253
16. Sumereder, C., & Muhr, M. (2005). Estimation of residual lifetime–Theory and practical problems. *Vol 1*, 1–6.
17. Hong, S., Hachiya, H., Balasubramanian, V. N., & Tsang, I., (2021). Multi-stream based marked point process. *proceedings.mlr.press*, 157:2021–2021
18. Tavakoli, H., Davoodi, N., Jalilian, A., & Zare, M. (2018). Aftershock forcasting experiment for Bushehr province of Iran using the epidemic-type aftershock sequence (ETAS) model. In *16th European Conference of Earthquake Engineering*, (pp. 18–21).
19. Motagi, S., Namilae, S., Gbaguidi, A., Parr, S., & Liu, D. (2021). Point-process modeling of secondary crashes. https://doi.org/10.21203/RS.3.RS-388055/V1.
20. Hirose, F., Tamaribuchi, K., & Maeda, K. (2021). Characteristics of foreshocks revealed by an earthquake forecasting method based on precursory swarm activity. *Journal of Geophysical Research: Solid Earth, 126*(9), e2021JB021673. https://doi.org/10.1029/2021JB021673
21. Altun, E., Bhati, D., & Khan, N. M. (2021). A new approach to model the counts of earthquakes: INARPQX(1) process. *SN Applied Science, 3*(2), 1–17. https://doi.org/10.1007/S42452-020-04109-8/FIGURES/9
22. Davoudi, N., Tavakoli, H. R., Zare, M., & Jalilian, A. (2020). Aftershock probabilistic seismic hazard analysis for Bushehr province in Iran using ETAS model. *Natural Hazards, 100*(3), 1159–1170. https://doi.org/10.1007/S11069-020-03854-8/METRICS
23. Nekrasova, A., & Kossobokov, V. (2022). Seismic risk assessment for the infrastructure in the regions adjacent to the Russian Federation Baikal-Amur Mainline based on the Unified Scaling Law for Earthquakes. *Natural Hazards*. https://doi.org/10.1007/s11069-022-05750-9
24. Syamsundar, A., Naikan, V. N. A., & Wu, S. (2020). Alternative scales in reliability models for a repairable system. *ReliabEngSystSaf*. https://doi.org/10.1016/j.ress.2019.106599
25. Lawless, J. F., & Crowder, M. J. (2010). Models and estimation for systems with recurrent events and usage processes. *Lifetime Data Analysis, 16*(4), 547–570. https://doi.org/10.1007/S10985-010-9157-X/METRICS

Chapter 7
Web Services Security Tools: A Brief Review

Amal Mohammed Albalawi, Enas Yahya Alqulaity, and Omar H. Alhazmi

Abstract Web services are self-descriptive building blocks that speed up the creation of distributed software applications. Web services employ web technologies like HTTP to convey communications between programs or electronic gadgets. Web services security refers to the managerial and technological controls put in place on the network to ensure the confidentiality, availability, and integrity of the information transferred. In this study, we describe a number of threats that pose a security risk to web services and investigate the most pressing challenges facing web services today. The complexity of web servers and applications makes attacks more difficult and harder to notice. Therefore, a web service intrusion detection system must be capable of processing the massive data set and the noise it creates efficiently and precisely. We discussed and compared IAPF, SoapUI pro, SOAPSonar, SOAtest, Strom, and Jmeater WS testing tools in terms of security testing. Modern distributed applications are built on Web Services. Therefore, it is essential that developers thoroughly test and certify their Web Services before making them available to users.

Keywords Web service · WS security · WS security tools

A. M. Albalawi
Department of Computer Science, College of Computer and Cyber Sciences, University of Prince Mugrin, Madinah, Saudi Arabia
e-mail: A.Albalawi@upm.edu.sa

E. Y. Alqulaity · O. H. Alhazmi (✉)
Department of Computer Science, College of Computer Science and Engineering, Taibah University, Madinah, Saudi Arabia
e-mail: ohhazmi@taibahu.edu.sa

E. Y. Alqulaity
e-mail: starinsky008@gmail.com

P. K. Kapur et al. (eds.), *Data-centric Approaches to Industrial Decisions*, Asset Analytics, https://doi.org/10.1007/978-981-96-7556-2_7

7.1 Introduction

A computing paradigm known as "service-oriented computing" makes use of Web services when developing software applications. Web services are self-descriptive components that facilitate the quick and inexpensive composition of distributed software applications [1]. The purpose of the Web services technology is to enable distributed applications to connect and cooperate using common Internet protocols and languages without direct human involvement [2]. SOC provides the ability to assemble distributed Web services. Currently, e-businesses use Web services for information exchange as a popular option. To achieve their intended goals, all business transactions today rely on Web services. Web technologies like HTTP are used by Web services to transfer messages between applications or electronic devices. These messages mostly come in XML and JSON file formats. A Web service is built on the Request-Response method and can be implemented using either of the two protocols, SOAP (Simple Object Access Protocol) or REST (Representational State Transfer). XML data can be transferred using either method [3]. XML and REST are used in many deployed systems and protocols [4, 5].

The technological and managerial measures implemented on the system to guarantee the privacy, availability, and integrity of the information exchanged by the Web service are known as Web services security. This work explains the threats and testing processes while examining security concerns related to Web services. Techniques are needed to guarantee security in Web services. The Web services must be protected if a successful deployment is to be achieved. True security requires awareness of potential security risks and proactive action to mitigate them.

In this work, Sect. 7.2 will discuss the security challenges in Web services. Section 7.3 explains the common threats to Web services in terms of service level threat and message level threat. Section 7.4 will discuss the techniques used to protect Web services against attacks. Furthermore, Sect. 7.5 will compare and discuss many Web service testing tools. Finally, Sect. 7.6 concludes the findings of this work.

7.2 The Security Challenges in Web Services

Using platform-neutral standards, such as XML, and internet-based protocols, web services are emerging as a new model for sharing data over the internet. Among their features are improved data accessibility, dynamic connections between services and communication channels, interactions between applications, and data exchange on multiple platforms located in different locations with minimal human intervention, as well as a high degree of autonomy of the services. Almost all of these aspects are incompatible with traditional security models and controls, leading to new security concerns that differ from traditional security issues and introducing new security challenges that differ from traditional security threats [6].

Data that was previously trapped behind firewalls is now accessible via web services, allowing businesses to take advantage of valuable business opportunities they never had access to before. In addition to their great potential, web services also have a huge problem: too much information is available. Patient records, credit card numbers, or manufacturing designs are all valuable resources that need protection from unauthorized access. Collaboration and information sharing are a company priority, but not at the expense of handing over all their resources. Businesses must remain vigilant and wary of the people they come in contact with. It is best to provide just enough detail, but not too much [7]. Current challenges include the need to isolate security from the core network. Security also depends on the end user while navigating between different web services. Databases are vulnerable to threats as well. Database security solely relies on the security of applications that directly access them, especially when the database application is available as a service or accessed by a web service. The exposed application should verify the applicant's right to access the DB. Before forwarding queries to the DB, the application must also filter them. Depending on the nature of the incoming requests, the database may be targeted by SQL injection attacks or other sensitive data mining efforts. To prevent inadvertent leakage, the application must filter the data before submitting it to the applicant [8].

We have two questions: What makes web services so hard to secure? How are web services different from other types of services? As mentioned, the architecture of web services is designed to be open and accessible. Firewalls allow HTTP traffic, and web service requests pass over HTTP. Internal networks can be easily exposed by firewalls. An application's threads are connected by web services. The client agent communicates with a database on the Internet to authorize requests to others. Technology based on PKI is point-to-point. Client/server connections work great but are not sufficient to secure applied themes. The security of web services requires many techniques, and different safety products and technologies are used by all companies involved in web service transactions.

As a result, companies have three options: turning off security, which is high-risk; agreeing to use one security technology, which is too expensive; or building and using a bridge between techniques, not always easy, but the ideal solution [7]. All organizations participating in online service transactions use a variety of security products and solutions.

7.3 Common Threats to Web Services

In this section, we will review the many types of attacks that threaten web services. Injection attacks and Denial of Service attacks are common attacks on web services [9].

7.3.1 SQL Injection Attack

Using SQL injection to target web services is a common attack method. Many web services fail to filter non-validated inputs because of improperly coded blocks. By injecting and embedding themselves as parameters in SQL statements, attackers attempt to execute non-administrative commands.

7.3.2 XML Injection Attack

XML injection vulnerabilities arise when a service fails to validate malicious XML content. When malicious XML content is injected into any service, its working logic can be altered.

7.3.3 XPath Injection Attack

Services/applications that use XPath to transform user input into XML documents are targets for XPath injection attacks. An XML document's structure can be obtained without authorization by submitting incorrect user input.

7.3.4 Denial of Service Attack

A DoS attack disables the target machine by depriving it of its resources [9]. By flooding the user authentication service with authentication attempts.

Web services present many security challenges that include threats, attacks, and security problems [6]. According to Salas et al. there are two types of attack: the attack at the service level and the attack at the message level [8, 10].

7.3.5 Service Level Threats

WSDL and UDDI Attack In WSDL files and UDDI registries, service level information is available. Any publicly accessible WSDL file may be altered by an attacker. There are two types of WSDL attacks: WSDL Scanning and WSDL Tampering. In WSDL scanning, operational data, ports, and other information are extracted from the WSDL file. Data can be altered as a result of WSDL tampering, and sensitive information may be accessed. Protection against these attacks is difficult using standard authentication and authorization procedures. Service level information is accessible via the Web Service Data Locator (WSDL) and UDDI registries.

Malicious Code Injection and Identity Spoofing Attackers can inject malicious code and compromise the service's functioning. Identity spoofing happens when a hacker assumes the identity of the service requester. Such assaults are difficult to detect. An attacker can create a well-formed XML request message and submit it to the service provider, deceiving the service provider into believing that it is responding to a legitimate request. In the second scenario, the attacker will convince real service requesters to submit messages to the fake service provider. Such attacks are challenging to detect.

The Tampering of XML Schemas An attacker can alter the XML schema to make it incorrect. Such attacks result in service provider failures at the end points.

Session Hijacking An attacker can obtain unauthorized access to the offered resources by stealing a user's session token. This results in the transmission of bogus requests or responses, and as a result, the session between the service requester and provider has been hijacked.

7.3.6 Message Level Threats

Modification or Injection of Messages Altered messages or malicious messages can be added between the client and the server. An XML file can be hacked in this way and gives the hacker many privileges.

Message Confidentiality and Espionage Web services are always at risk of being intercepted. Web services cannot be protected against such threats by traditional mechanisms such as VPN or SSL.

Replay of Messages Through unauthorized access to the services, the attacker captures and replies to a valid message. Hackers usually start off by hijacking sessions and tampering with the services to hack into web services. It is possible to detect patterns more easily with the right tools if the same or similar payload is sent over several mediums, such as HTTP, HTTPS, SMTP, and different interfaces [8].

7.4 Protecting Against Attacks

As an alternative to an attack or unusual behavior, an anomaly detection approach identifies traffic that deviates from a predefined rule of typical behavior. Although the anomaly detection model is an excellent way to identify new types of assaults, it is impossible to describe in advance all typical activity norms and network protocols [11].

7.4.1 Intrusion Detection System

An Intrusion Detection System (IDS) aims to detect malicious acts that pose a threat to the integrity of computer systems [12]. Current intrusion detection systems use either signature-based or anomaly-based analysis. The signature-based analysis technique identifies a specific pattern of a known attack threat by comparing the relevant string to the regular expression and analyzing the list of signatures already registered. This technique is very effective provided that it maintains an updated database of signature patterns, but it cannot detect unknown attacks or new malware, such as zero-day attacks.

Unlike signature-based analysis, abnormal-based analysis can detect assaults that differ from conventional traffic patterns, even if their signatures are not provided, and can swiftly identify novel types of attacks [11]. However, a system's performance depends on how well it is implemented and tested across all protocols. Since conventional traffic patterns are determined by the vendor's protocols, it is hard to define detection rules.

7.4.2 AI-Based Intrusion Detection

Recent studies on AI algorithms have extensively addressed issues with signature-based and anomaly-based intrusion detection systems. Machine learning-based methods allow the detection of new and hidden threats that occur in real-time without significant human screening or intervention [13]. Decision tree, Bayes network, SVM (support vector machine), GA (genetic algorithm), and KNN are examples of machine learning models. Although several studies in this field have shown over 90% accuracy, the majority of them have used known test datasets or were designed to detect certain types of attacks [14].

Intrusion detection systems based on deep neural networks have gained great interest from both industry and academia. Kim et al. trained an intrusion detection system using the KDDCup'99 dataset and the LSTM (Long-Term and Short-Term Memories) architecture of RNN [15]. The LSTM-RNN achieved a detection rate of 98.99% and an accuracy of 96.93% compared to other intrusion detection classifiers. In their impressive study of deep neural network-based intrusion detection models, Chang et al. showed the capacity of deep neural network models to evaluate real-time internet traffic data with high computational accuracy and speed. They also have outstanding performance for detecting web application intrusions, something signature-based systems cannot do [14].

Web servers and applications, unlike at the network layer, are very complex systems, which increases the potential for vulnerability and makes cyber-attacks difficult to identify. Also, a web service intrusion detection system should be able to process data quickly and with high computational accuracy, as the noise level in the dataset grows with the size of the dataset [11]. Given these limitations with web service attacks, a single approach cannot address these challenges. It is necessary to evaluate and examine the intrusion detection capabilities of hybrid deep neural network models with unstructured, letter- and number-based synthetic structures [14].

7.5 Web Service Testing Tools

Modern distributed applications are built on web services, allowing for rapid application development across platforms, languages, operating systems, and devices. Thanks to the simplicity and flexibility of integrating dissimilar components, it is now possible. However, because of this flexibility, it is essential that developers and security experts thoroughly test and certify their web services before making them available to users. In this section, we will discuss and compare IAPF, SoapUI Pro, SOAPSonar, SOAtest, Strom, and Jmeater WS testing tools in terms of security tests.

7.5.1 Iapf

An Integrated Application and Protocol-based Framework (IAPF) was proposed by Navya and Jigang [16] to tackle existing WS security problems and empower application-level web service developers to design and implement SOA producers to the IAPF standard. This can prevent DoS and DDoS based attacks and mitigate the effects of these attacks.

Figure 7.1 depicts the system architecture of IAPF, which involves four phases of protection. The first phase addresses the vulnerability of the UDDI protocol, while the second phase focuses on preventing vulnerabilities in the WSDL protocol. In the third phase, WS Security and other measures are used to safeguard against end-to-end vulnerabilities in the SOAP protocol. The fourth phase involves creating protection mechanisms to secure SOAP web services that are publicly accessible. These comprehensive solutions can be utilized by both web services security implementers and web services application developers, ensuring secure and safe web service access [16]. Additionally, Jigang and Liu explore the security vulnerabilities of SOAP messaging in a service-oriented architecture (SOA) environment and describe the core implementation of IAPF methods in providing intrusion-resistant SOAP messaging. They also discuss how IAPF aids in the early detection of XML injection and parameter altering attacks [17].

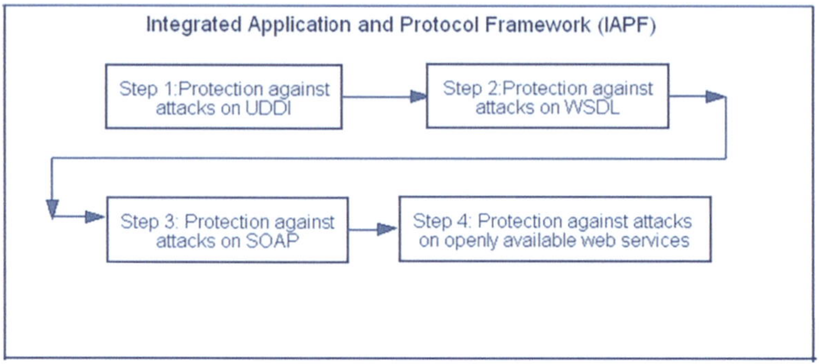

Fig. 7.1 IAPF system architecture [16]

7.5.2 SoapUI Pro

SoapUI Pro is a popular and effective testing software for SOAP web services. The SQL Injection Security Scan features attempts to attack web services that use SQL servers by substituting malicious strings for the Test Step's original arguments, revealing any potential vulnerabilities. To ensure the attack did not disclose sensitive information or return a session ID, assertions can be used. XPath Injection, Invalid Types, and Malformed XML security scans are also employed to identify and exploit subpar XML processing, incorrect input data, and improper handling of invalid XML in web services or web servers [18, 19].

7.5.3 SOAPSonar

SOAPSonar was created by Crosscheck Networks to provide thorough, code-free testing that is incredibly simple to set up and use. In a matter of minutes, you can generate functional, performance, interoperability, and vulnerability reports. Crosscheck Network is a leader in the field of active Web Services Vulnerability Assessment thanks to SOAPSonar Vulnerability assessment. Based on the WSDL, SOAPSonar generates a thorough test suite of Mutant requests using patent-pending XSD-Mutation technology. SOAPSonar attacks the Web Services using Automated Attack Functions (AAF), records the responses, and then employs Automated Vulnerability Detection (AVD) to identify vulnerabilities [20].

7.5.4 SOAtest

SOAtest was announced by Parasoft on April 24, 2002, with the goal of automating function, load, and regression testing within the enterprise. The objective of SOAPTest is to increase the dependability of Web service-related applications. It makes use of web service standards like Web Services Description Language (WSDL) and Simple Object Access Protocol (SOAP) (WSDL) [21]. It is a toolkit for testing and analyzing APIs and API-driven applications [22]. Functional unit tests, integration tests, regression tests, system tests, security tests, simulation and mocking tests, runtime error detection tests, web UI tests, interoperability tests, WS compliance tests, and load tests are all examples of basic testing functionality. SOATest Supported different technologies including Web services. Offering Free Access to the Full Suite of Enterprise-Class Test Automation Software from Parasoft, Parasoft [21].

7.5.5 Storm

Storm is a free and open-source tool for testing web services created by Erik Araojo. Storm was created in the F# programming language and is offered under the New BSD license. Testing web services created with any technology is possible thanks to Storm (.Net, Java, etc.). Storm enables editing and manipulating of raw soap requests and supports dynamic invocation of web service methods, even those with input parameters of complicated data kinds. The GUI is incredibly user-friendly and easy to use. It is possible to test multiple web services at once, which shortens testing times [23].

7.5.6 JMeter

JMeter is an open-source testing tool created by the Apache Software Foundation (ASF) and is offered with the Apache License. Although it was first intended to evaluate Web applications, it has since been expanded to include other test procedures. JMeter's primary use is for performance monitoring, but it may also be used to load test client/server applications. Additionally, JMeter aids in the development of test scripts with assertions, which is useful for regression testing. In this manner, we can confirm that the application produces the anticipated outcomes. Full multithreading is supported by JMeter, enabling simultaneous sampling of various functions by various threads as well as by various thread groups. JMeter's high level of extensibility is a result of the utilization of pluggable components. Timer, sampler, and visualization plugins are examples of these pluggable elements. JMeter provides an intuitive Graphical User Interface (GUI). A test plan's configuration and setup are

Table 7.1 WS Security testing tools

Tool	Supported technology	First release	Platforms supported	Programming language	Ease of use	License	Pricing
SoapUI pro	HTTPS SOAP Database, Web-HTTP	2005	Windows, Mac and	Java	Complex	Licensed version and open source	$599 per year and Free Trial
SOAPSonar	HTML, SOAP, REST, XML and JSON	2005	Windows, Mac and	Java	Easy	Licensed	Free Version
SOAtest	HTTP, XML Encryption, XML Signer and Signature Verifier, SOAP	2002	Windows, Linux, Mac	JavaScript	Easy	Licensed and supports the Open-Source Development Community	Free Version
Storm	SOAP	2008	Windows	F#	Easy	Open source	Free Version
Jmeater	HTTPS SOAP Database, Web-HTTP, and POP3(S) and IMAP(S)	2001	Windows, Linux, Mac	Java	Easy	HTTPS SOAP Database, Web-HTTP, and POP3(S) and IMAP(S)	2001

really simple. JMeter provides both graphical analysis and a variety of statistical outputs [24]. Table 7.1 illustrates WS security testing tools in terms of supported technology, first release, platforms supported, programming language used to build it, ease of use, license, and pricing [21, 25–27].

7.6 Conclusion

Web services are self-descriptive components that facilitate the quick composition of distributed software applications. Web technologies such as HTTP are used by web services to transfer messages between applications or electronic devices. The technological and managerial measures implemented in the system to guarantee the privacy, availability, and integrity of the information exchanged are known as web services security. This paper has discussed the most important challenges facing web services and reviewed many types of attacks that threaten web services security. Web servers and applications are complex systems, which increase sensitivity and make cyber-attacks difficult to detect. A web service intrusion detection system must be able to oversee the data quickly and accurately, as the size of the dataset increases the noise. Moreover, different web service testing tools are discussed and compared in this work. Modern distributed applications are built on web services, and it is essential that developers thoroughly test and certify their web services confidentiality, availability, and integrity before making them available to users.

References

1. Huhns, M. N., & Singh, M. P. (2005). Service-oriented computing: Key concepts and principles. *IEEE Internet Computing, 9*(1), 75–81.
2. Papazoglou, M. (2008). *Web services principles and technology*. Pearson Education.
3. Top 15+ most popular web service testing tools in 2023. (2023). *Software Testing Help*. Available at: https://www.softwaretestinghelp.com/web-services-testing-tools/ last accessed 2022/10/20.
4. Jagruti, B., Nidhi, P., Pandya, D. (2018). A survey on Webservice security techniques. In *2018 4th International Conference on Computing Communication and Automation (ICCCA)*. https://doi.org/10.1109/ccaa.2018.8777462.
5. Saravanaguru, R., Abraham, G., Venkatasubramanian, K., & Borasia, K. (2013). *Securing web services using XML signature and XML encryption*. (pp. 2–7). arXiv.
6. Ladan, M. I. (2011) Web services: Security challenges. In *2011 World Congress on Internet Security (WorldCIS-2011)*, (pp. 160–163). https://doi.org/10.1109/WorldCIS17046.2011.5749903.
7. Hartman, B. (2003). *Mastering web services security*. Wiley Technology Pub.
8. Aruna, S., & V. I. T. Vellore. (2016). Security in web services-issues and challenges. *International Journal of Engineering Reserach 5*, 09
9. Mouli, V. R., & Jevitha, K. P. (2016). Web services attacks and security—a systematic literature review. *Procedia Computer Science, 93*, 870–877. https://doi.org/10.1016/j.procs.2016.07.265
10. Salas, M. I. P., & Martins, E. (2014). Security testing methodology for vulnerabilities detection of XSS in web services and WS-security. *Electronic Notes in Theoretical Computer Science, 302*, 133–154. https://doi.org/10.1016/j.entcs.2014.01.024
11. Veeramreddy, J., & Prasad, K. M. (2011). A review of anomaly based intrusion detection systems A review of anomaly based intrusion detection systems. *International Journal of Computer Applications, 28*(7), 975–8887. https://doi.org/10.5120/3399-4730
12. Chandola, V., Banerjee, A., & Kumar, V. (2009). Anomaly Detection: A Survey. *ACM Computer Survey., 4*, 4. https://doi.org/10.1145/1541880.1541882
13. Hodo, E. et al. (2017) Shallow and deep networks intrusion detection system: A taxonomy and survey. arXiv preprint arXiv:1701.02145.
14. Chang, H., & Park, S. (2020). Developing the security threat detection model for the web service using deep neural network. *Journal of Theoretical and Applied Information Technology, 31*(6), 532–541.
15. Kim, J. et al. (2016) Long short term memory recurrent neural network classifier for intrusion detection. In *2016 International Conference on Platform Technology and Service (PlatCon)*, (pp. 1–5).
16. Sidharth, N., & Liu, J. (2007) A framework for enhancing web services security. In: *31st Annual International Computer Software and Applications Conference, COMPSAC 2007*, (pp. 23–30).
17. Sidharth, N., & Liu, J. (2008) Intrusion Resistant SOAP Messaging with IAPF. In *2008 IEEE Asia-Pacific Services Computing Conference*, (pp. 856–862).

18. Javatpoint: Soapui Security Test, https://www.javatpoint.com/soapui-security-test, last accessed 2022/10/20.
19. SoapUI: Working with Security Tests, https://www.soapui.org/docs/security-testing/working-with-security-tests, last accessed 2022/10/20.
20. WebServices Testing: Soap Sonar - A Quick Overview, http://webservices-testing.blogspot.com/2012/10/soap-sonar-quick-overview.html, last accessed 2022/10/21.
21. Wayback Machine: Parasoft's Web Service Testing Tool Should Aid Development, https://web.archive.org/web/20160306133135/http://www.gartner.com/id=356258, last accessed 2022/10/22.
22. Parasoft: Parasoft SOAtest - The Best API Testing Tool, https://www.parasoft.com/products/parasoft-soatest/api-testing, last accessed 2022/10/24.
23. Softpedia: Download Storm 1.1.0.0, https://www.softpedia.com/get/Internet/Servers/Server-Tools/Storm.shtml, last accessed 2022/10/20.
24. Apache JMeter™: Apache JMeter™, http://jmeter.apache.org, last accessed 2022/10/22.
25. SourceForge: JMeter vs. Soapsonar vs. SOAtest vs. SoapUI Comparison Chart, https://sourceforge.net/software/compare/JMeter-vs-SOAPSonar-vs-SOAtest-vs-SoapUI, last accessed 2022/10/25.
26. Softpedia: Download Storm 1.1.0.0, https://www.softpedia.com/get/Internet/Servers/Server-Tools/Storm.shtml, last accessed 2022/10/25.
27. Crosscheck Networks, Inc.: SOAPSonar Enterprise User Guide. Version 9.0, https://docplayer.net/165215686-Soapsonar-enterprise-user-guide-version-9-0.html, last accessed 2022/10/25.

Chapter 8
IoT in Transportation Systems

Prasad Dattaram Dessai and Milind Sakhardande

Abstract The smart technology that has emerged in the current generation has captured day-to-day life activities and promotes a fast pace in every process. The older systems are constantly being updated with rapid technological changes. More eco-friendly systems are being developed with greener technologies for more efficient, sustainable, reliable, and safe processes. In this paper, a SWOC analysis has been conducted to assess the feasibility of adapting IoT in transportation through a detailed research methodology to find proper literature and execute a literature review of selected work in the field of IoT and IoT in transportation. Addressing issues that have proven to be multidimensional due to strong vehicle growth and challenges to traffic management systems. Some unconventional and innovative solutions are needed to address the problems at a scale that is likely in the near future. The integration of the Internet-of-Things into the transportation system is leading to the notion of an intelligent transportation system to help address issues such as safety concerns and resource availability, traffic congestion, and transport overload. Relieving the burden on humans by integrating the physical world with software-based systems.

Keywords IoT – Internet-of-things · Intelligent transportation systems · Smart city · SWOC analysis

8.1 Introduction

Since the dawn of the industrial revolutions, smart, new, and secure technologies have evolved around us, updating the systems of daily life based on demands, needs, and research findings. The devices of the Internet-of-Things have evolved over the years according to the increasing requirements and research, the decision of the system,

P. D. Dessai (✉) · M. Sakhardande
Goa College of Engineering (Affiliated to Goa University), Goa 403401, India
e-mail: prasaddessai@gmail.com

M. Sakhardande
e-mail: milind@gec.ac.in

P. K. Kapur et al. (eds.), *Data-centric Approaches to Industrial Decisions*, Asset Analytics, https://doi.org/10.1007/978-981-96-7556-2_8

and its requirements that contribute to its efficient use, over the cost and other factors. Since 2010, Systems for identification based on RFID technology have been used in pharmaceuticals, retail, and logistics with advanced smart sensors in the last 2 decades. The key to IoT's success lies in standardization, ensuring interoperability, compatibility, reliability, and efficiency of operations worldwide [1].

The just-survived Covid 19 pandemic placed a major strain on supply chain management, or transportation, which helped move the bulk of supplies around the world. The growth of transportation systems depends on non-renewable resources to run the system, which leads to depletion and also impacts the environment after it is consumed. The solution for large transportation systems lies in adapting to new technologies that are smarter, safer, and bring performance to an optimal level with the help of the Internet-of-Things, which is now used in most aspects of day-to-day life [2].

8.1.1 Internet-Of-Things (IoT)

Internet-of-Things is a system or physical device connected over a network to enable or form the exchange of data over the network. Physical devices or systems are electronic devices or hardware integrated or not with the software required for communication. Individual or combined IoT devices are operated to achieve more results [3] (Fig. 8.1).

The concept of IoT revolves around the idea of making everyday objects "smart" by embedding them with sensors and connectivity, enabling them to communicate, analyze data, and performs tasks autonomously or semi-autonomously.

8.1.2 The IoT Architecture

The Internet-of-Things (IoT) architecture is a blueprint detailing the necessary components for facilitating smooth communication and data sharing among IoT devices, networks, and applications. The IoT architecture typically consists of several layers, each serving specific functions and interacting with one another. Here is a simplified overview of the key layers in the IoT architecture:

The application layer is the first layer that serves interaction, storage, and access by providing data in the proper format for use in software programs and establishing communication. The benefits are flexibility and mobility, scalability of the project, security aspects, disaster recovery, and cost savings.

The network layer is the intermediate layer that ensures that different networks can be interconnected. It forwards packets to the network routers via defined algorithms and is primarily responsible for connecting the IoT with the cloud server. Advantages include the transfer of files in smaller volumes, data security, the ability to run applications on different systems, user-controlled access, and scalability.

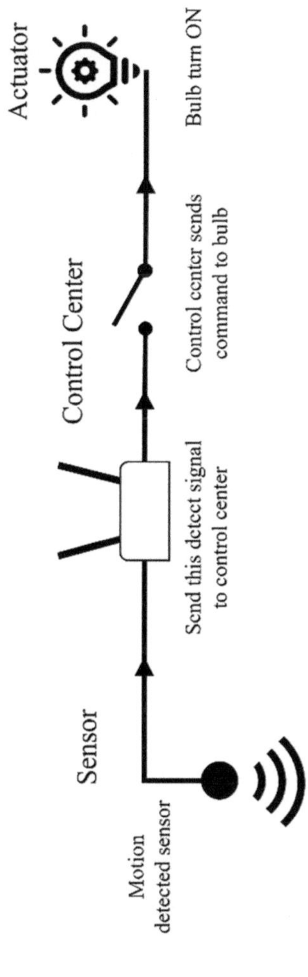

Fig. 8.1 Example IoT - sensor to actuator [3]

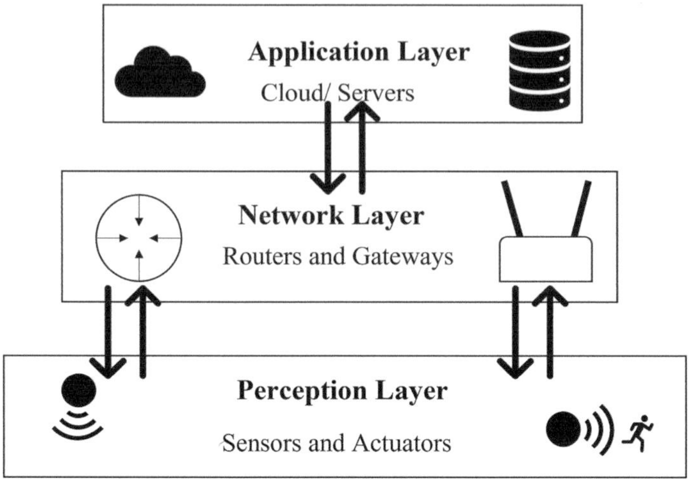

Fig. 8.2 IoT Architecture [5]

The perception layer is the final layer involving all IoT devices that are primarily responsible for acquiring, perceiving, or recognizing data and/or information from the delineated areas according to their ability to do so. The benefits are time and labor savings, error-free data or information, cost-efficient operation, increased productivity of processes and systems, improved occupational safety, and optimal use of resources and assets [4]

IoT architecture is formed with 3 Layers as below (Fig. 8.2).

8.1.3 Capabilities and Major Components of IoT Devices

Data acquisition and control to capture analog data or information at a specified time interval, data processing and storage as internal or external data processing or storage activities, connectivity devices connected to the network layer via wired or wireless modes, and power management for IoT devices that require a power source [6] (Fig. 8.3).

8.1.4 Transportation Systems

Transportation services to meet the demand for travel and related elements form a transportation system. The main function of transportation systems is to move goods, people, and information.

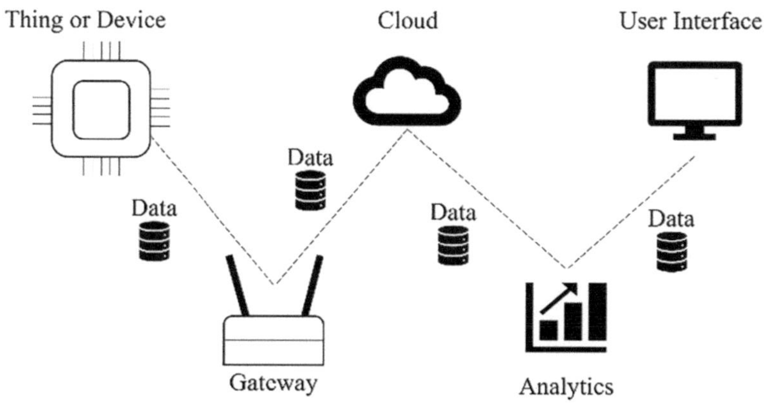

Fig. 8.3 Major components of IoT [3]

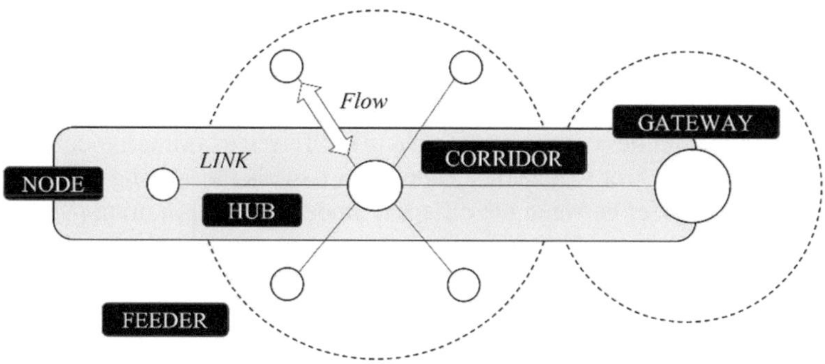

Fig. 8.4 Components of the transportation system [7]

8.1.4.1 Components of the Transportation System

(See Fig. 8.4).

8.1.4.2 Modes of Transportation & Issues

Various means of transportation are used to carry goods, people, and information [28, 29].

Roadways transport, where vehicles travel across the country via a road, has important aspects such as the combination of location, use, and route of facilities, and consumers as support.

Railways are also called train transportation. It is a vehicle that runs on rails used for transportation. It involves important aspects such as planning to improve service and/or reduce delays, off-track delays in pickup and delivery, yard activities, and irregular delivery schedules.

Package carriers like FedEx which transport tiny things, including letters and freight, with important aspects such as costly, prompt, and reliable delivery, small shipments with tight deadlines, the method of choice for e-businesses (e.g., Amazon, Dell), and grouping shipments (especially for air mode).

Waterway transportation is which movement of goods and people on waterways by using a vehicle that is compatible to drive true waterways, having key issues which are restricted to specific geographic regions, large loads for relatively little money, slowest, strong in international commerce (clothes, food, automobiles, etc.).

Airways transportation is a mobile means of transportation that operates in airspace but takes into account important aspects such as location and number of hubs, base locations for the fleet and crew, allocation of the fleet, crew planning, management of yield, and optimization of flight schedules.

Pipeline transportation for liquid/gas transfers, which pose important problems and significant fixed costs, especially for natural gas, petroleum products, and crude oil, would be used to transport crude oil to ports or refineries, but not to transport refined gasoline to service stations.

Intermodal transportation is the movement of a shipment to its destination through the use of multiple modes of transportation and is referred to as intermodal. Common examples are rail/truck or water/truck. The main issue is the sharing of information to speed up the transfer between the different modes of transportation [8].

8.1.4.3 Transportation Infrastructure

Adequate infrastructure is required to operate any transportation system in day-to-day life for the movement of goods, people, and information and to meet the demand of travel Roadways which require roads, bridges, power and energy, fuel, etc., Railways which require stations/terminals, rails, bridges, energy, fuel, etc., Waterways which require ports, canals, passways, etc., Airways which requires Airports, energy, fuel, etc. [9].

8.2 Research Methodology

The main objective of a systematic literature review is to locate, assess, analyze, and organize all pertinent articles related to one or more research inquiries and subjects. In contrast to a narrative literature review that primarily examines descriptive findings of a particular field and is prone to selection bias, a systematic literature review can offer a thorough summary of research in that area [7].

The process of conducting a systematic literature review involves the following stages:

1. Defining research inquiries: Creating the questions to be addressed in the research.
2. Conducting a literature review involves creating a search plan using a wide range of important terms to gain a thorough understanding of the topic being studied. Next, appropriate filters are established to request the most suitable sample of articles.
3. Conducting a bibliometric analysis involves quantitatively analyzing and visually representing a chosen set of articles to grasp important aspects of the topic such as publication patterns, journal sources, references, partnerships, and keyword emphasis.
4. Conducting a thorough content analysis of the chosen articles in order to outline the key findings in various interconnected subject areas. Given this information, we can grasp the current research environment and pinpoint potential areas for future research.

The research questions are designed to mirror the goal and extent. This article examines the connection between IoT and Transportation Systems, with a specific focus on their interactions in the existing literature. Regarding these ideas, three research questions are suggested to comprehend the level of understanding in utilizing IoT in Transportation Systems.

- Research Question 1: What written works discuss IoT in Transportation Systems and how can they be grouped?
- Research Question 2: What are the consequences of implementing IoT in Transportation Systems?
- Research Question 3: What future research areas should be explored to address the gaps?

Drawing from the research queries mentioned earlier, Fig. 8.5 outlines the process for searching documents, encompassing five stages:

(1) Conducting keyword searches,
(2) The configuration of filters,
(3) Examination of the titles and abstracts,
(4) Analysis of the complete text, and
(5) Analysis of findings.

(1) Keyword search: In this study, a keyword search was conducted on Google Scholar databases. The search took place in December 2022, with a focus on two sets of keywords associated with IoT and Transportation Systems. The initial keywords are linked to IoT and Transportation Systems, including "IoT," "Transportation Systems," and "IoT and Transportation Systems". In addition, as numerous IoT applications were examined within the framework of transportation systems, "Transportation network" was included in this category. Another group of keywords in the Transportation realm

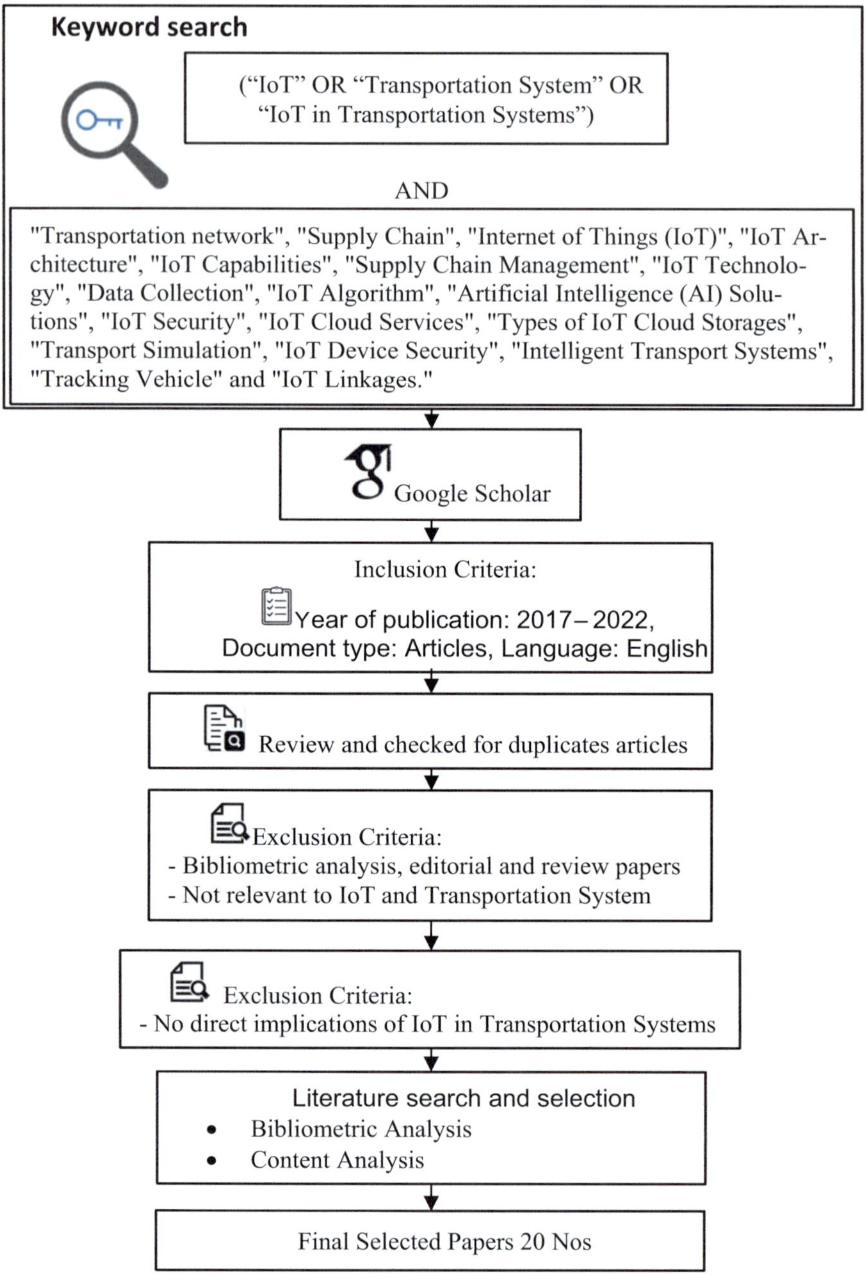

Fig. 8.5 Research method flow

is "Supply Chain," "Internet of Things (IoT)," "IoT Architecture," "IoT Capabilities," "Supply Chain Management," "IoT Technology," "Data Collection," "IoT Algorithm," "Artificial Intelligence (AI) Solutions," "IoT Security," "IoT Cloud Services," "Types of IoT Cloud Storages," "Transport Simulation," "IoT Device Security," "Intelligent Transport Systems," "Tracking Vehicle," and "IoT Linkages." The "OR" Boolean operator was utilized to link keywords within one group, while "AND" was used to connect the main groups of keywords concerning IoT and Transportation Systems. The first search produced 218 outcomes on Google Scholar.

(2) The configuration of filters: The next step involves configuring multiple filters to choose the most appropriate articles, and any papers that do not fall within the research scope or are unrelated to answering the research questions are excluded. Taking into account the quality and thoroughness of chosen papers, the search results were narrowed down to journal articles that had undergone the peer-review process. The language for publication was limited to English. Hence, this study did not include conference proceedings, book chapters, pre-prints, and papers published in a different language. Following the introduction of the new filters, the search yielded 96 relevant articles in Google Scholar, correspondingly.

(3) Examination of the titles and abstracts: Initially, we analyzed the nature of the papers in the refined sample, which consisted of 5 bibliometric analysis papers, excluding editorial and review articles. Next, we examined the thematic importance of these articles; papers that lack significance in utilizing IoT and Transportation Systems. Additionally, articles discussing the use of IoT in Transportation Systems were excluded if they did not have a specific focus on transportation. During this stage, 24 papers were eliminated in total.

(4) Analysis of the complete text: Next, we performed a thorough reading of the full text during the second round of paper selection. Special attention was given to the papers in this phase, particularly those that do not directly relate to the research questions. Although both papers include keywords IoT and Transportation System, they do not thoroughly explore the application of IoT in Transportation Systems, so they are deemed irrelevant for addressing the research queries. During this phase, an additional 26 papers were deemed unsuitable for the topic and subsequently eliminated. Afterwards, 20 papers were chosen in total. Wherein 1 set on specific to IoT and the other set is specific to Transportation Systems. 10 in each category. These articles were Both IoT and Transportation considered in 10 articles to cover the most available transportation systems.

(5) Analysis of findings: Using the chosen sample, a bibliometric analysis was carried out to present the outcomes of publication trends, source distribution, co-authorship, citations, and keyword co-occurrence. Afterwards, a

content analysis was conducted to explore ways in which various Transportation Systems can be enhanced through different types of IoT integration and to outline the potential opportunities, challenges, and future areas of research.

8.2.1 Bibliometric Analysis

8.2.1.1 Publication Trend

Figure 8.6 shows the quantity of articles released from 2017 to 2022. The increased focus on integrating IoT into Transportation Systems is clearly noticeable, with this movement gaining momentum notably from 2017 onwards. In the year 2022, 127 articles were released in global publications, representing 23.78% of all publications over the past six years. The current publication trend indicates that the increase in IoT research is offering new possibilities is unlocking numerous possibilities across various sectors by leveraging connected devices, data analytics, and automation, as recognized by industry professionals and academia. IoT research is continuously expanding the horizons of what is possible in various domains. The integration of IoT with AI, machine learning, and big data analytics is expected to drive further innovations, making systems smarter, more efficient, and more responsive to human needs.

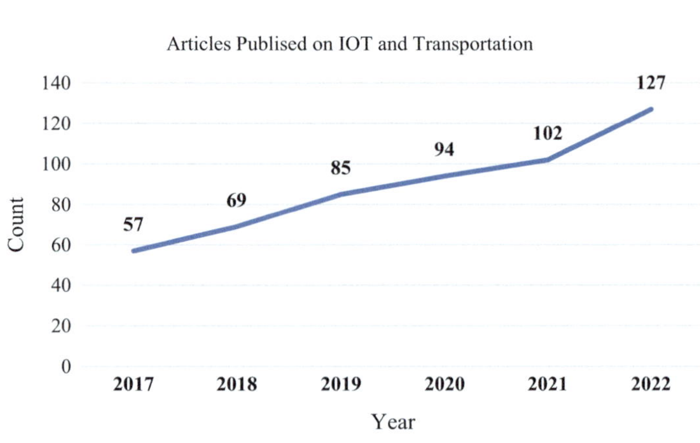

Fig. 8.6 Articles published trend from 2017 to 2022

8.2.1.2 Research Highlights and Keywords

Key findings and important terms from the study

An occurrence analysis of popular keywords linked to IoT and Transportation Systems was carried out to pinpoint the research highlights. To provide a comprehensive snapshot of the existing research environment, we utilized the "all keywords" and "full counting" functions to list every keyword from past studies and determine the total number of times they appeared together. The clusters, and instances of these keywords can be found in (Table 8.1).

The 71 common keywords are categorized into six clusters, which help identify the main research areas in IoT. The six groups of keywords have uncovered the various areas of research focus. Cluster 1 consists of 12 keywords that primarily concentrate on the utilization of IoT, such as IoT devices, in smart Transportation, smart Vehicle systems, and other supply chain functions. Cluster 2 consists of 12 items that mainly focus on utilizing IoT to enhance Transportation and Supply Chain. Cluster 3 includes 14 nodes that concentrate on smart and secure transportation systems. Cluster 4 is made up of 11 important words that are involved in communication, monitoring, and live monitoring. Cluster 5 comprises 11 nodes that prioritize enhancing decision-making through AI and other intelligent technologies. Cluster 6 includes 11 terms connected to the field of literature research, indicating a focus on consolidating recent findings.

8.2.1.3 Content analysis

The analysis of keyword search results has revealed the significance of IoT, technology, and data in the Transportation System, while content analysis is conducted to grasp the impact of smart technologies and data analytics on both IoT and the Transportation System. Analyzing content is a crucial process for systematically studying the research progress in various subject areas. In this part, we provide an intricate examination of four key subjects concerning transportation systems across different phases. Smart production initially alters demand patterns and the methods of meeting such demands, subsequently reshaping purchasing and logistics service requirements. Therefore, the discussion of IoT in transportation systems began. Next, innovative solutions for the top two crucial transportation systems, supply chain and transportation, were presented. Finally, the overall digitalization and system integration challenges were addressed to improve efficiency in various transportation processes. In Fig. 8.5, the distribution of articles published trend from 2017 to 2022 is displayed, with 23.78% of papers addressing multiple topical areas covering IoT and Transportation systems.

Table 8.1 The Clusters, of the keywords for search

Clusters	Keywords
Cluster 1	IoT Utilization
	IoT Devices
	Smart Transportation
	Smart Vehicle Systems
	Supply Chain functions
	IoT Functions
	Internet-of-Things
	IoT
	Transportation Systems
	Types of Transportation System
	IoT modules
	IoT functions
Cluster 2	IoT to Enhance Transportation
	IoT to enhance Supply Chain
	IoT to Tackle Transportation
	Transportation using IoT
	Advancement in Transportation
	Updated IoT modules
	Enhanced IoT devices
	IoT device sustainability
	IoT in supply chain
	Internet-of-Things and Transportation
	IoT in Public Transportation
	IoT Traffic Management
Cluster 3	Smart Transportation System
	Secure Transportation System
	Intelligent Transportation Systems (ITS)
	Connected Vehicles IoT
	IoT for Urban Mobility
	Vehicle-to-Infrastructure (V2I) IoT
	Vehicle-to-Vehicle (V2V) Communication IoT
	Smart Traffic Lights IoT
	IoT-enabled Autonomous Vehicles
	IoT and Smart Cities Transportation
	IoT-based Transportation Infrastructure
	IoT and Sustainable Transportation
	IoT for Traffic Flow Optimization

(continued)

Table 8.1 (continued)

Clusters	Keywords
	IoT in Transportation Security
Cluster 4	IoT communication
	System Monitoring
	Live Tracking
	Wireless Communication IoT Transportation
	IoT Networks in Transportation Systems
	M2M Communication in Transportation IoT (Machine-to-Machine)
	IoT-based Communication for Autonomous Vehicles
	Smart Transportation Communication Networks IoT
	IoT for Real-time Communication in Transportation
	IoT Communication for Connected Vehicles
	IoT Data Communication in Transportation
Cluster 5	Enhanced decision-making
	Intelligent technologies
	Machine Learning IoT Transportation
	AI in Transportation IoT
	Predictive Analytics IoT Transportation
	Intelligent Traffic Management IoT
	Big Data IoT Transportation
	IoT for Intelligent Mobility Solutions
	Real-time Decision Making IoT Transportation
	Cognitive IoT Transportation Systems
	IoT-enabled Smart Infrastructure Decision Making
Cluster 6	Literature Research
	Surveys
	Analytical hierarchy
	Research Article
	Systematic Review
	Recent Developments
	Comprehensive Review
	Current Trends
	Meta-Analysis
	State-of-the-Art
	Advances and studies

8.3 Literature Review

The final selected papers based on the research methodology were used to conduct a literature search for scientific articles that were derived from published or newly researched articles.

8.4 SWOC of Using IoT in Transportation Systems

Based on the literature review, a SWOC analysis has been conducted to understand the strengths, weaknesses, opportunities, and challenges of using IoT in transportation systems. This method is used to identify opportunities and future scope for improvement and success of IoT in transportation systems (Tables 8.2, 8.3).

8.4.1 Strengths of Using IoT in Transportation Systems

- **Visibility in the system:** by gaining more clarity about the system and the benefits of analyzing data and focusing on performance, the organization responsible for managing the transportation system can improve performance.
- **Enhanced safety and security**: IoT can improve road safety through connected vehicle technologies. Vehicles have the ability to communicate with one another as well as with infrastructure, allowing for collision avoidance systems and giving drivers instant warnings.
- **Improved user experience**: IoT applications in transportation enhance the overall user experience for passengers.
- **Predictive Maintenance**: IoT sensors can monitor the health of vehicles and transportation infrastructure in real-time. Predictive maintenance algorithms use this data to identify potential issues before they cause breakdowns, reducing maintenance costs and improving service reliability.
- **Sustainability and environmental impact**: IoT helps optimize transportation routes and operations, reducing unnecessary travel and carbon emissions. This contributes to more sustainable transportation systems and promotes environmentally friendly practices.
- **Efficient fleet management**: IoT-powered telematics and asset tracking enable efficient fleet management in public transportation and logistics. Fleet managers can monitor vehicle locations, fuel consumption, and maintenance needs, leading to better resource allocation and reduced downtime.
- **Traffic management**: providing signals in real-time can help manage traffic.
- **Ticketing and toll management**: Recent advances in India in toll management via the Internet-of-Things (IoT) are bringing many advantages such as reduced queues at toll booths, and time and fuel savings.

Table 8.2 Literature on the Internet-of-Things (IoT)

Author (Refs.)	Broad topic	Methods used
Phase Avani, Mhetre Nalini – [3]	IoT Applications for Supply Chain Management	Detailed on 1. IoT & Supply chain, 2. IoT Technologies & Algorithm
Tsang Y.P., Wu CH, Koo PS, and K.L. Choy [4]	A method for mitigating cold supply chain hazards based on the Internet-of-Things	IoTRMS. Wireless sensor network, Cloud databases, Fuzzy logic method
Mladen Jardas, K Ivanić, Marko Gulić, & Čedomir D [10]	The Impact of the Internet-of-Things on Port Development as a Supply Chain Holder	IoT Architecture in Smart Ports
Haddud, A, DeSouza, A Lee, H, AR, & Khare, [10]	Analyzing the possible advantages and difficulties of supply chain blend with the IoT	The survey by 87 university professors across six distinct continents
Tu Mengru [11]	Internet-of-Things deployment intent in supply chain and logistics management	Mixed methods research
Arun Aryal, Bo Li, Ying Liao & Prasnna Nattuthurai [12]	A systematic study of the use of IoT and analytics in Big data for the SCM	The Leximancer analysis for analyzing pertinent issues and trends in the literature
Zhi Li, Xinjun Lai, Guo Liu, Gangyan Xu & Layne Liu [13]	An IoT-based platform for tagging and chasing the supply chain of prepared foods	QR code & RFID to tagging and chasing. XML to share information
Tharaka de V, Shah J. Miah & Himanshu Shee [14]	IoT in management of supply chains A story about the sustainability of the retail industry	IoT technology usage by sampling in retail companies
Pishdar Mahsa, Jonas S, Fatemeh G, & Jurgita A [15]	Internet-of-Things and Supply Change Management Challenges	DEMATEL (Decision Making Trial & Evaluation Laboratory technique)
Tu Mengru, Ming-Fang Yang & Ming K. Lim [16]	IoT-based cyber-physical architecture: IoT-based production supply chain and logistics system	A research study, emulation experiment approach, and cost–benefit evaluation

Table 8.3 Literature on IoT and transportation systems

Author (Refs.)	Broad topic	Methods used
Hong Z a and Lu Xinxin b [17]	Network for vehicle communications in an IoT-based intelligent transportation system	The layered network simulation approach is a modeler for OPNET
Priyanka E, M Ponni, Thangavela and Maheswaria [18]	IoT & LQR-PID for online monitoring, pressure & flow in the fluid transport network	Embedded IoT-based framework, the local controller by the SCADA-integrated LQR-PID
Al-Turjmana Fadi and b, Joel Poncha L [19]	Overview of intelligence, security, and vehicle sensor networks in IoT-enabled smart cities	Vehicle sensor systems development in intelligent transport systems for confidentiality
Guie Li [20]	Construction of a 5G network and IoT-based logistics cold chain transportation system	Predictive Data Transfer paradigm technique for data transfer via Internet-of-Things
Rajesh A, Anita G, Dharam, Praveen M & Rajeev A [21]	Using machine learning and intelligent IoT sensors to digitally transform Highways	Deep learning approaches and IoT for vulnerable road users safety
S. Vanitha, J. Karthikeyan, Radhika, and G. Kavitha [22]	Using IoT to model an intelligent transport system for public health	The mockup uses AutoCAD and an integrated controller oversees the transport network
Mastalerza M, Bartosz, Sławomir, Aleksander M, & Anna Ś [23]	The intelligent transportation system, person BIBO recognition with IoT and ML	Techniques developed of a hands-free, pay-as-you-go, and BIBO smart fare collection
Al-Turjmana Fadi & Arman Malekloo [24]	A report on parking system in Internet - of - things cities	Examining emerging parking trends. Driver assistance systems, mixed parking paradigm
R Geetanjali, Varun M, Sujesh P. Lal, Razi I and Naveen J [25]	An IoT-focused trust analysis framework for automotive networks	A System for Intrusion Detection to the flow of data utilizing the MATLAB simulator
Lakshmi S I [26]	Applications for transport systems using AI	AI solutions to address problems in the global & Indian state transport industries

8.4.2 *Weakness of Using IoT in Transportation Systems*

- **Security and confidentiality**: managing data collected and transmitted over the Internet-of-Things is a challenge, and protecting data from cyberattacks and physical compromise through network or hardware attack.
- **Integration complexity**: With larger devices, encryption, and security protocols become a challenge. Although it may seem that IoT devices are doing simple tasks like measuring temperature, there is a lot of complex technology behind them. Moreover, if they send important data to another workflow or system, they can negatively impact all related processes.
- **Power dependability**: Many devices rely on connectivity and power to operate successfully. They need the internet and constant power. Both the devices and everything connected to them will stop working if either fails.
- **Reliability and Connectivity**: IoT devices rely on stable and robust connectivity to function effectively. In locations lacking good network connectivity or intermittent connectivity, IoT solutions may experience disruptions and reduce the accuracy of real-time data.
- **Higher costs (money and time)**: The deployment of IoT devices typically demands a substantial investment in both time and money. There are many devices to buy and configure, along with personnel to set up, others to incorporate into the network, and calls to vendor support.
- **Over-reliance on technology**: Relying heavily on IoT devices and systems may create dependencies that leave transportation systems vulnerable in the event of system failures, cyberattacks, or technical glitches.

8.4.3 *Opportunities for Using IoT in Transportation Systems*

- **Emergency response through IoT technology**: where the IoT can accelerate interactions between various emergency responders such as fire, police, and hospitals so they can leverage data and make faster, more informed decisions.
- **Smart public transit**: optimized commutes with apps where IoT GPS & sensors collect and analyze data on every road a user travels. They can provide advice, determine the best routes, and warn users of accidents or traffic jams. Less time in the car means less pollution.
- **Safety and security enhancements**: safe fleet management and truck driving, where IoT sensors can keep track of key vehicle safety components such as tire pressure and notify drivers of problems.
- **Autonomous public transit**: The future will see the integration of autonomous technologies with public transit systems, enabled by IoT.

- **Data-driven decision-making**: Intelligent traffic signs, where smart traffic signals resemble regular traffic lights, use a variety of sensors to continuously monitor traffic. The numerous signals are interconnected and instantly adjust to changing traffic conditions to reduce vehicle idling time.
- **Smart infrastructure**: Innovative parking technologies, where IoT applications and smart parking meters can alert drivers when a space is available, allow them to reserve the space, and provide location instructions.
- **Predictive maintenance**: for vehicles, that includes a vehicle's sensors keep track of a variety of variables like vibration, temperature, oil or fuel consumption, and light. This way, drivers can fix the problem before it gets worse, saving time and money by not having to inspect a vehicle that does not need it.

8.4.4 Challenges of Using IoT in Transportation Systems

- **Security risks**: Limited integration of security systems can be difficult or impossible due to their diversity and size. Insecure Communications Protocols where attackers and hackers can readily access your data and even remote controls due to insecure protocols.
- **Data privacy and compliance**: Data and identity theft in places like parking lots, electric vehicle charging stations, and video surveillance systems are just a few examples of insecure smart city structures that provide a lot of targeted personal information that hackers could use for identity theft and fraudulent transactions.
- **Software and firmware vulnerabilities**: insufficient computing power for effective integrated security. Access control of the IoT system is inadequate. Due to the technological limitations of IoT devices and budget constraints, there are no regular patches or upgrades.
- **Resistance to change**: Implementing IoT in transportation may face resistance from various stakeholders, including transportation authorities, employees, and passengers, who may be apprehensive about adopting new technologies or changing established practices.
- **Power consumption**: IoT devices often require continuous power supply, leading to increased energy consumption. Optimizing power usage and exploring energy harvesting techniques are necessary to ensure long-term sustainability.
- **Lack of staff awareness and training**: many transportation companies fail to adequately train and educate their employees. For example, employees might connect devices to the network without taking the necessary security measures [27].

To summarize on Table 8.4 and Fig. 8.7, IoT is transforming transportation by enabling smarter, more efficient, and safer systems. Key achievements include automation and control of fluid transport, improved traffic and safety management, enhanced security in smart cities, better vehicle communications, and optimized logistics and parking systems. These advancements contribute to more sustainable and resilient transportation networks by enhancing efficiency, reducing emissions, and improving safety and reliability.

Table 8.4 Area of improvement by using IoT in transportation

Area on improvement	Topic	IoT's Used	Applied sections	Achievements
Intelligent transportation system	IoT & LQR-PID for online monitoring, pressure, and flow control in the fluid transport	IoT module	Fluid transport networks	System for automation and control
	Applications for transport systems using AI	AI-enabled	Public transport	Traffic & Safety management
	Intelligent, security & vehicle sensor networks in IoT-enabled smart cities	Vehicular sensors, Sensors	Smart city	Security
	An IoT-focused trust analysis framework for automotive networks	Malicious devices	Smart & Green City	Security
	A vehicle communications network in an IoT-based intelligent transportation	Modeler for OPNET	Vehicle communications	Vehicle communications
	Using IoT to model an intelligent transport system for public health	Embedded control, GSM, Sensors	Transport	Security & Safety
Smart city	Using ML and intelligent IoT sensors to digitally transform roadways to improve vulnerable road safety	IoT sensors & machine intelligence	Highway	Vulnerable Road Safety & Renewable Energy
	A report on parking system in Internet-of-things cities	IoT-enabled Modules	Smart parking system	Planned Parking
	Intelligent transportation system person BIBO recognition with IoT assistance and machine learning approaches	Smartphones & a subset of machine intelligence	Automated Toll Payment mechanism	Prediction Analytics
Logistics	Construction of a 5G network and Internet-of-Things-based logistics cold chain transportation system	FPGA & Technique for Predictive Data Transfer	Cold Chain Logistics	Real-time Tracking

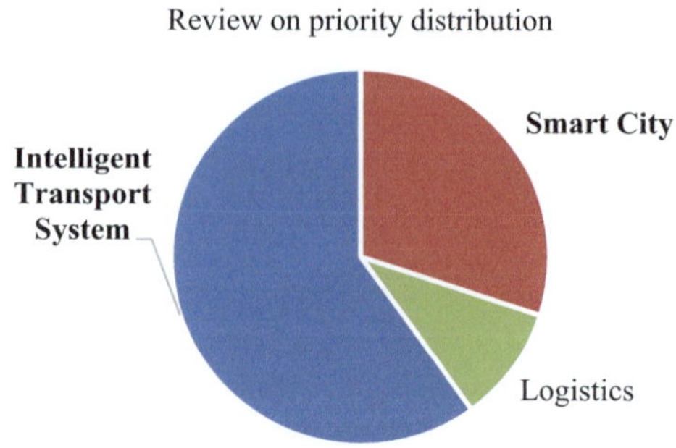

Fig. 8.7 Area of improvement by using IoT in transportation

8.5 Conclusion

Various IoT modules are being explored, such as vehicle trackers, RFID tags, IoT-based risk management systems, wireless sensor networks, cloud databases, QR codes, etc. When it comes to IoT applications in transportation systems, great improvements are needed in the field of IoT in intelligent transportation systems, smart cities, and logistics.

Vehicles equipped with older technology need to be updated with IoT devices such as GSM, real-time trackers, coded FPGAs, OPNET modelers, and AI. OPNET modeler can help by targeting vehicle communication networks to generate data and then using AI methodology for further technology improvements.

Key developments are taking place in public transportation, including smart cities & green cities, highways, smart parking, toll collection, and cold chain logistics. The IoT helps save fuel and resources by optimizing vehicle movements on the road, tracking vehicles properly, and also preventing major breakdowns through regular preventive maintenance. The versatility of IoT is also evident in its application in various areas of fluid transportation.

8.6 Future Scope

Through the literature review and proceeding with SWOC analysis, we found that the future scope of IoT in public transportation systems holds great potential for transforming urban mobility and enhancing the overall passenger experience. As cities become more connected and technology advances, IoT will play a crucial role in making public transportation more efficient, accessible, and sustainable. Some key areas of future research and growth for IoT in public transportation are.

- Real-time passenger information: IoT can enable the delivery of real-time information to passengers, including accurate arrival times, service disruptions, and alternative routes, and allow public transportation systems to dynamically adjust routes and schedules based on real-time demand data.

- Predictive maintenance: IoT can facilitate predictive maintenance of public transport vehicles and infrastructure.
- Smart fare collection: IoT-powered smart fare collection systems can streamline payment processes, enabling contactless and mobile-based payments.
- Safety and security enhancements: IoT can be used for enhancing public transportation safety and security. For instance, video surveillance systems can be integrated with analytics to detect suspicious behaviors.
- Autonomous public transit: The future can see the integration of autonomous technologies with public transit systems, enabled by IoT.
- Environmental monitoring and sustainability: IoT sensors can be deployed to monitor air quality, noise levels, and other environmental factors to understand the environmental impact of transportation systems.
- Urban traffic optimization: IoT-based traffic management systems will continue to evolve, incorporating real-time data from various sources, including connected vehicles, sensors, and smartphones. Advanced algorithms and AI-driven analytics will be used to optimize traffic flow, reduce congestion, and minimize travel times.

The future of IoT in transportation holds immense potential to create smarter, more efficient, and safer transportation systems. These advancements will not only contribute to sustainability by reducing emissions and energy consumption but also enhance the resilience of transportation networks by improving real-time monitoring, predictive maintenance, and communication among vehicles and infrastructure. As technology continues to evolve, the integration of IoT with AI, machine learning, and big data analytics will further drive innovations in transportation.

References

1. Shancang, L., Li Da, X., & Shanshan Z (2014). *The internet of things: A survey*. Published online: 26 April 2014 Springer Science Business Media New York 2014
2. Bhardwaj, K. K., et al. (2019). Designing energy-efficient IoT-based intelligent transport system: Need, architecture, characteristics, challenges, and applications. In M. Mittal, et al. (Eds.), *Energy conservation for IoT devices, studies in systems, decision and control 206*. Springer Nature Singapore Pte Ltd.
3. Phase, A., & Nalini, M. (2018). Using IoT in supply chain management. *International Journal of Engineering and Techniques, 4*(2), 973–979.
4. Tsang, Y. P., et al. (2018). An Internet of Things (IoT)-based risk monitoring system for managing cold supply chain risks. *Industrial Management & Data Systems, 118*(7), 1432–1462.
5. Haddud, A. et al. (2017). Examining potential benefits and challenges associated with the Internet of Things integration in supply chains. *Journal of Manufacturing Technology Management*
6. Rodrigue, J.-P. et al. (2020). *The geography of transport systems*. Hofstra University, Department of Global Studies & Geography, https://transportgeography.org. Assessed 25 Oct 2022.
7. Xu, S. et al. (2022). *The application of Industry 4.0 technologies in sustainable logistics: A systematic literature review (2012–2020) to explore future research opportunities*. Published online: 10 December 2021 Environmental Science and Pollution Research.

8. Sunil, C., Peter M., Kalra, D. V. (2017) *Supply chain management, strategy, planning and operation*. 6th Edition
9. Rodrigue, J.-P. (2020). *The geography of transport systems*. Fifth Edition, New York: snapRoutledge. Assessed 25 Oct. 2022.
10. Jardas, M., et al. (2018). The role of internet of things on the development of ports as a holder in the supply chain. *Pomorski zbornik, 54*(1), 61–73.
11. Tu, M. (2018). An exploratory study of Internet of Things (IoT) adoption intention in logistics and supply chain management: A mixed research approach. *The International Journal of Logistics Management*
12. Aryal, A. et al. (2018) The emerging big data analytics and IoT in supply chain management: A systematic review. *Supply Chain Management: An International Journal*
13. Li, Z., et al. (2017). IoT-based tracking and tracing platform for prepackaged food supply chain. *Industrial Management & Data Systems, 117*(9), 1906–1916.
14. De Vass, T., Himanshu, S., & Shah, J. M. (2021). Iot in supply chain management: A narrative on retail sector sustainability. *International Journal of Logistics Research and Applications, 24*(6), 605–624.
15. Pishdar, M. et al. (2018). *Internet of things and its challenges in supply chain management: A rough strength-relation analysis method*
16. Tu, M., Ming, K. L., & Ming-Fang, Y. (2018). IoT-based production logistics and supply chain system–Part 2: IoT-based cyber-physical system: a framework and evaluation. *Industrial Management & Data Systems*
17. Zhang, H., & Xinxin, L. (2020). Vehicle communication network in intelligent transportation system based on Internet of Things. *Computer Communications, 160*, 799–806.
18. Priyanka, E. B., et al. (2020). Integrating IoT with LQR-PID controller for online surveillance and control of flow and pressure in fluid transportation system. *Journal of Industrial Information Integration, 17*, Article 100127.
19. Al-Turjman, F., & Joel, P. L. (2020). Intelligence, security, and vehicular sensor networks in internet of things (IoT)-enabled smart-cities: An overview. *Computers & Electrical Engineering, 87*, Article 106776.
20. Li, G. (2021). Development of cold chain logistics transportation system based on 5G network and Internet of things system. *Microprocessors and Microsystems, 80*, Article 103565.
21. Singh, R., et al. (2021). Highway 4.0: Digitalization of highways for vulnerable road safety development with intelligent IoT sensors and machine learning. *Safety Science, 143*, Article 105407.
22. Vanitha, N. S., et al. (2020). Modelling of intelligent transportation system for human safety using IoT. *Materials Today: Proceedings, 33*, 4026–4029.
23. Mastalerz, M. W., et al. (2020). Passenger BIBO detection with IoT support and machine learning techniques for intelligent transport systems. *Procedia Computer Science, 176*, 3780–3793.
24. Al-Turjman, F., & Arman, M. (2019). Smart parking in IoT-enabled cities: A survey.". *Sustainable Cities and Society, 49*, Article 101608.
25. Rathee, G., et al. (2020). A trust analysis scheme for vehicular networks within IoT-oriented green city. *Environmental Technology & Innovation, 20*, Article 101144.
26. Iyer, L. S. (2021). AI enabled applications towards intelligent transportation. *Transportation Engineering, 5*, Article 100083.
27. Frue, K. (2022) SWOT analysis of the internet of things. In *PESTLE Analysis*, 13 May 2019, pestleanalysis.com/swot-analysis-of-the-internet-of-things. Accessed 25 Oct 2022.
28. Philpott, M. (2022). *What are the different modes of transportation in logistics*? Philburn Logistics Inc., 16 Mar. 2017, philburn.com/what-are-the-different-modes-of-transportation-in-logistics. Assessed 25 Oct. 2022.
29. Sanders, L. (2022). Top Four Modes of Transport | Logistics | Supply Chain. Top Four Modes of Transport | Logistics | Supply Chain, 25 July 2022, wearedg.com/news/article/modes-of-transport-logistics. Assessed 25 Oct. 2022.

Chapter 9
Contact to Customer Lead Generation Predictive Model Using Machine Learning Techniques

V. L. Helen Josephine, Chandana Sai, Justin Joy, and Ajay A. Kashyup

Abstract Lead generation is the process of turning an outside person or business into a customer of the business. Traditionally, marketing personnel must conduct significant follow-ups in order to convert even one potential consumer. In an effort to convert bad client leads, businesses have burned through the majority of their cash reserves. Due to this, it is now necessary to develop an automated system that can correctly anticipate whether or not a lead should be explored (convert to customer or not). In this study, we attempt to evaluate historical data regarding leads produced by other businesses in order to train and validate an ML/DL model and test it against real-world characteristics to categorize them as hot leads (convert to customers) or cold leads (failed leads). This can be achieved by employing ML algorithms, low code-no code libraries like pycaret in python and make predictions regarding probable lead creation, propensity to convert generated leads, optimal communications team's actions on the leads. The Supervised ML Algorithms like Logistic Regression, Decision Trees, Random Forests and other models using python library were built to score leads for identifying potential conversions. With good and broad lead scoring models in place they can optimize their CTI actions on the basis of lead prioritization and let go of non-prospect leads at the right time to cut costs and enable efficiency. The result of the study finding reveals that 52% of the sample of 74,779 leads are cold leads and 48% are hot leads that are sales qualified. The leads are qualified using the lead score matrix. As a result, it aids digital businesses in removing unqualified leads and managing the leads better, which raises the caliber of the leads sent to clients. This will improve conversion rates for individual customers. The increased conversion rates support the business strategy of digital marketing firms.

Keywords Machine learning · Supervised ML algorithms · Logistic regression · Decision tree · Hot leads · Cold leads

---

V. L. H. Josephine (✉) · C. Sai · J. Joy · A. A. Kashyup
School of Business and Management, Christ University, Bangalore, India
e-mail: helenjose.cbe@gmail.com

J. Joy
e-mail: justin.joy@christuniversity.in

P. K. Kapur et al. (eds.), *Data-centric Approaches to Industrial Decisions*, Asset Analytics, https://doi.org/10.1007/978-981-96-7556-2_9

9.1 Introduction

The practice of finding and cultivating potential customers for a company's goods or services with the intention of turning them into paying clients is known as lead generation. It entails the gathering and analysis of information on people who are probably interested in a company's products or services, as well as the development of targeted marketing campaigns to connect with and engage these people. A subfield of artificial intelligence called machine learning uses statistical models and algorithms to carry out tasks devoid of explicit instructions. Machine learning can be used to automate and improve a number of lead generation processes, including lead scoring, data collecting, and data analysis. Lead generation and machine learning are two topics that are interconnected and can be combined to forge a potent solution for sales and marketing teams. Data gathering, lead scoring, predictive modeling, customer segmentation, and personalization are just a few of the lead generation processes that may be automated and optimized using machine learning. Machine learning (ML), which employs predictive modelling algorithms to assess customer data from prior interactions and project future customer journeys and outcomes, is one of the most efficient methods for lead scoring. The marketing automation sector increasingly favors predictive lead scoring since it increases total conversion and aids teams in coordinating acquisition and sales goals.

The actions and communications that led to a conversion are also identified by machine learning. A business can use lead scoring to find "hot leads," or targets or leads that may convert, when it correctly identifies trends and common behaviors. Businesses may automate lead creation processes including data collecting, analysis, and lead scoring using machine learning algorithms, which can enhance accuracy and save time. To identify the people who are most likely to become consumers, machine learning algorithms can examine customer data such as demographics, behaviors, and purchase history. Using this data, personalized marketing strategies may be developed to reach and interact with these high-potential leads. Machine learning can also be used to optimize lead nurturing by personalizing marketing messages based on a lead's unique characteristics and behaviors. This can help increase engagement and conversion rates and lead to higher customer lifetime value.

9.2 Motivation

Organizations often spend a lot of money and effort trying to convert all the leads they get for their products or services, hoping to generate revenue and recoup their investment. However, they frequently do not distinguish between leads that are likely to convert and those that are not during the customer acquisition process. This can result in the company wasting its financial resources on pursuing unproductive leads that would not generate any revenue. This can also increase the organization's expenses, including marketing materials, sales team salaries, and other overhead costs, without any financial gain. As a result, the organization may miss out on potential leads that could have resulted in successful conversions and sales if they had been identified and pursued properly. To prevent this, organizations should use effective lead scoring and qualification processes to differentiate between the promising leads and non-promising ones. This can help them focus their resources and efforts on the leads that are most likely to result in successful conversions and sales, thus reducing costs and maximizing returns.

The Objectives of the research are (1) To identify the generated leads and categorize them into hot leads or cold leads through various predictive models and scoring. (2) To identify prospectful leads to help organizations focus on the high-quality leads based on their dependence on various related characteristics of the leads generated. (3) To identify and compare the best performing predictive lead scoring models the organizations can employ for setting up automated lead scoring systems in their customer acquisition cycles.

9.3 Literature Survey

The upcoming wave of digital disruption will be fueled by machine learning. Companies are required to get ready for it. Companies that started using it early benefit from it. Computer vision, natural language processing, virtual agents, robots and autonomous vehicles, deep learning, and machine learning are examples of future technologies. The cornerstone for the upcoming wave of machine learning applications is digitization. Very often, industries that have embraced digitization are also leading the way in machine learning. It is also believed that they will promote growth. It is anticipated that machine learning will hasten changes in market share, income, and profit margin. All of these are trademarks of companies, products, and services affected by digital technology [1].

Demand Response (DR) has seen a rise in popularity recently as a low-cost method of enhancing the flexibility and dependability of the energy system. The necessity for frequent near-real-time judgements, the utilization of enormous amounts of data, and the high complexity of disaster recovery (DR) operations have all fueled recent developments in artificial intelligence (AI) and machine learning (ML), a subset of AI [2].

They have developed as a vital technology for providing demand-side response. The best group of consumers to respond to, learning their attributes and preferences, dynamic pricing, scheduling, device control, and learning how to fairly incentivize and reward participants in DR schemes are just a few of the issues that can be addressed using artificial intelligence-based methods. Based on a thorough analysis of over 160 papers, 40 businesses and commercial initiatives, and 21 large-scale projects, this research offers an overview of AI technology utilized for disaster relief applications. Based on the AI/ML algorithm(s) utilized and the energy DR application domain, the studies are categorized. Then, large-scale innovation projects leveraging

AI for energy DR as well as commercial activities (from both start-ups and established businesses) are emphasized. The merits and potential drawbacks of the investigated AI techniques for various DR applications are discussed in the paper's conclusion, along with prospective directions for future research in this rapidly developing field. With the purpose of identifying and discussing trends for AI-based techniques in the energy DR sector, the authors of this study reviewed more than 160 papers that were published between 2009 and 2019, in addition to 40 businesses, commercial efforts, and 21 large-scale projects. The examination of the existing literature in this article demonstrates that AI methodologies are a viable technology for DR applications. The general success of DR initiatives in the future will depend heavily on the adoption of AI. Although AI-based techniques offer tools for addressing a variety of DR difficulties, they also have some disadvantages and restrictions. For successful use in a DR situation, a greater comprehension of the procedures and their constraints is necessary [3].

Our analysis revealed that a wide variety of AI techniques are employed, but it is evident that some of them are more effective for particular tasks than others. Indeed, it is demonstrated that supervised learning is commonly employed in short-term load and price forecasting by ANNs, which are typically used for multivariable function approximation and regression. On the other hand, RL-based algorithms are frequently employed to record human feedback, which makes them perfect for control tasks in HEMS that contain a DR solution. Contrarily, unsupervised learning is typically employed for clustering where there is no prior knowledge of the categories, as is the situation with DR consumers' aggregator level clustering tasks [4].

After classifying DR clients and predicting their consumption, aggregates organize the activation of DR participants and determine their rewards and penalties. For these challenges, a number of approaches have been highlighted, including optimization, which may involve the use of nature-inspired optimization techniques (such as swarm intelligence) when standard, deterministic optimization methods are less accurate. To find the most effective pricing and scheduling strategy, some techniques employ multi-agent systems in game-theoretic scenarios [5].

9.4 Methodology

The study uses the overall recommended procedure for information systems research using predictive analytics. With the research's primary aim being data understanding focuses on analyzing the data and finding and resolving potential issues contained in it, whereas data mining focuses on the building and evaluation of prospective predictive machine learning models for automated lead scoring. Figure 9.1 depicts the lead generation process overview. As part of the data preparation process, the data is processed to cope with missing values and outliers as well as to produce a variable structure suitable for future machine learning model development using feature extraction, filtering, and feature selection [6, 7]. Several models are developed and assessed using machine learning techniques in the following processes. After determining the best model, visualization tools are used to analyze the key findings.

9.4.1 Data Collection

The datasets consisted of both quantitative and qualitative data comprising of lead details and their corresponding campaign and CTI details. Additionally, the quantitative data was computed by accounting for derived variables while data preparation. The datasets included categorical/nominal data extensively. Some important variables are LeadID, CompaignID, Source, Status, Location, vechicle type, tyre brand, size, call segment disposition, call status, campaign, Contact, Lead's email, etc.

9.4.2 Data Pre-Processing

Lead data and communications and technical team data sets from various months were combined into separate data frames called lead data dataframe and communications and technical team dataframe. These dataframes were then added to the leads' Lead IDs after being filtered to include just "Closed-Won" and "Closed-Lost" Lead IDs, respectively. The campaign member dataset and the lead dataset underwent a similar integration procedure, with the dataframes being appended to the lead IDs of the leads filtered to "Closed-Won" and "Closed-Lost" Lead IDs as needed.

This involves cleaning up the data, which involved giving the relevant columns new names that made sense, removing the irrelevant columns from the dataframes, computing the required derived columns using the relevant columns, performing missing analysis, making relevant data imputations, then performing outlier detection and keeping the relevant bands of quartile ranges while discarding the rest in accordance with standard procedures.

9.4.3 Model Building

The "no free lunch theorem" states that no single classifier can perform optimally across different datasets. As a result, it is advised to test a variety of classifiers on a master feature vector to see which one yields the best outcomes. Consequently, four distinct categories of algorithms/models were chosen (Table 9.1).

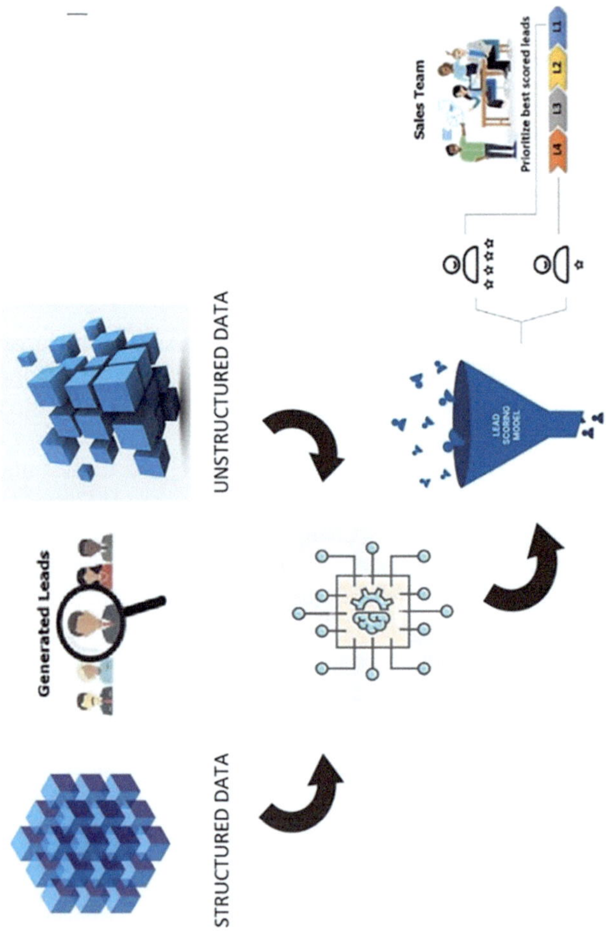

Fig. 9.1 Lead geenration process overview

Table 9.1 Categorization of Machine learning and ensemble techniques

Traditional ML and statistical algorithms	Regression algorithms	Bagging algorithms	Boosting algorithms
1. Decision Tree 2. K Nearest Neighbors 3. Naïve Bayes 4. SVM-Linear Kernel 5. Linear Discriminant Analysis 6. Quadratic Discriminant Analysis	1. Logistic Regression 2. Ridge Classifier	1. Random Forest 2. Extra Trees Classifier	1. Light Gradient Boosting Machine 2. Gradient Boosting Classifier 3. Ada Boost Classifier

9.4.3.1 Traditional Machine Learning and Statistical Algorithms

Traditional ML models refer to single, standalone models that are trained on a dataset to make predictions on new data. These models are typically built using a single algorithm or a combination of a few algorithms, and the training data is used to optimize the model's parameters.

1. **Decision Tree:** This algorithm is supervised. It creates classification rules in the form of a tree, with each internal node signifying attribute conditions, each branch representing result conditions, and the leaf node indicating the class label.

   Gain (T, X) = Entropy (T) — Entropy (T, X).

2. **K Nearest Neighbor:** In pattern classification, the K-nearest neighbors (KNN) algorithm is a non-parametric approach that is frequently used for classification and regression applications [8, 9].
3. **Naïve Bayes:** Naive Bayes is a machine learning algorithm that is based on Bayes' theorem, which states that the probability of a hypothesis (target variable) is proportional to the conditional probability of the evidence (input variables) given that hy-pothesis, multiplied by the prior probability of that hypothesis.
4. **SVM-Linear Kernel:** Using the theory that the optimum decision boundary is the one that has the most distance and margin from both classes of data, SVM determines the best hyperplane that divides the data [10].
5. **Linear Discriminant Analysis:** A well-liked approach for supervised feature extraction and classification is LDA. The goal of LDA is to reduce within-class distance while maximizing distance across classes [11, 12].
6. **Quadratic Discriminant Analysis:** Quadratic discriminant analysis (QDA) is a statistical model used for classification tasks. It is a type of discriminant analysis that assumes that the data is normally distributed and uses the covariance matrix of each class to model the decision boundaries between classes [13].

9.4.3.2 Regression Algorithms

Regression algorithms are primarily used for predicting continuous values. However, some regression algorithms can be adapted for classification prediction by applying a threshold or boundary to the predicted continuous values.

Logistic Regression: The logistic regression model estimates the coefficients using maximum likelihood estimation. The goal is to find the values of the coefficients that maximize the likelihood of observing the binary outcome given the predictor variables [14].

Ridge Classifier: In Ridge classification, the goal is to find a hyperplane that separates the data points into different classes. The hyperplane is defined by a set of weights that are learned during the training process. The regularization parameter in Ridge classification controls the complexity of the learned weights, preventing overfitting, and improving generalization performance.

9.4.3.3 Bagging Algorithms

Bagging stands for bootstrap aggregating, which involves creating multiple models by resampling the original dataset with replacement, and then aggregating the results. Bagging models use a parallel ensemble learning approach, meaning that each model in the ensemble is trained independently and in parallel with the others. The final result is obtained by aggregating the output of all models in the ensemble.

Random Forest: It is a particular ensemble classifier made out of several decision trees. An instance is classified using the voting results of each decision tree's class predictions.

Extra Trees Classifier: ETC builds multiple decision trees on random subsets of the training data and features. However, instead of using the traditional decision tree algorithm, which chooses the best split point based on a measure of information gain, ETC selects the split point at random for each feature.

9.4.3.4 Boosting Algorithms

Boosting, on the other hand, involves iteratively training a series of models, with each subsequent model learning from the errors of the previous model. Boosting models use a sequential ensemble learning approach, meaning that each model in the ensemble is trained sequentially, with each subsequent model focusing on improving the performance of the previous model.

Light Gradient Boosting Machine: Gradient boosting is an ensemble learning technique that combines multiple weak learners, typically decision trees, to create a strong learner. LightGBM employs a specific type of gradient boosting called Gradient-based One-Side Sampling (GOSS) and Exclusive Feature Bundling (EFB) to achieve high efficiency and accuracy [15].

Gradient Boosting Classifier: The GBC algorithm works by iteratively training decision trees on the errors of the previous trees. During each iteration, the algorithm fits a new tree to the residual errors of the previous tree, and then adds it to the ensemble of trees. This process continues until the algorithm has built a predetermined number of trees or until the error rate has converged to a minimum level [16].

Ada Boost Classifier: It is one of the best-boosting algorithms, strengthening weak learning algorithms.

$$H(x) = sign\left(\sum_{t=1}^{T} a_t h_t(x)\right)$$

9.4.4 Result and Discussions

Different machine learning algorithms were chosen to be tested based on the final dataset mentioned in the preceding section, prompted by the results of our literature review on the most popular algorithms.

The Fig. 9.2 displays the multiple machine learning models are built using one liner command and are being compared/based on their AOC(Area Under the Curve of ROC Values) in Descending Order, as we can see from above, this order is the order of highest to lowest scores of Accuracy, Precision, Recall, F1 Score, Kappa Value, MCC Score, and TT Score. The best model for prediction pertaining to this scenario is Random Forest Classifier. The Random Forest model, which achieved an accuracy score of 83.27%, clearly outperforms all other models in all metrics, followed by the Extra Trees Classifier and Decision Tree models with corresponding accuracy scores of 83.22% and 83.16%. Meanwhile, due to their simplicity, KNN and Gradient Boost achieved 76.89% and 74.08%, respectively.

For binary classification issues, a graphic performance assessment known as the Receiver Operator Characteristic (ROC) curve is used. The area under the curve (AUC), a measure of a classifier's ability to distinguish between classes, is used in data analysis. Figure 9.2–9.3 shows the AUC curve and Fig. 9.4 depicts the confusion matrix for random classifier.

The performance of the models was evaluated using more complex ratings. These indicators are calculated using the first-level indicators and go by the name secondary indicators. Secondary indications that are frequently employed are accuracy (percentage of correctly predicted events among all predicted events), precision (percentage of accurate positive forecasts among all positive predictions), sensitivity or recall (percentage of accurate positive predictions among all positives).Metrics Computation Results are as follow with True Positives being 16,863, True Negatives

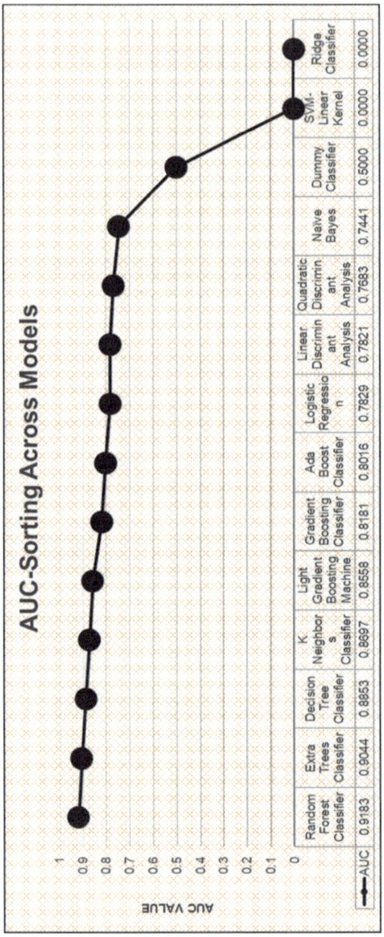

Fig. 9.2 AUC sorting across models

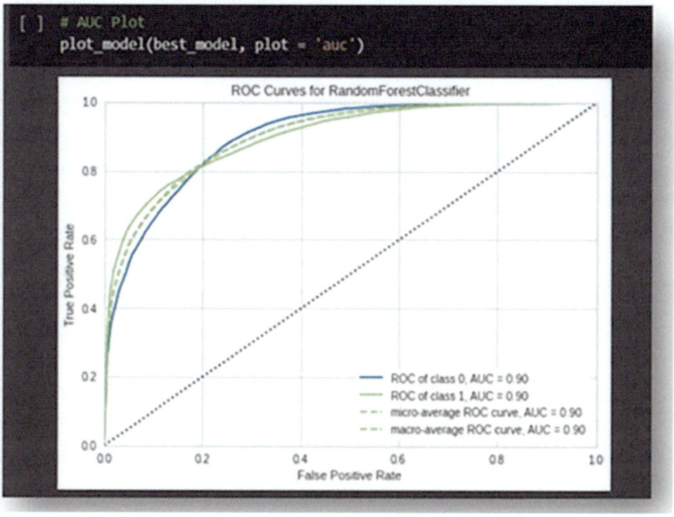

Fig. 9.3 AUC Curve for random classifer

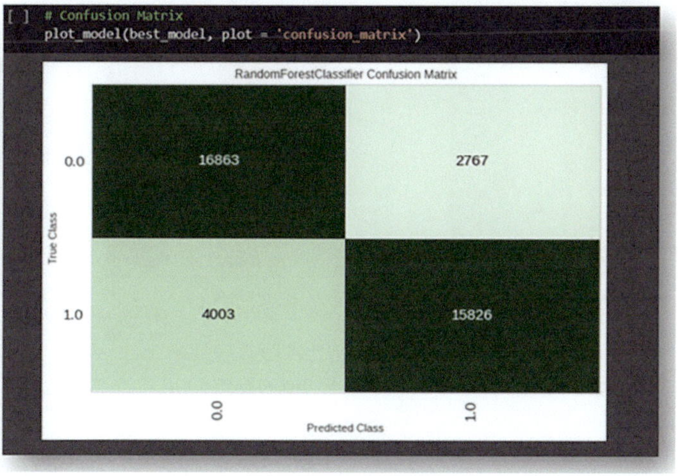

Fig. 9.4 Confusion matrix for random classifer

being 1582, False Positives as 2767, False Negatives as 4003, Precision(True Positives)/(True Positives + False Positives) being 0. 8590 and Recall(True Positives)/(True Positives + False Negatives) being 0.8081.

The F1 score is a performance metric used in binary classification tasks, combining precision and recall. It helps assess a model's ability to balance both true positive and false negative predictions. Ranging from 0 to 1, higher values indicate better overall model performance. The F1 score is widely employed in machine learning evaluations. Figure 9.5 the bar graph represents the F1-Score measure with the Decision Tree model, which is 0.8250, followed by the K Neighbors Classifier(0.7838), LDA (0.6861), SVM(0.6710), QDA(0.6670) and the Naïve Bayes (0.56) models in terms of their capacity to combine accuracy and recall into a single score to measure the best performing models among the traditional machine learning algorithms to classify leads into hot and cold. Figure 9.6, the bar graph represents the F1-Score measure with the Ridge Classifier model, which is 0.6860, followed by the Logistic Regression Classifier (0.6855) in terms of their capacity to combine accuracy and recall into a single score to measure the best performing models among the traditional machine learning algorithms to classify leads into hot and cold.

In Fig. 9.8 Comparison of Boosting models F1 Scores. The Fig. 9.7 bar graph signifies the F1-Score measure with the Random Forest Classifier model, which is 0.8286, followed by the Extra Trees Classifier (0.8284) in terms of their capacity to combine accuracy and recall into a single score to measure the best performing models among the traditional machine learning algorithms to classify leads into hot and cold. The Fig. 9.8 bar graph denotes the F1-Score measure with the Light Gradient

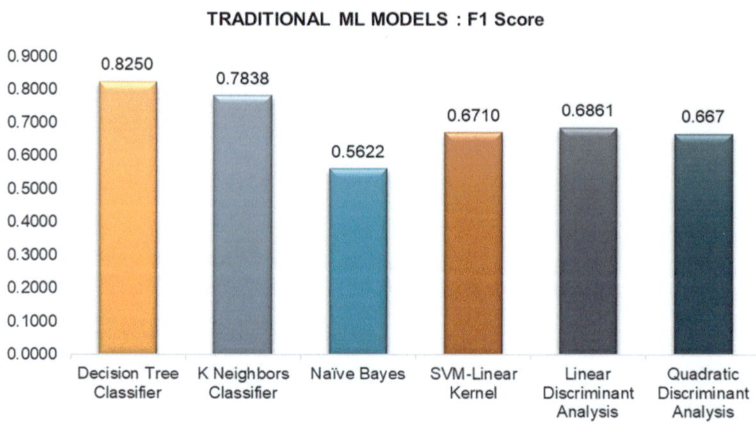

Fig. 9.5 Comparison of Traditional ML models F1 Scores

Fig. 9.6 Comparison of Regression models F1 Score

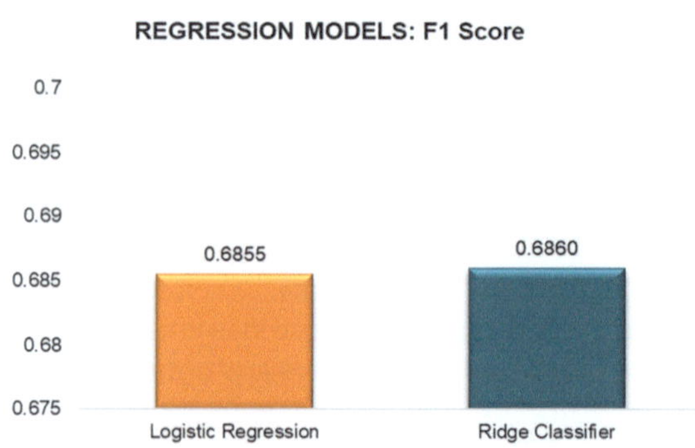

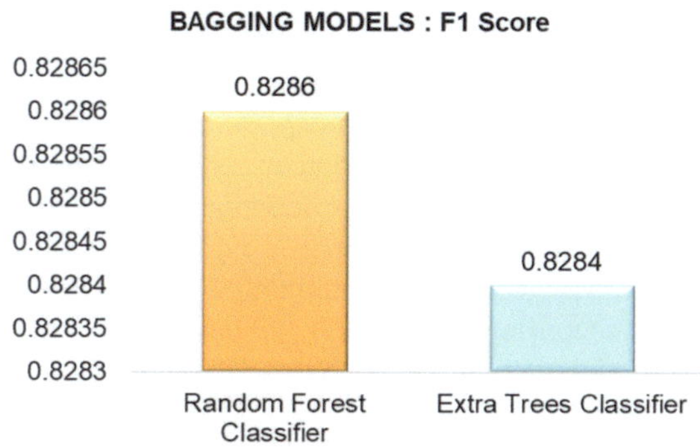

Fig. 9.7 Comparison of Bagging models F1 Scores

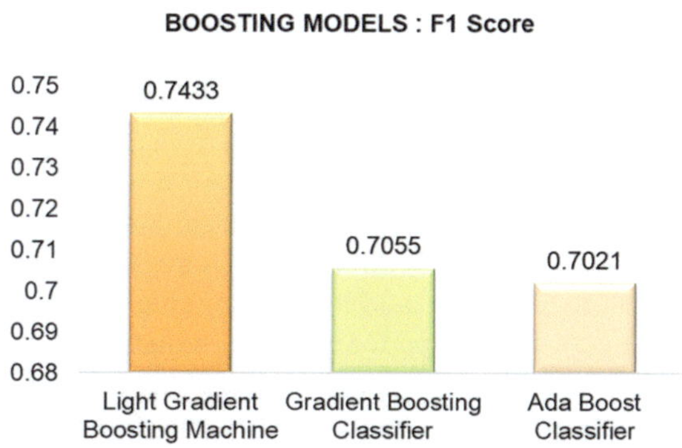

Fig. 9.8 Comparison of Boosting models F1 Scores

Boost Machine model, which is 0.7433, followed by the Gradient Boosting Classifier (0.7055) and Ada Boost Classifier (0.7021) models in terms of their capacity to combine accuracy and recall into a single score to measure the best performing models among the traditional machine learning algorithms to classify leads into hot and cold.

According to the findings of the research, the use of a random forest classifier produced the most favorable outcome. This method was able to successfully categorize 52% of the 74,779 leads as cold or bad leads, while identifying the remaining 48% as potential good leads which the organization should prioritize its efforts on. When dealing with data that has high variability and the risk of overfitting, the bagging models were found to be more effective than the boosting models. In terms of accurately predicting cold and hot leads, the boosting models and traditional machine learning models were less effective, and therefore, not the primary method used to support predictive lead scoring. The study suggests that using a random forest classifier is the best approach for lead scoring. However, the choice between bagging and boosting models will depend on the specific characteristics of the data. It is also important to note that traditional machine learning models may not provide the highest level of accuracy for predictive lead scoring.

9.4.5 Conclusion and Future Work

For firms looking for new prospects, predictive lead scoring is a popular study topic. In order to replace the outdated scoring system, this article focuses on researching the application of machine learning techniques to automate prospect identification. Potential leads have been divided into qualified and unqualified leads using several ML techniques, including Logistic Regression (LR) and Random Forest (RF). According to historical and behavioral data on the company website, the two top models (RF and Extra Trees Classifier) have both achieved good performance and can predict, with an accuracy of 83.27% and 83.22%, respectively, whether visitors would purchase. From the study, it is understood that the sound usage of right fit predictive models according to the pertained organization's domain and relevant independent variables, they can cut down on costs incurred through concentrating their efforts only on good leads and not the bad ones. The models preferred according to the study are bagging models followed by boosting, traditional ml and regression models. These might assist companies in differentiating high-quality prospects from regular website users, enhancing their return on investment (ROI). As manual lead scoring and lead scoring utilizing machine learning were not compared to one another, it is impossible to tell for sure which method is superior.

Future research could look into multiclass classifiers as opposed to binary ones. If the conversion type for example, requesting a call back, contacting a chatbot for assistance, etc., is used as a class label, it is also possible to predict the sort of action a user may take in the future. When using sales data, the names or types of products may be used as labels. This might produce a model that is far more insightful. They can incorporate factors such as adding customer value to lead scoring, which would produce a monetary value that might appear more real than a mere purchase likelihood. One could, for instance, just multiply the purchase likelihood by the client value duration. Using refined regression rather than categorization to calculate the customer lifetime value of leads is another example. Identifying various lead types would also be helpful for businesses. By nurturing campaigns, they could handle the various lead kinds with various sorts of marketing content. As it is uncertain how many distinct kinds of leads there are, this could be accomplished using unsupervised learning. The process of developing a machine learning model could be further refined to perhaps improve the final classification outcomes.

References

1. Kashyap, P. (2017). *Industrial applications of machine learning*. Apress. https://doi.org/10.1007/978-1-4842-2988-0_5
2. Mohs, R. C., & Greig, N. H. (2017). Drug discovery and development: Role of basic biological research. *Alzheimers Dement (N Y).*, *3*(4), 651–657.
3. David, L., Arús-Pous, J., Karlsson, J., et al. (2019). Applications of deep-learning in exploiting large-scale and heterogeneous compound data in industrial pharmaceutical research. *Frontiers in Pharmacology, 10*, 1303.

4. Han, Y., Yang, J., Qian, X., et al. (2019). DriverML: A machine learning algorithm for identifying driver genes in cancer sequencing studies. *Nucleic Acids Research, 47*(8), Article e45.
5. Asif, M., Martiniano, H., Vicente, A. M., Couto, F. M. (2018). Identifying disease genes using machine learning and gene functional similarities, assessed through Gene Ontology. PLoS ONE.
6. Krishnadas, G., & Kiprakis, A. (2020). A machine learning pipeline for demand response capacity scheduling. *Energies, 13*(7), 1848.
7. Tokui, S., Oono, K., Hido, S., & Clayton, J. (2015). Chainer: a next-generation open source framework for deep learning. In *Proceedings of workshop on machine learning systems (LearningSys) in the twenty-ninth annual conference on neural information processing systems (NIPS)* (Vol. 5, pp. 1–6).
8. Jain, K., Duin, R. P. W., & Mao, J. (2000). 'Statistical pattern recognition: A review. *IEEE Transactions on Pattern Analysis and Machine Intelligence, 22*(1), 4–37.
9. Wang, N. A., Chen, G., Li, L., & Alterovitz, G. (2014) Accelerating incremental wrapper based gene selection with K-Nearest-Neighbor. In *Proceedings of the IEEE International Conference on Bioinformatics and Biomedicine (BIBM)*, (pp. 21–23). Belfast
10. Zhang, H. F., Jia, M. M., Xiang, B. B., & Ma, C. (2020). Predicting missing links in complex networks via an extended local naïve Bayes model. *EPL Europhysics Letters, 130*(3), 38002. https://doi.org/10.1209/0295-5075/130/38002
11. Ye, Q., Yang, J., Liu, F., Zhao, C., Ye, N., & Yin, T. (2018). L1-norm distance linear discriminant analysis based on an effective iterative algorithm. *IEEE Transactions on Circuits and Systems for Video Technology, 28*(1), 114–129. https://doi.org/10.1109/tcsvt.2016.2596158
12. Wang, J., Wang, D., Zhu, Z., Zheng, T. F., & Soong, F. (2014). Discriminative scoring for speaker recognition based on I-vectors. In *Signal and Information Processing Association Annual Summit and Conference (APSIPA), Asia-Pacific*. https://doi.org/10.1109/apsipa.2014.7041619
13. Wang, J., Plataniotis, K., Lu, J., & Venetsanopoulos, A. (2008). Kernel quadratic discriminant analysis for small sample size problem. *Pattern Recognition, 41*(5), 1528–1538. https://doi.org/10.1016/j.patcog.2007.10.024
14. Ariza-Garzon, M. J., Arroyo, J., Caparrini, A., & Segovia-Vargas, M. J. (2020). Explainability of a machine learning granting scoring model in peer-to-peer lending. *IEEE Access, 8*, 64873–64890. https://doi.org/10.1109/access.2020.2984412
15. Pamir, J. N., Akbar, M., Aldegheishem, A., Alrajeh, N., & Mohammed, E. A. (2022). Employing a machine learning boosting classifiers based stacking ensemble model for detecting non technical losses in smart grids. *IEEE Access, 10*, 121886–121899. https://doi.org/10.1109/access.2022.3222883
16. Umutcan Uguz, D., Berief, F., Leonhardt, S., & Hoog Antink, C. (2020). Classification of 12-lead ECGs using gradient boosting on features acquired with domain-specific and domain-agnostic methods. In *2020 Computing in Cardiology Conference (CinC)*. https://doi.org/10.22489/cinc.2020.363.

Chapter 10
Effect of Perceived CSR on Customer Loyalty: Exploring Mediating Role of Corporate Image and Consumer Trust with Parallel Mediation Approach

Divya Sharma and Rakesh Kumar

Abstract Corporate social responsibility (CSR) is a concept which is beyond the managerial implications and contains standardizing suggestion that lack clarity and specificity. CSR plays a vital role in promoting and improving the public image of many companies. This study aims to analyze the effect of perceived CSR (legal CSR, economic CSR, societal CSR, and environmental CSR) on customer loyalty taking the corporate image and consumer trust as a mediator. A total of 199 valid questionnaires from the customers were collected using purposive sampling and analyzed using structural equation modelling in Amos 22.0 and SPSS (Process). The results indicated that legal CSR, and economic CSR have a significant impact on a corporate image while societal CSR, and environmental CSR exhibits no significant impact on corporate image. On other hand, economic CSR and environmental CSR have a significant impact on consumer trust but other perceived CSR has no significant impact on consumer trust. Also, corporate image and consumer trust have a significant impact on customer loyalty. In addition, the results of parallel mediation analysis demonstrated that corporate image and customer trust mediate the effect of perceived CSR (economic, social, legal, and environmental) on customer loyalty. Finally, the conclusion of our findings are discussed.

Keywords Corporate social responsibility (CSR) · Perceived CSR · Corporate image · Consumer trust · Customer loyalty

---

D. Sharma (✉) · R. Kumar
School of Management Studies, Motilal Nehru National Institute of Technology Allahabad, Allahabad, India
e-mail: ds2818ds@gmail.com

R. Kumar
e-mail: mmmec.rakesh@gmail.com

P. K. Kapur et al. (eds.), *Data-centric Approaches to Industrial Decisions*, Asset Analytics, https://doi.org/10.1007/978-981-96-7556-2_10

10.1 Introduction

Since the early 90 s, sustainable development and CSR of organisations remains an issue. As community and organisation depend on each other, they need to care about the well-being of both [1]. Today, the world is more aware and concerned about corporate social responsibility (CSR) and sustainable development. Numerous organizations work together to ensure that sustainable development is not overlooked as they pursue their objectives—profit-making, providing social services, etc. Thus, in an attempt to execute CSR, businesses must incorporate social and environmental issues into all aspects of their operations, including contacts with stakeholders [2]. This interconnection between CSR, the organisation's reputation and trust leads to long-term profitability. Organisations are acknowledging the need for CSR in determining the perception of customers and fostering their loyalty [3].

CSR is a dynamic concept that evolves the culture of an organization and relies on the transformation of cultural norms and behaviour [4]. According to Arnold, a key advantage of CSR is strengthening the interaction between organisations and society [5]. CSR provides several benefits to an organisation such as increased profitability, attracting investors, improving quality, and avoiding government regulation and legal action. Additionally, it benefits society by contributing to natural environment protection, raising the standard of living and better employment opportunities [6]. For businesses, maintaining loyalty is an important concern, especially in times of economic hardship like the present global recession [7]. It can be comprehended that customer loyalty enhances their satisfaction with products/services offered by the organisation which leads to trust-based relations, customer retention and subsequently promoting the organisation [8]. Customers are considered an important resource of the organisation and influence their performance directly [9]. The relationship between CSR and the evaluation of an organisation is mediated by the relationship between CSR, organisation features and the customers [10, 11]. Customers' trust in an organisation and its products is increased when CSR initiatives are carried out successfully [11]. Corporation believes that considering CSR strategies helps them in determining objectives for improving corporate image and profitability [12]. CSR practices of the organisation help in positively improving the corporate image and achieving market share [13]. CSR is essential for projecting a positive image of the organisation. Based on the popular organisational case argument for CSR [14], an effective CSR approach leads to improved organisations' (non-) financial performance and fosters organisation value. Hence, an appropriate corporate government framework is required to minimise greenwashing, and data saturation and stimulate the organisation's image [15].

It is necessary to understand the customer's perception of social activity performed by the organisation and whether such activities impact its image. Moreover, previous research studies have conceptualized both corporate image [16–18] and customer trust [19, 20] as mediators but little research has been done on perceived CSR taking the corporate image and customer trust as a mediator, especially in the Indian context. Hence, the current study aims to analyse the effect of perceived CSR namely, legal

CSR, economic CSR, societal CSR, and environmental CSR on customer loyalty. In addition, the study conceptualizes that perceived CSR activities exhibit direct as well as indirect effects (mediated through corporate image and customer trust) on customer loyalty.

Following is the remaining structure of the research paper. Section 10.2 contains a brief description of the variable along with the hypotheses. Section 10.3 deals with the methodological aspect including sampling description and data analysis. Section 10.4 presents a brief discussion of the study's result and Sect. 10.5 presents the implications of the study. The last section covers the conclusion and future recommendations.

10.2 Literature Review

Due to the downfall of the economy during the pandemic and its dawdling recovery, organizations require reassessing their priorities and also aim to manage resources carefully [21]. Organizations are needed to anticipate consumer trends ahead of ever as consumer situations are changing rapidly. In today's time, the need for movement towards social responsibility is unforgotten, but the sustainability issue grabs more attention. Organizations need to encourage firm and strong CSR strategies to draw customers' attention, and trust and bring out constructive changes [22].

CSR literature is highly influenced by stakeholder theory [23]. It states that a stakeholder is a person, group or institute with which an organisation engages while seeking its objective [24]. The stakeholders are concerned about society and cooperate with organizations that are more concerned about social, economic, and environmental issues. Conscious customers are well-informed and are sensitive about organizations that have more impact on society. Nowadays, it is very difficult for an organization to grow unless it is not involved in society [25]. Hence, focusing on stakeholders' anticipation and understanding is crucial as they have a significant impact on the organisation directly and indirectly. All components of perceived CSR consider various stakeholders and their expectations. Stakeholders view the overall image of customers before engaging with them, as positive perception can enhance favourable attitudes and beliefs. Additionally, customers, being important stakeholders, include their perceptions (trust and loyalty) as part of stakeholder theory.

The study has taken perceived CSR, corporate image, customer trust and customer loyalty as crucial constructs. The perceived CSR include four elements which have been measured separately to comprehend the impact of each variable on image and trust. Due to the obligatory nature of CSR in India, organisations are needed to put some per cent of their profit into CSR. Thus, understanding which component affects customers and society the most helps them to optimise cost and efficiency along with changes in corporate image and perception of customers. Moreover, the study has considered customer loyalty as the dependent variable to identify the impact of CSR, corporate image and trust on it. The analysis will able to help in determining customer attitude and perception towards a particular organisation.

10.2.1 Corporate Social Responsibility (CSR)

Corporate social responsibility (CSR) is described as an organization's attempt to take social responsibility and stand regarding its stakeholder obligations, to escalate its positive consequence and alleviate its negative consequence on society [26]. The influence of CSR activities on consumer behavioural responses results in affecting consumer's perception change and change in process of inference based on phenomena and information, creating a process that is not easily visible.

In today's competitive world, the price and quality of products are not the only preferential advantage that the organisation has, so they are moving towards CSR which is recognized as an important quality in the market because customers' perceptions and responses are affected by the social and ethical activity done by them [26, 27]. Customer regards, if the organisation is engaged in CSR show its customer concern and satisfy their needs and preference [28–30].

The most prominent aspect of a customer's behavioural intent to maintain a long-lasting and fixed engagement with a company is said to be customer loyalty [31]. Customer loyalty is vital for any business not only because it helps them to grow and survive, but also becomes a major marketing goal and develops a sustainable advantage basis [32].

Based on the social contract reasoning the organizations have moral and ethical obligations towards society and are considered the foundation for the stakeholder's theory [33]. For formatting and operating organizational relationships in different circumstances, the stakeholder's theory supports a practical and ethical approach [34]. Carroll propounded four types of CSR namely, legal, economic, ethical, and discretionary [35].

10.2.2 Legal CSR

A company is required to conduct business under the legal obligations of society. Managers are needed to develop policies that are under legal constraints. A belief that all trade partners' conduct will be credible regardless of any contractual or legal limits is the foundation of trust-based relationships. This assumption is fuelled by the idea that a firm is moral and responsible [36]. Legal CSR can be defined as the action which is performed by the company to improve living standards of employee, customers or other stakeholders [37]. It includes provide safe working environment with fair wages, environmental protection and community philanthropy, lastly meeting legal requirements and improve society [38]. For the conformity and obligation of CSR legal norms, legislation, jurisdiction, and ordinance are crucial operators, and due to this reason organizations are expected to perform in a social-legal system parallel to aiming for their economic purpose [39–42]. According to Conchius [43], legal CSR involve adhering environmental, product and employment laws, which are essential for company. However, it does not cover all aspects of the company. Legal aspect requires following rules and law such as payment of taxes in order to meet basic legal requirements [37]. A company's legal requirements might influence the functioning of product and quality directly, influencing customer's behaviour [44]. Some authors argue that legal CSR is necessary as it helps build trust and goodwill for the company, subsequently enhancing sales [18, 45, 46]. While other authors view legal CSR as ineffective and can backfire, causing customers to doubt the company's intentions. Moreover, prior research shows significant impact of Legal CSR on corporate Image [20, 47].

H1: Legal CSR will positively affect corporate image.

H2: Legal CSR will positively affect consumer trust.

10.2.3 Economic CSR

In a world where a competitive edge is necessary for any business, it is necessary to invest and innovate to meet the customer's needs and gain profit and sustain in long run [48]. For businesses to succeed and stay in operation, society believes that they must fulfill their economic obligations to society by producing what they want [48]. Economic CSR refers to the degree to which company meets its responsibilities and commitment associated with business activities to maintain economic development [49]. According to Yassin and Backmann (2023), economic dimension of CSR is defined as the obligation of company to yield profit for shareholders and investors. It is considered as the most crucial component of CSR which must be complied with legal CSR. It is not possible to implement other dimension without economic dimension of CSR (Caroll, 1991). A company is perceived as just and fair when they are ethically commitment (Virvilaite and Daubaraite, 2011). However, economic activities benefits company directly, but not the society [49]. Costa and Menichini [50], economic CSR effects the company's economic activity indirectly, while impacting all the stakeholders. Economic practices are defined as the policies and procedure to prevent legal exploitation and affirms adherence with law. Businesses should incorporate economic operations with other CSR aspects for value creation, maximizing profit, and sustainability, based on an argument regarding the relationship between enterprises' economic duties and performance [51–53]. It has empirically proven that corporations' economic responsibilities improve performance in form of corporate image [20, 47].

H3: Economic CSR will positively affect corporate image.

H4: Economic CSR will positively affect consumer trust.

10.2.4 Social CSR

Based on social identity theory, the bond between customer and company is developed by customer trust. It is widely established as an association in social events and pursuits and initiatives to assist the welfare of the community and enhance their living standard for individuals and social level [54, 55]. Social CSR pursuits such as consideration of community affairs, employees' healthcare, and support for children's education, etc., affects customers' rational assessment of the organization [56]. External stakeholders willing to corporate with company who focuses more towards well-being of society. Consequently, a company's efforts are aimed at benefiting society as it tends to gain more favour from customers [57]. However, Husted and Salazar [58] reasoned these actions to be self-centred as they ultimately aim to enhance reputation. Companies are required to invest strategically to gain corporate image and competent staff. It will further enhance value and profitability of the company. As highlighted by Huster et al. [59, 60], strategically investing in social welfare and practices leads to enhance in value and provide competitive edge. Social CSR charitable social conducted behaviour of company leading to favourable social image. For example, sponsoring non-profit organisation and endorsing community activity. Moreover, charitable activities not only display social obligation and honesty of company but also assist in sustainable development of society [61].

H5: Social CSR will positively affect corporate image.

H6: Social CSR will positively affect consumer trust.

10.2.5 Environmental CSR

Environmental CSR is characterized as a company's contribution to society's advancement of environmental consciousness. These initiatives fall under the umbrella of sustainable development and are consistent with long-term strategies for protecting the environment. Companies are getting concern about CSR and potential risk of degrading environmental footprints [62, 63]. The conservation and safeguarding of nature have reached one of the concerns of the organization's strategic goals due to the increasing concern about the environmental impact of the organization on the planet (e.g.[64, 65]). To fulfil the needs of stakeholders, companies integrate environmental endeavours in their strategies [66, 67]. Environmental CSR can be said a combination of environmental administration cognizance in an organization's functioning products, pursuits, and implements to shorten effect on the environment [68]. These endeavours are undertaken to reduce harmful impact of company's activities [69]. The integration of environmental activities in company's operation foster company's performance improves leading to increase corporate image. Stakeholder especially customers attribute favourable attitude towards the company which are eco-friendly, depicting the concern for environment and its protection [70]. Environmental CSR include those activities which are done by company which are not part of profit-making and is conducted for purpose of benefiting the society and environment. A strict environmental regulation compels company to engage in environmental endeavours which focus basically on ecological services [71]. According to Morales-Raya et al. [72], company's meeting environmental expectations of its users, creates and uphold eco-friendly image. Graham and Potter [73] show positive impact of environmental activity on business, which is contrast to the study conducted by Chen and Ho [74] demonstrating no increment in sales due to such activities. However, Mijatovic et al. [75] highlighted the positive impact in form of corporate image and competitive edge.

H7: Environmental CSR will positively affect corporate image.

H8: Environmental CSR will positively affect consumer trust.

10.2.6 Corporate Image

It is portrayed as the response to the weight of faith, opinion, and the assumption that society has in describing an organization [76]. It includes construction of favourable personality of company, communicating through channels and providing feedback to the customers [77]. Corporate image is a multidimensional construct which can be described as intangible involvement with psychological condition, fostering the attitude and behaviour [78]. Worcester [79] defined "corporate image is the net result of all experiences, impressions, beliefs, feelings, and knowledge people have about a company". Bernstein [80] mentioned that corporate image should reflect actual experience, not be separate from it. The stakeholders are the crucial components of the organization and corporate image being an intangible element of the organization affects stakeholders' perceptions. Corporate image is segmented into organization competence and CSR initiative [37, 76]. For the organization's benefit corporate image is necessary, and social transparency is increased to establish an image of the organization [40]. CSR helps in connecting the good image of an organization to customers' perception leading to a good corporate image. So, it is a critical planned deviation to achieve feasible benefit [81]. Chattananon and Lizarranga [82] highlighted the positive CI leads to brand recognition and loyalty. Similarly, Chattananon and Lawley [82] shows that corporate image is stakeholder's perception about company and is determined by controlled and uncontrolled source of information. It helps in gaining competitive edge, as it is formed over long period of time, which foster trust [83].

H9: Corporate image will positively affect customer loyalty.

10.2.7 Consumer Trust

An organization can overcome the roughness of the market and sustain well with loyal customers which can be generated consumer trust [84]. Trust can be considered a crucial element for the organization which impacts the consumers [85]. Trust includes the expectations of customer that product is consistent and reliable [86]. According to Swean and chumpitaz [87] trust is the opinion, aspiration and gathering towards a brand, which is derived from "characteristics, motives, and intentions of exchange partner" along with the assumption customer make regarding a company product/brand [88]. There are three major viewpoint of trust- perceived creditability, perceived integrity and perceived benevolence [87]. Perceived creditability is the capability to satisfy consumer need, perceived integrity is the capability to hold their commitment and perceived benevolence includes the capability to emphasize customer's preference. CSR is viewed as a significant indicator of trust because it reflects the company's commitment to social well-being and its action influence trust of customer. Trust can be considered an antecedent of loyalty. It positively affects consumer loyalty [89]. If a consumer feels that they are deceived then the trust will diminish, so the organization should stick to those commitments that avail in advocating and keeping up their coherence [90]. To comprehend the relation between customer and company which leads to loyalty and commitment can be attained through customer trust in company and brand [91].

H10: Consumer trust will positively affect customer loyalty.

10.2.8 Customer Loyalty

Customer loyalty can be defined as re-purchase and—patronizing of the product or services subsequently despite conditional impact and trade attempts which can probably let switching conduct [92]. It makes a joint advantage to the organization and customers [93]. It is the strong dedication to steadily choose the identical product or services in the future, despite external stimuli that persuade customers to switch to another [92]. Loyalty is relied on the emotions of its customer thus reflecting the genuine feelings and relation with company (provider) leading to positive attitude [94, 95]. Customer company bonding is affected via social performance of company leading to increase loyalty [96]. Organization's profitability is supported by loyal consumers [hayes 2008], which shows rebuying dedication for long-term engagement that helps in sales enhancement without investing in promotional activities. Unlike non-loyal customers, they don't tend to switch as they are less price-sensitive [97]. Customer loyalty can be assumed as deliberate and intentional conduct concerning a service or organization [98]. CSR activities construct a positive attitude and trust between company and customer leading encouraging loyalty [99]. According to Du et al. [100], integration of CSR in company's strategy to compete foster creditability and long-term loyalty intention. Bhattacharya and Sen [101] stated that good CSR can generate loyal customer who endorse for the brand. Furthermore, it increases sales consequently the investors. Therefore, CSR leads to forming connection with stakeholders, gradually leading to loyalty [102].

Figure 10.1 shows the research model based on the literature review and hypothesis.

10.3 Methodology

10.3.1 Sample and Procedure

The current study used a descriptive cross-sectional research design to accomplish research objectives and test hypotheses. Data were collected through an online survey from consumers living in Prayagraj and neighbouring cities. A total of 199 responses were collected using a structured questionnaire developed from the previous literature [103–107]. The introductory part of the questionnaires discusses the objective of the study and categorically informs the respondents that the detailed information provided by them would be used only for academic purposes. The questionnaire was divided into two major sections. The first section, comprises the statement on the model constructs on a 5-point Likert scale (where "1 = strongly disagree", "2 = disagree", "3 = neutral", "4 = agree" and "5 = strongly agree"). Whereas the second section addresses demographics in general, such as age, gender, education level, and occupation.

As shown in Table 10.1, 50.8% of the sample were males, 48.7% were females, and 0.5% were others. The majority of the participants (85.5%) were between the age group of 18–25. More than half of the respondents (63.5%) were graduate students, followed by post-graduates (29.5%), then 12th pass or below respondents (7%). Student consists of 66.8%, employed persons were 23.1%, business owners were 5%, unemployed were 3.5% and others were 1.5% of the sample.

10.3.2 Data Analysis

Data were analyzed using structural equation modelling (SEM) in Amos 22.0 and SPSS 26.0. SEM was used to test the proposed model and examine the underlying relationships among the study constructs. It was conducted in two steps- first to identify the model fit, reliability and validity which was conducted through measurement model. Second was to identify the relationship between the variables which was done via structural model. In addition, parallel mediation approach was employed to test the mediating effects of corporate image and consumer trust in SPSS Process Macro [108]. Customer Loyalty was specified as the dependent variable, corporate image

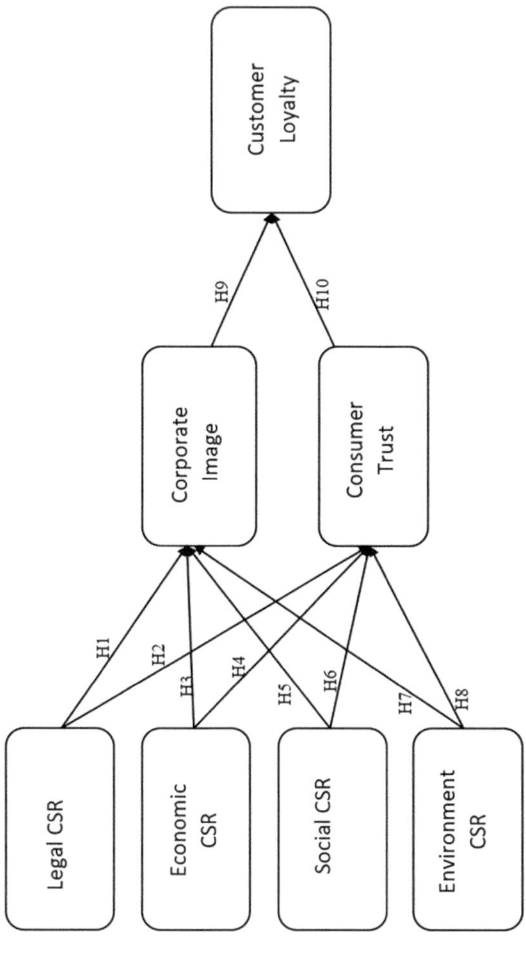

Fig. 10.1 Research model

Table 10.1 Demographic details of the respondent

S No	Variables	Categories	Frequency	Percentage (%)
1	Gender	Male	101	50.80
		Female	97	48.70
		Others	1	0.50
2	Age	18–25	170	85.40
		26–35	27	13.60
		36–45	0	0.00
		45 and above	2	1.00
3	Education	12th or below	14	7.00
		Graduation	127	63.80
		Post-graduation and above	58	29.10
4	Occupation	Student	133	66.80
		Employed	46	23.10
		Business Owner	10	5.00
		Other	3	1.50
		Unemployed	7	3.50

Source Author's Compilation

and consumer trust were specified as the mediators. Legal CSR, Economic CSR, Social CSR, and Environment CSR were specified as the independent variables.

10.3.2.1 Measurement Model

Before examining the structural relationships, the measurement model was assessed through confirmatory factor analysis (see Fig. 10.2). The results (Table 10.2) indicated that the measurement model produced satisfactory model fit [107] with fit indices including CMIN/df = 1.579 (<3); CFI = .968 (>0.95); RMSEA = .054 (<0.08); SRMR = 0.045 (<0.08) and PClose = 0.294 (>0.05).

In addition to goodness of fit, reliability and validity was also ascertained in confirmatory factor analysis (see Table 10.3). The value of composite reliability for each conduct were more than the critical value of 0.7 [109] which shows that constructs hold internal consistency. Also, values of average variance extracted (AVE) range between 0.605 and 0.745 which is higher than the accepted threshold of 0.5 confirming convergent validity [110]. In addition, square roots of AVE (values in diagonals) were found greater than the inter-construct correlation which confirms discriminant validity [110].

Fig. 10.2 Measurement model

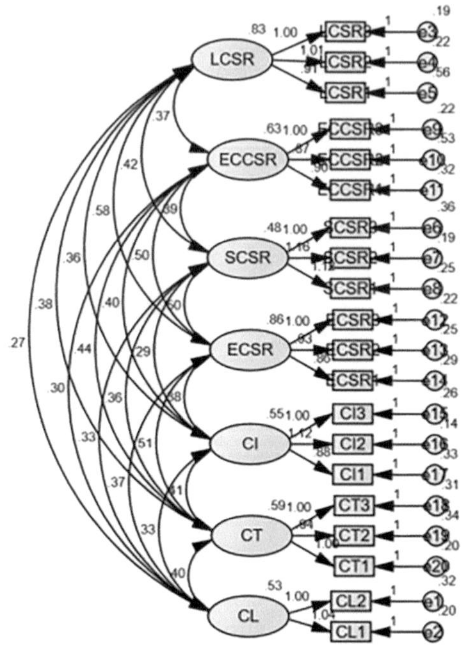

Table 10.2 Measurement model goodness of fit

Measure	Measure	Measure	Interpretation
CMIN	235,343	–	–
df	149	–	–
CMIN/df	1.579	Less than 3	Excellent
CFI	0.968	0.95	Excellent
SRMR	0.045	< 0.08	Excellent
RMSEA	0.054	< 0.06	Excellent
PClose	0.294	> 0.05	Excellent

Source Author's Compilation

Table 10.3 Reliability and validity measurement

	CR	AVE	Cl	CL	LCSR	SCSR	ECCSR	ECSR	CT
Cl	0.868	0.689	**0.830**						
CL	0.808	0.678	0.612	**0.823**					
LCSR	0.884	0.718	0.535	0.415	**0.848**				
SCSR	0.867	0.686	0.556	0.646	0.657	**0.828**			
ECCSR	0.82	0.605	0.684	0.53	0.517	0.712	**0.778**		
ECSR	0.897	0.745	0.552	0.544	0.691	0.776	0.683	**0.863**	
CT	0.858	0.668	0.720	0.713	0.543	0.677	0.730	0.712	**0.817**

Source Author's Compilation

Table 10.4 Structural model path analysis

Hypotheses	Path	Estimate	S.E	C.R	*P*- value	Std. estimates	Result
H1	LCSR → CI	0.191	0.077	2.494	0.013	0.237	Supported
H2	LCSR → CT	0.030	0.073	0.42	0.675	0.036	Non-supported
H3	ECCSR → CI	0.585	0.112	5.241	***	0.623	Supported
H4	ECCSR → CT	0.463	0.102	4.544	***	0.471	Supported
H5	SCSR → CI	– 0.044	0.138	– 0.323	0.747	– 0.042	Non-supported
H6	SCSR → CT	0.131	0.131	0.999	0.318	0.119	Non-supported
H7	ECSR → CI	– 0.002	0.1	– 0.017	0.986	– 0.002	Non-supported
H8	ECSR → CT	0.228	0.096	2.363	0.018	0.274	Supported
H9	CI → CL	0.226	0.085	2.644	0.008	0.229	Supported
H10	CT → CL	0.539	0.091	5.908	…	0.571	Supported

Source Author's Compilation

10.3.2.2 Structural Model

Structural equation modelling (SEM) was conducted to test the hypothesized relationships. The standardized path coefficients and their significance levels are presented in Table 10.4 and Fig. 10.3.

The findings of the path analysis showed that both legal CSR (S.E. = 0.077; p = 0.013) and economic CSR (S.E. = 0.112; p = ***) have positive effect on corporate image. Social CSR (S.E. = 0.138; p = 0.0.747) and environmental CSR (S.E. = 0.100; p = 0.986) does not significant influence corporate image. Moreover, economic CSR (S.E. = 0.102; p = ***) and environmental CSR (S.E. = 0.096; p = 0.018) have significant impact on customer trust. In contrast, legal CSR (S.E. = 0.073; p = 0.675) and social CSR (S.E. = 0.131; p = 0.318) does not impact customer trust. Additionally, corporate image (S.E. = 0.085; p = 0.008) and customer trust (S.E. = 0.091; p = ***) significantly influence customer loyalty.

This shows that hypotheses H1, H3, H4, H8, H9, H10 are accepted, while H2, H5, H6, H7 are rejected.

10.3.2.3 Parallel Mediation Analysis

With direct effect, the study also, estimated indirect effect of perceived CSR (legal, economic, social, and environmental) on customer loyalty through corporate image and consumer trust (see Table 10.5). The bootstrapping method was used to estimate the indirect effect, and the confidence interval was used to verify the significance of the influence. The findings indicate that Legal CSR has an indirect effect on customer loyalty both through corporate image (indirect effect = 0.0750; LLCI = 0.0031, ULCI = 0.1499) and consumer trust (indirect effect = 0.3160; LLCI = 0.1972, ULCI

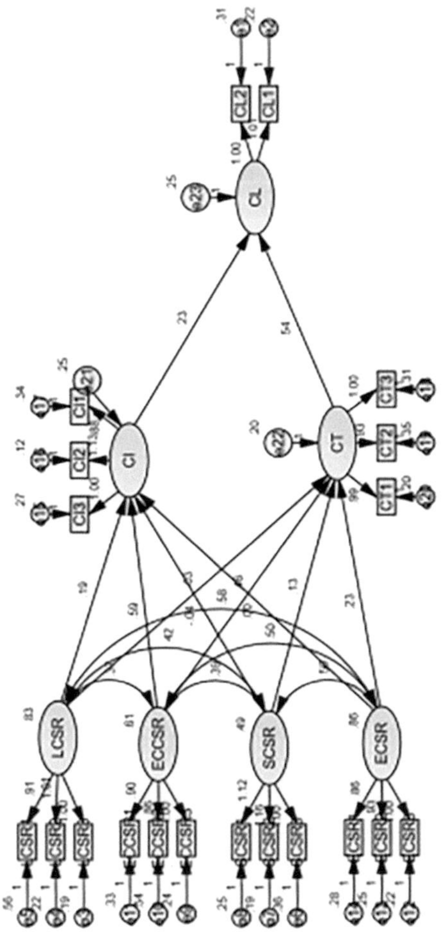

Fig. 10.3 Structural model

Table 10.5 Results of parallel mediation analysis

Mediator(s)	Indirect Effects	BootSE	Boot LLCI	Boot ULCI	Result(s)
LCSR → CI → CL	0.0750	0.0370	0.0031	0.1499	Partial mediation
LCSR → CT → CL	0.3160	0.0637	0.1972	0.4464	Partial mediation
ECCSR → CI → CL	0.0864	0.0488	– 0.0086	0.1858	No mediation
ECCSR → CT → CL	0.3656	0.0768	0.2251	0.5224	Partial mediation
SCSR → CI → CL	0.0718	0.0387	– 0.0032	0.1469	No mediation
SCSR → CT → CL	0.3921	0.0586	0.2753	0.5030	Full mediation
ECSR → CI → CL	0.0750	0.0377	0.0016	0.1514	Partial mediation
ECSR → CT → CL	0.3160	0.0636	0.1943	0.4430	Partial mediation

Source Author's Compilation

= 0.4464). Thus, both corporate image and consumer trust were found to partially mediate the impact of Legal CSR on customer loyalty. However, for Economic CSR, only consumer trust (indirect effect = 0.3656; LLCI = 0.2251, ULCI = 0.5224) was found to mediate the effect of Economic CSR on customer loyalty while corporate image didn't exhibit significant indirect effect (indirect effect = 0.0864; LLCI = – 0.0086, ULCI = 0.1858). Similarly, only consumer trust (indirect effect = 0.3921; LLCI = 0.2753, ULCI = 0.5030) was found to exhibit indirect effect of Social CSR on customer loyalty while corporate image (indirect effect = 0.0718; LLCI = – 0.0032, ULCI = 0.1469) was found to have no significant indirect effect on customer loyalty. It shows full mediating effect of trust in this relation. Finally, both corporate image (indirect effect = 0.0750; LLCI = 0.0016, ULCI = 0.1514) and consumer trust (indirect effect = 0.3160; LLCI = 0.1943, ULCI = 0.4430) were found to partially mediate the effect of Environmental CSR on customer loyalty.

10.4 Discussion

This study focused on finding antecedents of perceived CSR that effects corporate image and consumer trust and to what degree. These antecedents were legal CSR, economic CSR, societal CSR, and environmental CSR.

In this study, two antecedents were found to affect the corporate image i.e. legal CSR and economic CSR. While consumer trust was highly impacted by legal CSR, environmental CSR, and economic CSR. It was also found that consumer trust and corporate image have a significant effect on customer loyalty. Moreover, corporate image and consumer trust act as a mediator uplifting the effect of perceived CSR in customer loyalty.

The outcome is aligned with the general notion that CSR significantly impacts both corporate image and consumer trust. The economic CSR affecting consumer trust here implies that the higher the company works on the economic feasibility of consumer behaviour, the more they flourish. For this purpose, establishing a reasonable pricing strategy and generating jobs can be an example of economic CSR. In addition, companies' voluntary acts in favour of ecological sustainability and other rights lead to the formation of a positive attitude and a great corporate image. Furthermore, customers do have a favourable image of a company that conforms to rules and laws. Finally, a company's corporate image is highly impacted by the social endeavour the company takes in or supports social well-being. Further findings supporting the relationship between corporate image and consumer trust as a mediator on customer loyalty is because the company strives to enhance its marketing strategy focusing on these two factors to gain customer loyalty that can in turn be turned into customers. Apart from the antecedent's linkage to customer trust and corporate image, legal CSR and societal CSR have a mediating effect on corporate image and consumer trust. While economic CSR and environmental CSR have a mediating effect on customer trust. And in studying the mediating effect on customer loyalty, consumer trust plays a major role in corporate image.

10.5 Practical Implications

The study demonstrates the significant impact of legal and economic CSR on corporate image and trust. Managers need to adhere the rules and regulations with ethical standards and ensure proper and fair practices which will help in fostering image of company. Economic activity such as fair price and authentic report of financial status of CSR will boost trust by reflecting the sincerity and commitment. Managers should ensure the legal and economic activities of CSR are transparent and properly communicated to stakeholders while acknowledging their issues. Consistent involvement in CSR assists managers in maintaining customer loyalty and boost reputation. Economic and environmental CSR positively influence customer trust. Managers need to align economic practice (like fair price and transparent financial report) to ethical operations (like reducing waste or optimize energy use). For this transparency is crucial which can be done through regular and detailed CSR report, encouraging stakeholder to communicate openly and provide feedback, presenting actual contribution to environmental sustainability, and leveraging CSR as selling point can foster trust. This trust aid company in time of crisis and difficult time, as customers advocate and provide constructive feedback. Corporate image and customer trust influences customer loyalty. Managers should use proper CSR strategies to enhance its image by maintaining ethical practices, increasing transparency and engaging stakeholders.

Additionally, customers are less aware about the CSR initiatives performed by the company, especially in Indian economy. Thus, managers need to properly use its communication platform such as website and report to highlight their CSR activities. They should communicate with customers honestly and uphold the ethical standards to enhance loyalty which will build long-term relationship between them. Furthermore, the study finds that customer trust has a more significant impact on customer loyalty compared to corporate image. While corporate image can influence purchase intentions in the short term, building long-term customer trust provides a competitive edge. Overall managers need to integrate CSR strategies into their core business strategies. Moreover, they need to tailor these strategies aligning with customers' expectations. They can also provide training to employees on CSR in order to better comprehend it with company's goal.

10.6 Conclusion and Future Direction

The study aimed to analyse the link between the antecedents of CSR and its impact on corporate image and customer trust, subsequently effecting customer loyalty. With empirical data acquisition and analysis, the result showed impact of legal CSR, economic CSR and environmental CSR, but social CSR has no influence on either corporate reputation or customer trust. However, image and trust impact customer loyalty. Moreover, mediating impact was analysed between CSR variables and loyalty through corporate image and trust.

Certain limitations were identified which act as opportunity for future research direction. First, the study does not capture the cognitive measurement of perceived CSR initiatives as perception can be considered to be a dual construct (cognitive and affective). However, the study was conducted based on affective component while measuring perceived CSR. Thus, future study can consider both cognitive and affective dimension. Second, the study has identified several demographic data such as age, gender and other which has not been used in this study. Moreover, CSR awareness is important aspect of CSR literature especially in Indian context where people are less aware about company CSR performance. In future, moderation role of demographic profile along with CSR awareness in same study can be conducted.

References

1. Ramakrishnan, R. (2022). *CSR and Sustainable Development—A primer*. https://ssrn.com/abstract, 39, 04878.
2. Chaudhury, S. K., Das, S. K., & Sahoo, P. K. (2011). Practices of corporate social responsibility (CSR) in banking sector in India: An assessment. *Research Journal of Economics, Business and ICT*, 4.
3. Luo, X., & Bhattacharya, C. B. (2006). Corporate social responsibility, customer satisfaction, and market value. *Journal of Marketing, 70*(October), 1–18.

4. Szőcs, I., & Schlegelmilch, B. B. (2020). Embedding CSR in Corporate Strategies . In: *Rethinking Business Responsibility in a Global Context*, (pp. 45–60).
5. Arnold, M. F. (2010). Competitive advantage from CSR programmes. In C. Louche, S. O. Idowu, & W. L. Filho (Eds.), *Innovative CSR: From risk management to value creation.* Sheffield: Greenleaf Publishing Limited.
6. Księżak, P. (2016). The benefits from CSR for a company and society. *Journal of Corporate responsibility and leadership, 3*(4), 53–65.
7. Pérez, A., Salmones, M. d., & Bosque, I. (2013). The effect of corporate associations on consumer behaviour. *European Journal of Marketing, 47*, 218–238.
8. Arslan, I. K. (2020). The importance of creating customer loyalty in achieving sustainable competitive advantage. *Eurasian Journal of Business and Management, 8*(1), 11–20.
9. Edvardsson, B., Johnson, M. D., Gustafsson, A., & Strabdvik, T. (2000). The effect of satisfaction and loyalty on profits and growth: Products verses services. *Total quality management, 7*, 917–9247.
10. Sen, S., & Bhattacharya, C. B. (2018). Does Doing Good Always Lead to Doing Better? Consumer Reactions to Corporate Social Responsibility. *Journal of Marketing Research, 38.*
11. Castaldo, S., Perrini, F., Misani, N., & Tencati, A. (2009). The missing link between corporate social responsibility and consumer trust: The case of fair trade products. *Journal of Business Ethics, 84*, 1–15.
12. Pérez, A., & Bosque, I. R. (2015). Corporate social responsibility and customer loyalty: Exploring the role of identification, satisfaction and type of company. *Journal of Services Marketing, 29*, 15.
13. Kim, H. L., Rhou, Y., Uysal, M., & Kwon, N. (2017). An examination of the links between corporate social responsibility (CSR) and its internal consequences. *International Journal of Hospitality Management, 61*, 26–34.
14. Schaltegger, S., Hörisch, J., & Freeman, R. E. (2019). Business cases for sustainability: A stakeholder theory perspective. *Org Environ, 32*, 191–212.
15. Ortas, E., Alvarez, I., & Zubeltzu, E. (2017). Firms' board independence and corporate social performance: A meta analysis. *Sustainability, 9*, 1–26.
16. Wang, C. C. (2020). Corporate social responsibility on customer behaviour: The mediating role of corporate image and customer satisfaction. *Total Quality Management & Business Excellence, 31*, 7–8.
17. Luo, H., & Qu, X. (2023). Impact of environmental CSR on firm's environmental performance, mediating role of corporate image and pro-environmental behavior. *Current Psychology, 107*, 32255.
18. Ali, H. Y., Danish, R. Q., & Haq, M. A. (2020). How corporate social responsibility boosts firm financial performance: The mediating role of corporate image and customer satisfaction. *Corporate Social Responsibility and Environmental Management, 27*, 166–177.
19. Ahn, J., Shamim, A., & Park, J. (2021). Impacts of cruise industry corporate social responsibility reputation on customers' loyalty: Mediating role of trust and identification. *International Journal of Hospitality Management, 92*, 102706.
20. Chen, C. C., Khan, A., Hongsuchon, T., Ruangkanjanases, A., Chen, Y.-T., Sivarak, O., & Chen, S.-C. (2021). The role of corporate social responsibility and corporate image in times of crisis: The mediating role of customer trust. *International Journal of Environmental Research and Public Health, 18*, 8275.
21. Verghese, A. K. (2020). Post COVID-19: 5 Trends In Corporate Social Responsibility. *Retrieved from Business World*: https://www.businessworld.in/article/Post-COVID-19-5-Trends-In-Corporate-Social-Responsibility-/31-08-2020-314866/
22. Murray, J. (n.d.). Why CSR Is Essential in 2023. *Retrieved from Prowly*: https://prowly.com/magazine/why-csr-is-essential-in-2019/
23. Latif, K. F., & Sajjad, A. (2018). Measuring corporate social responsibility: A critical review of survey instruments. *Corporate Social Responsibility and Environmental Management, 25*(6), 1174–1197.

24. Wherther, W. B., & Chandler, D. (2006). *Responsibility: Stakeholders in a global environment*. SAGE Publications.
25. Apiday. (2023). *4 reasons companies should adopt CSR, Corporate Social Responsibility*. Retrieved from Apiday: https://www.apiday.com/blog-posts/4-reasons-companies-should-adopt-csr-corporate-social-responsibility
26. Bernal-Conesa, J. A., Nieto, C. d., & Briones-Peñalver, A. J. (2017). CSR strategy in technology companies: Its influence on performance, competitiveness and sustainability. *Corporate Social Responsibility and Environmental Management*, *24*(2), 96–107.
27. Al-Ghamdi, S. A., & Badawi, N. S. (2019). Do corporate social responsibility activities enhance customer satisfaction and customer loyalty? Evidence from the Saudi banking sector. *Cogent business and management*, *6*.
28. Ajina, A. S., Roy, S., Nguyen, B., Japutra, A., & Al-Hajla, A. H. (2020). Enhancing brand value using corporate social responsibility initiatives: Evidence from financial services brands in Saudi Arabia. *Qualitative Market Research an International Journal, 23*(4), 575–602.
29. McNamara, T. K., Carapinha, R., Marcie Pitt-Catsouphes, M.-C., Valcour, M., & Lobel, S. (2017). Corporate social responsibility and employee outcomes: The role of country context. *Business Ethics A European Review, 26*(4), 413–427.
30. Raza, A., Saeed, A., Iqbal, M. K., Saeed, U., Sadiq, I., & Faraz, N. A. (2020). Linking corporate social responsibility to customer loyalty through co-creation and customer company identification: Exploring sequential mediation mechanism. *Sustainability, 12*(6), 2525.
31. Ailawadi, K. L., Neslin, S. A., Luan, Y. J., & Taylor, G. A. (2014). Does retailer CSR enhance behavioral loyalty? A case for benefit segmentation. *International Journal of Research in Marketing, 31*, 156–167.
32. Dick, A., & Basu, K. (1994). Custromer loyalty: Towards an integrated conceptual framework. *Journal of Marketing Science, 22*(2), 99–113.
33. Khan, I., & Fatma, M. (2023). CSR influence on brand image and consumer word of mouth: Mediating role of brand trust. *Sustainability, 15*(4), 3409.
34. Harrison, J.S., Freeman, R.E., Abreu, M.C. S.d., (2015). Stakeholder theory as an ethical approach to effective management: applying the theory to multiple contexts. *Revista brasileira de gestao ˜ de negocios ´ 17*(55), 858–869.
35. Carroll, A. B. (1979). A three-dimensional conceptual model of corporate performance. *Academy of management review, 4*(4), 497–505.
36. Swaen, V., & Chumpitaz, R. (2008). Impact of corporate social responsibility on consumer trust. *Recherche et Applications en Marketing, 3*(4), 7–33.
37. Carroll, A. B. (1991). The pyramid of corporate social responsibility: Toward the moral management of organizational stakeholders. *Business Horizons, 34*(4), 39–48.
38. Mohr, L. A., Webb, D. J., & Harris, K. E. (2001). Do consumers expect companies to be socially responsible? The impact of corporate social responsibility on buying behavior. *Journal of Consumer Affairs, 35*(1), 45–72.
39. Huang, C. C., Yen, S. W., Liu, C. Y., & Huang, P. C. (2014). The relationship among corporate social responsibility, service quality, corporate image and purchase intention. *International Journal of Organizational Innovation, 6*(3).
40. Martínez, P., & Del Bosque, I. R. (2013). CSR and customer loyalty: The roles of trust, customer identification with the company and satisfaction. *International Journal of Hospitality Management, 35*, 89–99.
41. Martínez, P., Herrero, Á., & Gómez-López, R. (2019). Corporate images and customer behavioral intentions in an environmentally certified context: Promoting environmental sustainability in the hospitality industry. *Corporate Social Responsibility and Environmental Management, 26*(6), 1382–1391.
42. Afthanorhan, W. M. A. B. W., Ahmad, S., & Mamat, I. (2014). Pooled confirmatory factor analysis (PCFA) using structural equation modeling on volunteerism program: A step by step approach. *International Journal of Asian Social Science, 4*(5), 642–653.
43. Conchius, T. (2006). Corporate social responsibility in Dutch SME: motivations and CSR stakeholder. *Unpublished Masteral thesis, Maastricht University, Netherlands.*

44. He, Y., & Lai, K. K. (2014). The effect of corporate social responsibility on brand loyalty: The mediating role of brand image. *Total Quality Management & Business Excellence, 25*(3–4), 249–263.
45. Sharma, V., Poulose, J., Mohanta, S., & Antony, L. E. (2018). Influence of the dimensions of CSR activities on consumer purchase intention. *Innovative Marketing, 14*(1), 23–32.
46. Abdeen, A., Rajah, E., & Gaur, S. S. (2016). Consumers' beliefs about firm's CSR initiatives and their purchase behaviour. *Marketing Intelligence & Planning, 34*(1), 2–18.
47. Al Mubarak, Z., Ben Hamed, A., & Al Mubarak, M. (2019). Impact of corporate social responsibility on bank's corporate image. *Social Responsibility Journal, 15*(5), 710–722.
48. Kim, Y., Lee, S. S., & Roh, T. (2020). Taking another look at airline CSR: How required CSR and desired CSR affect customer loyalty in the airline industry. *Sustainability, 12*(10), 4281.
49. Zhao, L., Lee, J., & Moon, S. (2019). Employee response to CSR in China: The moderating effect of collectivism. *Personnel Review, 48*(3), 839–863.
50. Costa, R., & Menichini, T. (2013). A multidimensional approach for CSR assessment: The importance of the stakeholder perception. *Expert Systems with Applications, 40*(1), 150–161.
51. Bosch-Badia, M. T., Montlllor-Serrats, J., & Tarrazon, M. A. (2013). Corporate social responsibility from Friedman to Porter and Kramer. *Theoretical Economics Letters, 3*(3), 11–15.
52. Paramar, B. L., Freeman, R. E., Harrison, J. S., Wicks, A. C., Purnell, L., & Colle, S. D. (2010). Stakeholder theory: The state of the art. *Academy of Management Annals, 4*(1), 403–445.
53. Al-Shammari, A. M., Banerjee, S. N., & Rasheed, A. A. (2021). Corporate social responsibility and firm performance: a theory of dual responsibility. *Management Decision.*
54. Aguilera, R. V., Rupp, D. E., Williams, C. A., & Ganapathi, J. (2007). Putting the S back in corporate social responsibility: A multilevel theory of social change in organizations. *Academy of Management Review, 32*(3), 836–863.
55. Chen, X., Huang, R., Yang, Z., & Dube, L. (2018). CSR types and moderating role of corporate competence. *European Journal of Marketing, 52*(7/8), 1358–1386.
56. Venger, O., & Pomirleanu, N. (2018). Linking CSR communication activities to consumer brand evaluations: An examination of mediating and moderating factors linking CSR communication to brand evaluations. *Journal of Promotion Management, 24*(5), 675–693.
57. He, H., & Li, Y. (2011). CSR and service brand: The mediating effect of brand identification and moderating effect of service quality. *Journal of Business Ethics, 100*, 673–688.
58. Husted, B. W., & de Jesus Salazar, J. (2006). Taking Friedman seriously: Maximizing profits and social performance. *Journal of Management studies, 43*(1), 75–91.
59. Husted, B. W., & Allen, D. B. (2007). Strategic corporate social responsibility and value creation among large firms: Lessons from the Spanish experience. *Long Range Planning, 40*(6), 594–610.
60. Husted, B. W., Allen, D. B., & Kock, N. (2015). Value creation through social strategy. *Business & Society, 54*(2), 147–186.
61. Frank, A., Sheriff, O. A., & Mensah, H. K. (2015). The effect of CSR on brand equity in the IT solutions industry; The case of Alsale Service Limited. *The International Journal of Humanities & Social Studies, 3*(9), 312–336.
62. Henriques, I., & Sadorsky, P. (1996). The determinants of an environmentally responsive firm: An empirical approach. *Journal of Environmental Economics and Management, 30*, 381–395.
63. Buysse, K., & Verbeke, A. (2003). Proactive environmental strategies: A stakeholder management perspective. *Strategic Management Journal, 24*, 453–470.
64. Jeon, H. J., & Gleiberman, A. (2017). Examining the role of sustainability and green strategies in channels: Evidence from the franchise industry. *Journal of Marketing Theory and Practice, 25*(2), 189–199.
65. Menon, A., Menon, A., Chowdhury, J., & Jankovic, J. (1999). Evolving paradigm for environmental sensitivity in marketing programs: A synthesis of theory and practice. *Journal of Marketing Theory and Practice, 7*(2), 1–15.
66. Sharma, S., & Vredenburg, H. (1998). Proactive corporate environmental strategy and the development of competitively valuable organizational capabilities. *Strategic Management Journal, 19*, 729–753.

67. Dowell, G., Sastry, A., Hart, S., & Bernicke, J. (1997). Corporate environmental reputation: Comparing two industries. *Corporate Reputation Review, 1*, 140–146.
68. Rahman, N., & Post, C. (2012). Measurement issues in environmental corporate social responsibility (ECSR): Toward a transparent, reliable, and construct valid instrument. *Journal of Business Ethics, 105*, 307–319.
69. Gadenne, D. L., Kennedy, J., & McKeiver, C. (2009). An empirical study of environmental awareness and practices in SMEs. *Journal of Business Ethics, 84*, 45–63.
70. Wu, S. I., & Wang, W. H. (2014). Impact of CSR perception on brand image, brand attitude and buying willingness: A study of a global café. *International Journal of Marketing Studies, 6*(6), 43.
71. Hoogendoorn, B., Guerra, D., & Van Der Zwan, P. (2015). What drives environmental practices of SMEs? *Small Business Economics, 44*, 759–781.
72. Morales-Raya, M., Martín-Tapia, I., & Ortiz-de-Mandojana, N. (2019). To be or to seem: The role of environmental practices in corporate environmental reputation. *Organization & Environment, 32*(3), 309–330.
73. Graham, S., & Potter, A. (2015). Environmental operations management and its links with proactivity and performance: A study of the UK food industry. *International Journal of Production Economics, 170*, 146–159.
74. Chen, C. M., & Ho, H. (2019). Who pays you to be green? How customers' environmental practices affect the sales benefits of suppliers' environmental practices. *Journal of Operations Management, 65*(4), 333–352.
75. Mijatovic, I., Maricic, M., & Horvat, A. (2019). The factors affecting the environmental practices of companies: The case of Serbia. *Sustainability, 11*(21), 5960.
76. Adeniji, A. A., Osibanjo, A. O., Abiodun, A. J., & Oni-Ojo, E. E. (2015). Corporate image: A strategy for enhancing customer loyalty and profitability. *Journal of South African Business Research, 2015*(2015), 1–12.
77. Thøger Christensen, L., & Askegaard, S. (2001). Corporate identity and corporate image revisited—A semiotic perspective. *European Journal of Marketing, 35*(3/4), 292–315.
78. Tran, M. A., Nguyen, B., Melewar, T. C., & Bodoh, J. (2015). Exploring the corporate image formation process. *Qualitative Market Research: An International Journal, 18*(1), 86–114.
79. Worcester, R. (2009). Reflections on corporate reputations. *Management Decision, 47*(4), 573.
80. Bernstein, D. (1993). *Company image & reality: A critique of corporate communications*. Cassell Educational, London, Sage Publications.
81. Keh, H. T., & Xie, Y. (2009). Corporate reputation and customer behavioral intentions: The roles of trust, identification and commitment. *Industrial Marketing Management, 38*(7), 732–742.
82. Chattananon, A., & Lawley, M. (2007). Developing a model of the impact of societal marketing on corporate image. *Society and Business Review, 3*, 230–253.
83. Yeo, R. K., & Youssef, M. A. (2010). Communicating corporate image into existence: The case of the Saudi Banking Industry. *Corporate Communications: An International Journal, 3*, 263–280.
84. Mulaessa, N., & Wang, H. (2017). The effect of corporate social responsibility (CSR) activities on consumers purchase intention in China: Mediating role of consumer support for responsible business. *International Journal of Marketing Studies, 9*(1), 73–81.
85. Oney, E., Oksuzoglu Guven, G., & Hussain Rizvi, W. (2017). The determinants of electronic payment systems usage from consumers' perspective. *Economic Research-Ekonomska Istraživanja, 30*(1), 394–415.
86. Delgado-Ballester, E. (2004). Applicability of a brand trust scale across product categories: A multigroup invariance analysis. *European Journal of Marketing, 38*(5), 573–592.
87. Swaen, V., & Chumpitaz, R. C. (2008). Impact of corporate social responsibility on consumer trust. *Recherche et Applications en Marketing, 23*(4), 7–33.
88. Doney, P. M., & Cannon, J. P. (1997). An examination of the nature of trust in buyer-seller relationships. *Journal of Marketing, 61*(2), 35–51.

89. Stanaland, A. J., Lwin, M. O., & Murphy, P. E. (2011). Consumer perceptions of the antecedents and consequences of corporate social responsibility. *Journal of Business Ethics, 102*, 47–55.
90. Cazier, J. A. (2007). A framework and guide for understanding the creation of consumer trust. *Journal of International Technology and Information Management, 16*(2), 45–111.
91. Delgado-Ballester, E., Munuera-Aleman, J. L., & Yague-Guillen, M. J. (2003). Development and validation of a brand trust scale. *International Journal of Market Research, 45*(1), 35–54.
92. Oliver, R. L. (1999). Whence consumer loyalty?. *Journal of marketing, 63*(4_suppl1), 33–44.
93. Reichheld, B. Y. F. E., & Detrick, C. (2003). Want to know how to keep expenses low? Think loyalty. *American Banker, 168*(181), 6.
94. Kumar, N., Stern, L. W., & Steenkamp, J.-B.E.M. (1995). The effects of perceived interdependence on dealer attitudes. *Journal of Marketing Research, 32*, 348–356.
95. Gundlach, G. T., Archol, R. S., & Mentzer, J. T. (1995). The structure of commitment in exchange. *Journal of Marketing, 59*, 78–92.
96. Vlachos, P. A. (2012). Corporate social performance and consumer-retailer emotional attachment: The moderating role of individual traits. *European Journal of Marketing, 46*(11/12), 1559–1580.
97. Bowen, J. T., & Shoemaker, S. (2003). Loyalty: A strategic commitment. *Cornell Hotel and Restaurant Administration Quarterly, 44*(5/6), 31–46.
98. Hirschman and A.O. (1970). *Exit, voice, and loyalty: Responses to decline in firms. Organizations, and States*. Harvard University Press.
99. Oppewal, H., Alexander, A., & Sullivan, P. (2006). Consumer perceptions of corporate social responsibility in town shopping centres and their influence on shopping evaluations. *Journal of Retailing and Consumer Services, 13*, 261–274.
100. Du, S., Bhattacharya, C. B., & Sen, S. (2007). Reaping relational rewards from corporate social responsibility: The role of competitive positioning. *International Journal of Research in Marketing, 24*(3), 224–241.
101. Bhattacharya, C., & Sen, S. (2003). Consumer-Company Identification: A framework for understanding consumers' relationships with companies. *Journal of Marketing, 67*, 76–88.
102. O'Brien, I. M., Jarvis, W., & Soutar, G. N. (2015). Integrating social issues and customer engagement to drive loyalty in a service organisation. *Journal of Services Marketing, 29*(6/7), 547–559.
103. Salmones, M. D. M. G. D. L., Crespo, A. H., & Bosque, I. R. D. (2005). Influence of corporate social responsibility on loyalty and valuation of services. *Journal of Business Ethics, 61*, 369–385.
104. Currás-Pérez, R., Dolz-Dolz, C., Miquel-Romero, M. J., & Sánchez-García, I. (2018). How social, environmental, and economic CSR affects consumer-perceived value: Does perceived consumer effectiveness make a difference? *Corporate Social Responsibility and Environmental Management, 25*(5), 733–747.
105. Flavian, C., Guinaliu, M., & Torres, E. (2005). The influence of corporate image on consumer trust: A comparative analysis in traditional versus internet banking. *Internet Research, 15*(4), 447–470.
106. Kaur, H., & Soch, H. (2013). Mediating roles of commitment and corporate image in the formation of customer loyalty. *Journal of Indian Business Research, 5*, 33.
107. Gürlek, M., Düzgün, E., & Uygur, S. M. (2017). How does corporate social responsibility create customer loyalty? The role of corporate image. *Social Responsibility Journal, 13*(3), 409–427.
108. Hair, J. F., Black, W. C., Babin, B. J., & Anderson, R. E. (2010). *Multivariate data analysis* (7th ed.). Pearson.
109. Nunnally, J. C. (1978). An overview of psychological measurement. In *Clinical diagnosis of mental disorders: A handbook*, (pp. 97–146).
110. Hu, L. T., & Bentler, P. M. (1999). Cutoff criteria for fit indexes in covariance structure analysis: Conventional criteria versus new alternatives. *Structural Equation Modeling, 6*, 1–55.

Chapter 11
Analyzing the Impact of Freemium Concept and Modeling Subscription Growth for OTT Platforms

Adarsh Anand, Garima Babbar, and Niyati Aggrawal

Abstract The freemium model has witnessed a surge in popularity within the Over-the-top (OTT) industry in recent years. This model offers businesses and services a valuable opportunity to attract customers who might have otherwise been reluctant to try their products. By providing a limited version of the service for free, users can gain first-hand experience and assess the product's value prior to committing to a subscription. This research article aims to comprehensively explore the freemium policy in OTT platforms, encompassing its advantages, disadvantages, and impact on user behavior and preferences. To evaluate the effectiveness of the freemium model, two analytical models have been proposed to examine the influence on subscriber growth across four major OTT platforms, namely Amazon Prime, Disney + , Hulu, and HBO. The findings from these models showcase promising results, unequivocally establishing the freemium business model as an effective strategy. Consequently, this research strongly advocates for the wider adoption of the freemium model among other OTT platforms, highlighting its potential to significantly augment customer acquisition and retention rates. By embracing the freemium approach, OTT platforms can position themselves for increased success in the dynamic and competitive digital landscape.

Keywords Freemium · Over-the-top (OTT) · Streaming services · Subscription

---

A. Anand · G. Babbar (✉)
Department of Operational Research, University of Delhi, Delhi 110007, India
e-mail: garimababbar.du.or.20@gmail.com

N. Aggrawal
Department of Computer Science, JIIT, Noida, India
e-mail: niyati.aggrawal@jiit.ac.in

P. K. Kapur et al. (eds.), *Data-centric Approaches to Industrial Decisions*, Asset Analytics, https://doi.org/10.1007/978-981-96-7556-2_11

11.1 Introduction

Over the past thirty years, the freemium model has gained popularity as a business strategy. The term "freemium" merges the words "free" and "premium." This model offers a basic version of a product for free, while a premium version is available to paying customers. To motivate users to upgrade, the free version includes only limited introductory features. The Freemium approach is also utilized by providing free trials to customers for a limited time. This allows the business to attract customers who would otherwise have been apprehensive to try the product or service [1–4].

> *"The easiest way to get one million people paying is to get one billion people using."*—Phil Libin, Evernote

The concept of Freemium dates back to the 1980s when Adobe started releasing lighter versions of their software as free product for the users. The users would then be required to pay for using the premium features. Since then, hundreds and thousands of such services have emerged on the internet where the users are lured in using the free version and are encouraged to buy the premium version to remove advertisements, unlock new features, and enhance the quality or to upgrade the current available features. For instance, a free tier of a content service can be plagued with advertisements and the users can pay to get rid of them to enjoy uninterrupted viewing experiences [5]. In today's digital era, the freemium model is used in almost every available digital service. The freemium model provides a low-cost or free entry point for customers to experience the platform's content, which increases customer acquisition and brand recognition. For example, Google offers a free 15 GB free limit on each Google account, with this storage shared across its various platforms including Gmail, Google Photos, Google Drive. Its premium service Google One offers various annual as well as monthly plans to get up to 30 terabytes of storage (Fig. 11.1).

The freemium model operates on the assumption that users, after experiencing the basic service for free, will be inclined to pay for additional features or services. A commonly employed technique in the freemium model involves setting a restriction on the usage of an app's primary feature. Regular users may be content with the limited usage offered in the free version, but those who require more advanced or specialized usage may choose to pay for the removal of any restrictions hindering their ability to fully utilize the app. The gaming industry earns a major chunk of its revenue by allowing the users to download and play the games for free but then making them buy virtual currency to advance their gaming performance [6, 7]. According to a report by App Annie in 2018, the global mobile application market's total revenue was estimated at US$101 billion, and it is probable that over 50% of this revenue was generated through the freemium business model, as stated by Iqbal [8]. It can even be found in many games that it becomes impossible to advance some levels without the help of the virtual currency. Thus, the freemium model acts as a promising revenue model. Gupta et al. [9] discovered that individuals who frequently use a gaming app and excel at in-app challenges are less inclined to become super engagers. Among those who do become super engagers, some are motivated by social competition, as

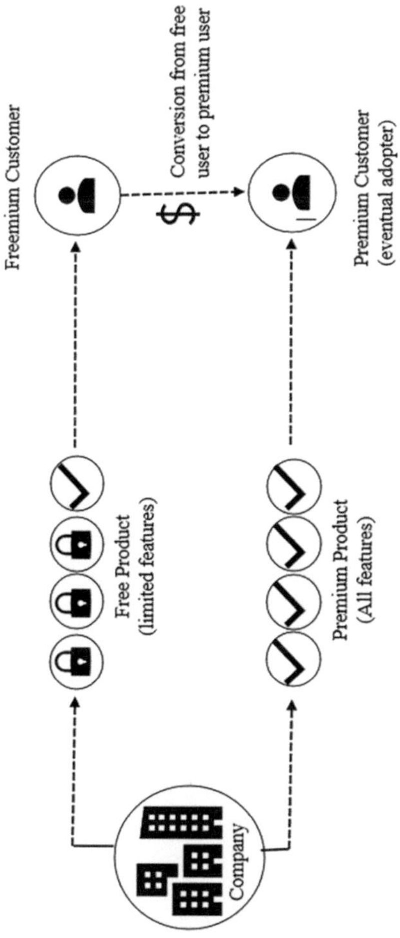

Fig. 11.1 The freemium model

evidenced by their tendency to make in-app purchases when unable to advance on the leader board.

It might be surprising that traditional news outlets, like The New York Times or The Washington Post, use the tactic of limiting access to their digital content. This involves offering a limited number of free articles to be read in a given month while providing unrestricted access to paid subscribers, as pointed out by Rivelo [10]. Therefore, the premium product must provide significant additional advantages over the free version to encourage customers to pay for it. If the premium product lacks sufficient extra features or benefits to warrant the cost, customers will remain with the free version, potentially causing the freemium model to fail. Wagner et al. [11] conducted a research study aimed at investigating the promotional effect of the free version of a freemium service on the premium version for music streaming services. Their findings suggest that service providers need to offer more than just basic functionality to promote their premium versions. They must deliver additional value that users find compelling enough to pay for. To motivate users to upgrade to the premium version, music service providers might need to create a sense of urgency by limiting the duration of the free trial and providing exclusive content and features to premium subscribers. Additionally, providers should focus on creating a high-quality premium product with desirable features and benefits to incentivize users to pay for the premium version.

Subramanian et al. [12] studied the factors that influence the success of the freemium model in the (Over-the-top) OTT market. They discovered that elements such as cost, content, website features, and trust can influence the effectiveness of the freemium model in attracting and retaining users. According to Wagner et al. [13], freemium service providers can improve the likelihood of converting users into paying customers by offering a strong functional fit between their free and premium services. Lui et al. [3], in their research analyzed the impact of implementing a freemium strategy on Google Play. Their research indicated that although providing a free version can enhance a product's visibility, the quality of the freemium product is more crucial in attracting customers. Furthermore, the study demonstrated that positive online reviews of the free version of a mobile application can also boost the market success of the paid version. In their study, Lambrecht and Misra [14] conducted a comparative analysis of a revenue model in which an online content provider offers free content versus one based on a traditional subscription fee. Surprisingly, their findings suggest that it would be more advantageous for a firm to provide free content during periods of high demand.

11.1.1 Advantages of the Freemium Model

A key benefit of the freemium model is its ability to rapidly build a large customer base. By providing a free version of the product or service, businesses can attract users who might not have considered trying the product otherwise. Once customers try the free version and like it, they are more likely to upgrade to the premium version, resulting in increased revenue for the business. Previous studies have confirmed the crucial role of social engagement [15] and peer influence [16] in driving the conversion of users to premium subscribers in freemium communities. Kurniawan and Alversia [17] discovered that online streaming freemium businesses can boost loyalty intentions among their current customers. The study revealed that satisfaction, brand familiarity, interpersonal relationships, and switching costs significantly affect both the intention to continue using and the intention to purchase. Conversely, free mentality and personal innovativeness impact only the intention to purchase. Additionally, value for money and relative advantage also significantly influence the intention to purchase.

An additional benefit of the freemium model is that it enables businesses to generate revenue from various sources, which is vital for long-term sustainability and financial stability. For example, companies can provide a free version of their product or service to attract a large user base and subsequently monetize this user base through advertising, data analytics, or partnerships. Concurrently, they can offer premium upgrades with enhanced features, thereby catering to users willing to pay for a superior experience. This dual revenue stream not only maximizes profit but also mitigates risks associated with relying on a single income source. For example, Spotify, a leading music streaming service, exemplifies this model. Its free version, supported by advertisements, attracts millions of users worldwide. Meanwhile, its premium version, which offers additional features such as ad-free listening, high-quality audio, and offline playback, generates substantial revenue from subscription fees [18]. This approach allows Spotify to leverage its large user base to attract advertisers while simultaneously providing premium users with a value-added service, thereby ensuring a steady and diversified revenue stream.

11.1.2 Challenges of the Freemium Model

The freemium model also presents some drawbacks. A primary challenge is achieving an optimal balance between the free and premium versions of a product or service. If the free version is overly restricted, it may fail to provide sufficient value, leading customers not to upgrade to the premium version. Conversely, if the free version is too accommodating, customers might not perceive a strong incentive to transition to the premium version. Upon realizing that only a small number of website visitors who did not subscribe converted to paying customers, The New York Times had to adjust its initial offer of 20 free online articles per month to 10 articles, as reported by Kumar [2]. Following this modification, the Times experienced a substantial improvement in its user conversion rate, with digital-only subscriptions increasing by 46% year on year and generating an additional $83 million in revenue, according to Ember [19]. Another example of a freemium model that failed is the game "Dungeon Keeper" by EA Games. The game was released as a free-to-play mobile game in 2014, but the freemium model was heavily criticized for being too aggressive with its microtransactions. Players were forced to wait long periods of time or pay real money to speed up the game's progress, leading to negative reviews and low player retention rates. The game was eventually pulled from app stores and considered a failure for EA Games.

Another challenge is that the freemium model can be expensive to implement. Offering a free version of the product or service requires resources and infrastructure, which can be costly for businesses. Furthermore, businesses need to invest in marketing and promotional efforts to attract customers to the free version of the product or service. Shi et al. [20] demonstrated that the freemium model is not always the most effective strategy, even when network effects are present. In cases of symmetric network effects, companies may be motivated to increase their network size and could find it advantageous to target low-end customers, but this does not necessarily lead to the freemium model being a stable equilibrium. Instead, firms are better off providing a low-end product with a positive price to attract customers. The freemium model can only become viable if the high-end and low-end products generate varying levels of marginal network effects. Contrary to traditional beliefs, companies using the freemium approach might enhance the quality of their low-end product beyond what is efficient to minimize differentiation.

An additional challenge for companies using the freemium business model is assessing whether they have reached an optimal conversion rate. This rate denotes the percentage of users who transition to the premium tier after interacting with the company's offering. Empirically, freemium firms can anticipate a conversion rate of around 3% to 5% of total users [21, 22]. This conversion rate range aligns with the generally accepted view that a freemium business should aim for a moderate conversion rate [2, 4]. A low or high conversion rate may signal that the freemium offer is suboptimal, as an exceedingly high conversion rate may indicate a lacklustre free tier, which could impede potential user acquisition. While some industry experts have advocated for a high conversion rate of up to 24% for freemium businesses [23], this is not a widely accepted view.

11.1.3 Over-The-Top (OTT) Platforms

Recent advancements have transformed the entertainment industry through the rise of Video on Demand (VoD) services, which have greatly enhanced the convenience and accessibility of watching movies and TV shows. These services utilize applications and websites, commonly referred to as Over-the-top (OTT) platforms, to provide video content directly to viewers. OTT platforms are a type of video streaming service that bypasses traditional cable and satellite TV providers and delivers video content directly to viewers over the internet. According to a report by Grand View Research [24], the global OTT market is expected to reach $220.54 billion by 2027, growing at a Compound Annual Growth Rate (CAGR) of 18.7% from 2020 to 2027. This growth can be ascribed to the expanding number of internet users and the growing demand for online video content. In recent years, OTT platforms have revolutionized the entertainment industry and are continuously altering global viewing habits. These platforms allow users to access content anytime and from any internet-connected device, offering enhanced control over their viewing experience. This has resulted in a shift towards on-demand and personalized video content consumption.

Anand et al. [25, 26] conducted a study that examined the key factors that users consider when choosing an OTT platform, and identified the most popular platform in the Indian market based on these attributes. In 2021, Puthiyakath and Goswami [27] conducted a study on the satisfaction levels of consumers using OTT and TV services, as well as their competition, coexistence, and competitive superiority. The researchers applied niche theory to assess the levels of gratification, similarity, and competitive advantage between OTT and traditional TV across seven specific gratification dimensions. The study's findings indicated that OTT services delivered higher satisfaction in each of these dimensions.

In the world of entertainment, especially in movies and TV shows, there existed a global imbalance where certain countries produced the majority of content while others consumed it. This led to a preference for certain cultures and types of entertainment over regional content. However, this trend appears to be changing with the rise of OTT video streaming platforms, both local and foreign. These platforms are increasingly acknowledging the importance of offering original content, along with international content that is subtitled, adapted, or dubbed. Golwalkar and Mishra [28] discovered that Indian regional language content is carving out its own niche and finding an audience, despite the abundance of international content available on OTT platforms. Their study also identified challenges facing regional content creators and the need for content evolution to widen the opportunities available to them.

11.1.4 Freemium Strategy for OTT Platforms

To stay competitive, many platforms have implemented a freemium model to attract new users and retain current ones. For example, Amazon Prime offers a 30-day free trial of its premium service, which includes access to Prime Video, ad-free music streaming on Amazon Music, as well as benefits like free unlimited fast delivery and exclusive deals. This extensive trial period allows users to experience a wide range of services, increasing the chances of converting them to paid memberships. The freemium model is widely adopted in the OTT industry, with platforms like Hulu, Amazon Prime, Voot, Disney + , Zee5, and HBO each customizing their approach to suit their content and audience. Hulu, for example, offers a basic ad-supported plan, while Disney + provides occasional free trials to attract new subscribers. Zee5 and Voot use extensive regional content to engage users, offering free content as a gateway to premium, ad-free experiences. The effectiveness of the freemium model in the OTT sector is clear from its widespread use and impact on user engagement and retention. By offering initial free access, these platforms can build a large user base and convert a significant portion into paying subscribers, driving both growth and long-term loyalty. Consequently, the freemium model has become a prevalent pricing strategy for OTT platforms, supporting their goals of expanding market presence and enhancing customer retention.

Identifying the specific user segments to focus on can be a valuable strategy for a company that is implementing a freemium business model and seeking to target its user acquisition efforts. This approach is particularly relevant in the current data-driven environment, as emphasized by Nieborg [7]. Shim et al. [29] conducted a research to identify a profitable niche in the Korean OTT market through the use of market segmentation theory. The study also examined strategies for content creation that would successfully convert free users into paying customers who belonged to the identified profitable niche segments.

However, there are concerns regarding the financial sustainability of OTT platforms using the freemium model. The model may not be universally effective across all types of OTT platforms. For instance, it often works better for platforms with a broad content range, such as Netflix or Amazon Prime, which can attract a wide audience with their diverse offerings. In contrast, platforms specializing in niche content might not benefit as much from this model. The success of the freemium model can also be influenced by the competitive landscape and the perceived value of the premium content or services offered. For example, a niche platform like Criterion Channel, which focuses on classic and indie films, may struggle more to convert free users into paying subscribers compared to a platform with a vast array of mainstream content. Additionally, the effectiveness of the freemium model is contingent upon how well the premium content differentiates itself from the free version and whether it provides sufficient additional value to justify the cost. A study by Lee and Shin [30] analyzed the financial performance of freemium-based OTT platforms and found that while the model can increase user engagement and retention, it can also lead to lower revenues and profits due to the costs of providing free services. The authors suggested that OTT platforms need to carefully evaluate the extent to which free content can be provided without incurring losses.

The freemium model can also lead to a cannibalization effect, where the free services reduce the sales of premium services. Cheng and Tang's [31] research centered on the utilization of a feature-limited free trial strategy, with particular focus on the tradeoff between consumer valuation gains resulting from positive network effects and lost sales attributable to the cannibalization effect. Holm and Jensen [32] discovered that providing a product or service for free does not guarantee business success, even if the offering is of excellent quality. They explained that for a new free product to succeed, it must be highly attractive to the mass market from its initial launch and fulfill an existing or potential need of potential users.

Thus, the freemium model has become a popular pricing strategy for OTT platforms seeking to attract and retain users. While some studies have shown positive results for the model, others have raised concerns about its effectiveness and financial sustainability.

The aim of the current study is to analyze the impact of the freemium model on the subscriber growth of top over-the-top (OTT) platforms. This research seeks to understand how offering a free tier, complemented by premium subscription options, influences the rate at which users subscribe to these services. By examining various datasets from leading OTT platforms, the study will evaluate the effectiveness of the freemium model in attracting and retaining subscribers.

The structure of the remaining sections is as follows: In Sect. 11.2, the fundamental components of the proposed modeling framework are explored. Section 11.3 delves into the specific models being proposed. Section 11.4 provides an illustration of the models. The discussion is presented in Sect. 11.5, and the conclusions are outlined in Sect. 11.6. The references are listed at the end.

11.2 Building Blocks of the Proposed Model

11.2.1 Assumptions

The article's proposed model relies on the following assumptions:

1. At any given time, the subscription rate is proportional to the number of potential subscribers yet to subscribe.
2. The subscriptions due to the effect of premium policy are considered.
3. The total number of potential subscribers "A" is fixed.

11.2.2 Notations

The following notations are used in the proposed modeling framework:

N	Potential market size
p	Number of additional purchases due to freemium
b	Rate of subscription
β	Learning rate
δ	Proportion of people joining the system due to freemium
g	Rate at which subscribers are added to the system due to freemium policy
$p(t)$	Time dependent number of subscribers which are added to the system due to freemium policy
$S(t)$	Expected number of subscriptions by time t

11.2.3 The Proposed Freemium Modeling Framework

In the field of social media, limited amount of research has been conducted for mathematically modeling the growth in the count or number of growing audience for a social media platform. Most of the studies conducted on Over-the-top (OTT) platforms are based on theoretical analysis of the trends and growth patterns in the entertainment industry due to emergence and rise of OTT platforms [33–35]. These studies analyze how the popularity of OTT platforms has affected the entertainment industry and its traditional models of content distribution, consumption, and revenue generation. These studies focus on factors such as consumer behavior, market competition, content production and licensing, pricing strategies, and technological advancements. They do not necessarily involve empirical research or data analysis of the performance or impact of specific OTT platforms.

However, in context of YouTube, numerous studies have been conducted to model the view-count of YouTube videos [25, 26, 36–40]. These studies model the growth pattern of these videos. In a similar fashion, it is possible to develop models for the growth of subscribers for an OTT platform.

This study draws on the work of Aggrawal et al. [41] in the context of YouTube video view-count modeling. They utilized a framework that applied literary theory from marketing science to analyze the growth in view-counts of YouTube videos. This approach can be adapted to OTT platforms for analyzing the count of subscribers joining an OTT platform.

Thus, the following equation can be used to determine the number of paying subscribers gained by an OTT platform:

$$\frac{dS(t)}{dt} = b(t)[N - S(t)] \tag{11.1}$$

where $b(t)$ represents the rate at which new subscribers are joining an OTT platform.

The utilization of any policy to expand the market by a business results in the acquisition of new customers, which in turn causes the market size of the total number of potential customers to become variable with time, rather than remaining fixed. By providing functional free version, the Freemium model aims to attract potential customers to the product or service and convert them into paying customers for the premium version. This can lead to an increase in the total number of potential customers over time, as more people become aware of the product and its offerings.

In the case of our model, the implementation of a Freemium policy by an OTT platform causes the market size of potential subscribers to become time-dependent. Therefore, to incorporate the effect of the Freemium model, we consider a time-varying form of the potential market size, denoted as $N(t)$, where the value of N changes with respect to time.

Therefore, in our model, we acknowledge the impact of the Freemium model by considering a time-varying form of the potential market size. This approach allows us to account for the changing dynamics of the market as a result of the implementation of the Freemium policy by an OTT platform.

Thus, the above Eq. (11.1) can be modified as:

$$\frac{dS(t)}{dt} = b(t)[N(t) - S(t)] \tag{11.2}$$

To incorporate the impact of the expansion of the number of subscribers, we adopt the following form of $N(t)$:

$$\frac{dS(t)}{dt} = b(t)\big[(N + \delta p(t)) - S(t)\big] \tag{11.3}$$

where, δ represents the proportion of people joining the platform due to the freemium policy and $p(t)$ represents the time-dependent number of subscribers which are added to the platform due to the freemium policy. Together $\delta p(t)$ represents the subscribers who join the platform due to the freemium model. In other words, it captures the impact of the Freemium policy on the conversion rate of potential subscribers to actual subscribers.

Thus, the following equation is used for modeling the subscription growth:

$$\frac{dS(t)}{dt} = \frac{b}{1 + \beta e^{-bt}}\big[(N + \delta p(t)) - S(t)\big] \tag{11.4}$$

where, $b(t) = \frac{b}{1+\beta e^{-bt}}$

Thus, the above Eq. (11.4) can be used to determine the number of premium subscribers obtained at time 't' due to the effect of the freemium business model.

11.3 Proposed Models

The authors have utilised the above approach given by Eq. (11.4) and considered various forms of $p(t)$ to cater to different subscription patterns due to the freemium policy given as follows:

11.3.1 Model 1: Exponential Growth

In this model, we consider $p(t) = pe^{gt}$ i.e., the subscribers joining the OTT platform due to the freemium policy are following an exponential growth pattern.

Thus, Eq. (11.4) can be written as:

$$\frac{dS(t)}{dt} = \frac{b}{1 + \beta e^{-bt}}\big[\big(N + \delta pe^{gt}\big) - S(t)\big] \tag{11.5}$$

On solving the above equation using $S(0) = 0$, the following equation can be obtained:

$$S(t) = \frac{1}{1 + \beta e^{-bt}}\left[N\left(1 - \frac{1}{e^{bt}}\right) + \frac{\delta pb}{(b + g)}\left(e^{gt} - \frac{1}{e^{bt}}\right)\right] \tag{11.6}$$

Equation (11.6) represents the cumulative number of subscribers by time 't' following an exponential distribution, where the rate of subscription inculcates the effect of the freemium policy.

11.3.2 Model II: Linear Growth

In this model we consider, $p(t) = p(1 + gt)$ i.e., the number of customers joining the premium membership of an OTT platform due to the freemium model follow the linear growth pattern.

Thus, Eq. (11.4) can be written as:

$$\frac{dS(t)}{dt} = \frac{b}{1 + \beta e^{-bt}}\left[(N + \delta p(1 + gt)) - S(t)\right] \tag{11.7}$$

On solving the above equation using $S(0) = 0$, the following equation can be obtained:

$$S(t) = \frac{N\left(1 - e^{-bt}\right)}{1 + \beta e^{-bt}} + \frac{p\delta}{1 + \beta e^{-bt}}\left[1 + \frac{g}{b}(tb - 1) - \frac{e^{-bt}}{b}(b - g)\right] \tag{11.8}$$

Equation (11.8) depicts of the cumulative number of subscribers by time 't', where the rate of subscription due to the freemium policy follows the linear pattern.

11.4 Model Illustration

Four datasets were used to validate the models proposed in this research.

Dataset I – Amazon Prime

The first dataset includes quarterly data on the number of Amazon Prime subscribers (in millions) in the United States, spanning from December 2013 to December 2019. Amazon Prime Video, often known as Prime Video, is a subscription-based streaming and rental service available either as a standalone option or as part of the Amazon Prime membership. This platform primarily features films and television series produced by Amazon Studios and MGM Holdings, as well as licensed content under the Amazon Originals label. It also offers content from third-party providers, additional content options, live sports events, and services for video rental and purchase. A free 30-day trial of Amazon Prime provides access to Prime Video, ad-free music streaming on Amazon Music, and additional benefits such as free unlimited fast delivery.

Dataset II - Disney +

The second dataset encompasses the global count of Disney + subscribers (in millions) from the first quarter of 2020 through the third quarter of 2022. Disney

+ is an American subscription-based streaming service managed by Disney Entertainment, a division of The Walt Disney Company. In India, Disney + Hotstar, the country's largest OTT platform, boasts a subscriber base of 300 million, with the platform's partnership with Disney + in 2020 contributing to 30% of the latter's overall subscription base. It offers both subscription and free tiers, with the latter including select films, news content, and TV shows for Hindi-speaking audiences, available only in standard definition and integrated with advertisements. The platform generates 70% of its revenue from advertising, making it a successful "freemium" OTT business model with high returns, particularly in India. Its unique business model has been effective in expanding its reach and attracting diverse audiences, demonstrating the potential for success with the freemium model in the OTT industry.

Dataset III - Hulu

The third dataset includes the count of paying subscribers (in millions) for Hulu in the US from Q4 2010 to Q3 2022. Hulu is a subscription-based streaming service in the United States, predominantly owned by The Walt Disney Company, with NBCUniversal of Comcast also holding a stake. Launched on October 29, 2007, Hulu offers a diverse selection of films and television shows from multiple studios, including 20th Century Studios, Searchlight Pictures, Disney Television Studios, ABC, Freeform, and FX Networks, in addition to its own original programming. New users can access a 30-day free trial.

Dataset IV - HBO

The fourth dataset consists of the count of subscribers (in millions) for HBO and HBO Max in the US, starting from 2015 until the first quarter of 2022. HBO Max is a subscription-based, over-the-top streaming service owned by WarnerMedia Direct, LLC, a subsidiary of Warner Bros.

The datasets cover a wide range of time periods, allowing for an in-depth analysis of the impact of the freemium model on subscriber growth. The datasets were collected from Statista, ensuring their validity and reliability for use in this research.

11.4.1 Model Parameters

In order to evaluate the proposed models, this study has employed the nonlinear least square (NLLS) method (Srinivasan & Mason, 1986) to estimate the unknown parameters. To evaluate the effectiveness of the proposed models, they have been compared to the existing information diffusion model (Aggrawal et al. [41]). Tables 11.1, 11.2, 11.3, and 11.4 contain the estimated parameters for each of the four datasets.

Table 11.1 Estimates values of model parameters for Amazon Prime

Amazon Prime	Model I	Model II	I.D.P. Aggrawal et al. [41]
N	52.265	45.071	116.483
p	30.473	25.54	
b	0.188	0.344	0.188
β	0.001	0.001	1
δ	0.99	0.273	
g	0.06	0.623	

Table 11.2 Estimates values of model parameters for Disney +

Disney +	Model I	Model II	I.D.P. Aggrawal et al. [41]
N	57.207	241.888	241.967
p	28.345	77.118	
b	0.205	0.101	0.101
β	0.001	1	1
δ	0.99	0.001	
g	0.124	0.001	

Table 11.3 Estimates values of model parameters for Hulu

Hulu	Model I	Model II	I.D.P. Aggrawal et al. [41]
N	60.024	29.852	61.243
p	6.39	10.005	
b	0.091	0.156	0.091
β	1.131	1.003	1.131
δ	0.185	0.247	
g	0.001	0.42	

Table 11.4 Estimates values of model parameters for HBO

HBO	Model I	Model II	I.D.P. Aggrawal et al. [41]
N	44.145	29.163	45.963
p	1.6	1.458	
b	0.99	0.989	0.99
β	266.61	124.35	290.279
δ	0.015	0.99	
g	0.395	0.99	

11.4.2 Model Validation

The performance of the different models was assessed using various comparison metrics, such as Mean Square Error (MSE), Root Mean Square Error (RMSE), variance, bias, and R-square. The results from the goodness of fit analysis are detailed in Tables 11.5, 11.6, 11.7, and 11.8.

Table 11.5 Values of comparison parameters on the proposed models for Amazon Prime

Amazon Prime	Model I	Model II	I.D.P. Aggrawal et al. [41]
R^2	0.98	0.982	0.965
Bias	0.584	0.439	1.22
Variance	3.977	3.763	5.111
RMSE	3.902	3.677	5.106
RMPSE	2.078	1.989	2.569

Table 11.6 Values of comparison parameters on the proposed models for Disney +

Disney +	Model I	Model II	I.D.P. Aggrawal et al. [41]
R^2	0.976	0.965	0.965
Bias	1.151	1.803	1.803
Variance	6.213	7.42	7.42
RMSE	6.097	7.374	7.374
RMPSE	2.745	3.267	3.267

Table 11.7 Values of comparison parameters on the proposed models for Hulu

Hulu	Model I	Model II	I.D.P. Aggrawal et al. [41]
R^2	0.993	0.993	0.993
Bias	0.042	0.009	0.042
Variance	1.1	1.115	1.1
RMSE	1.077	1.092	1.077
RMPSE	1.049	1.056	1.049

Table 11.8 Values of comparison parameters on the proposed models for HBO

HBO	Model I	Model II	I.D.P. Aggrawal et al. [41]
R^2	0.972	0.96	0.97
Bias	-0.23	-0.785	-0.213
Variance	3.344	3.906	3.472
RMSE	3.23	3.845	3.352
RMPSE	1.843	2.126	1.875

The Tables 11.5, 11.6, 11.7, and 11.8 display the performance of the predictive models across various OTT platforms. The results show that Model I is the most suitable for three out of the four datasets, highlighting its broad applicability. Conversely, Model II is the best fit for one particular dataset.

11.4.3 Graphical Analysis

Figures 11.2, 11.3, 11.4 and 11.5 show how well the predictive models align with the actual data, highlighting the real-world implications of the freemium model. The closeness of the graphs indicates that the proposed model effectively captures the influence of the freemium approach, validating its applicability in reflecting actual subscriber growth scenarios.

11.5 Discussion

Tables 11.1, 11.2, 11.3, and 11.4 present the evaluation and comparison of the proposed models, showing a strong correlation between them. Tables 11.5, 11.6, 11.7, and 11.8 depict the performance of these models across various OTT platforms, indicating that Model I is the most suitable for three out of the four datasets studied, while Model II is the best fit for one dataset, as evident from a straightforward analysis of the comparison parameters. Therefore, the addition of subscribers resulting from the freemium policy follows an exponential pattern for Disney + , Hulu, and HBO. In contrast, for Amazon Prime, it follows a linear pattern. The effectiveness of the proposed models has been further evaluated by comparing them to the widely used information diffusion model in marketing science. The results consistently demonstrated that the proposed models outperformed the existing model in terms of accuracy and reliability. The performance of the two proposed models varies across the datasets. Model II demonstrates superior performance for the Amazon Prime dataset, while Model I outperforms for the remaining three datasets: Disney + , HBO, and Hulu. Notably, both proposed models surpass the already established Information Diffusion Model by Aggrawal et al. [41] in terms of performance for the Amazon Prime, Disney + , and HBO datasets. However, for the Hulu dataset, Model I still exhibits better performance compared to the Information Diffusion Model Aggrawal et al. [41].

11.5.1 Research Contributions

This research significantly contributes to the existing literature on promotional broadcasting strategies and their impact on YouTube video view-counts. By proposing and

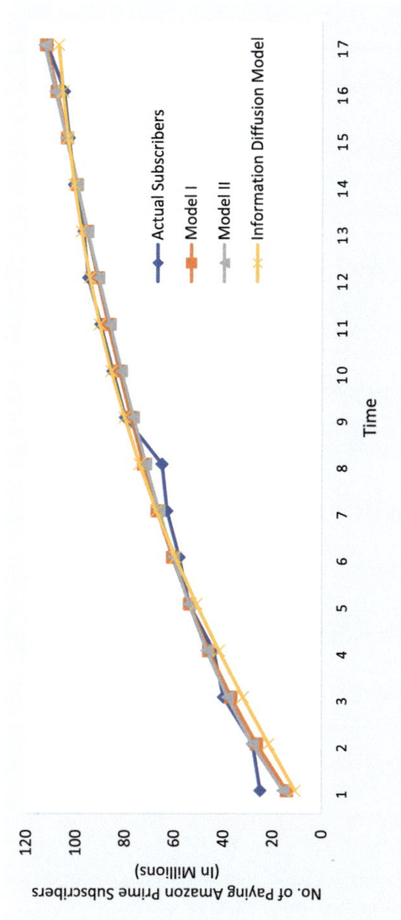

Fig. 11.2 Graphical analysis for Amazon Prime

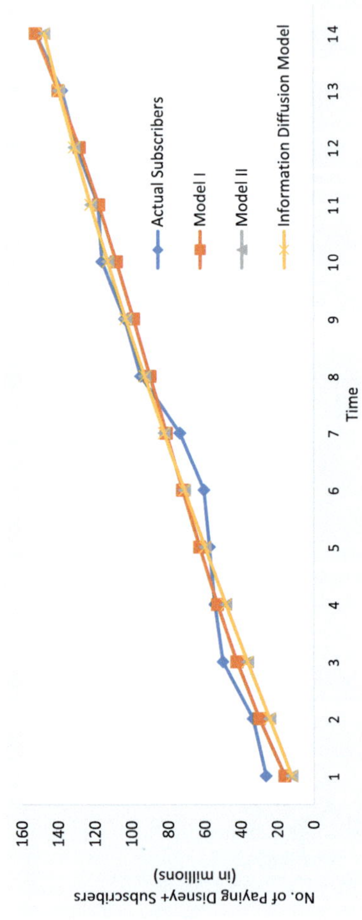

Fig. 11.3 Graphical analysis for Disney +

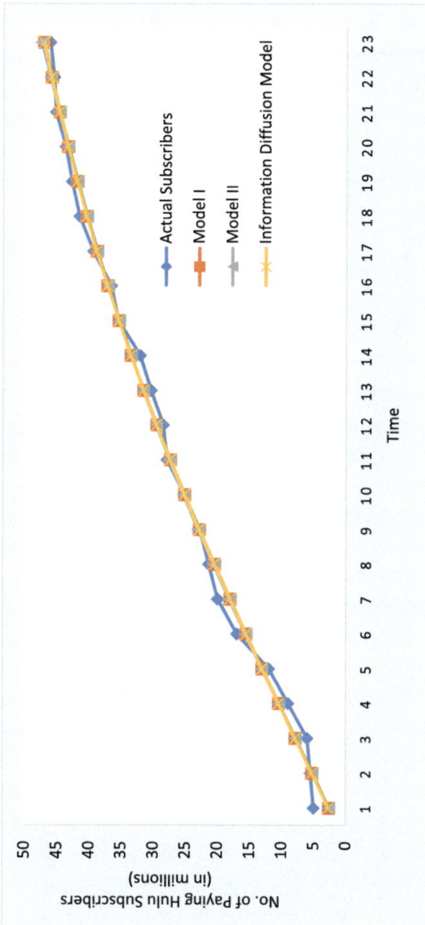

Fig. 11.4 Graphical analysis for Hulu

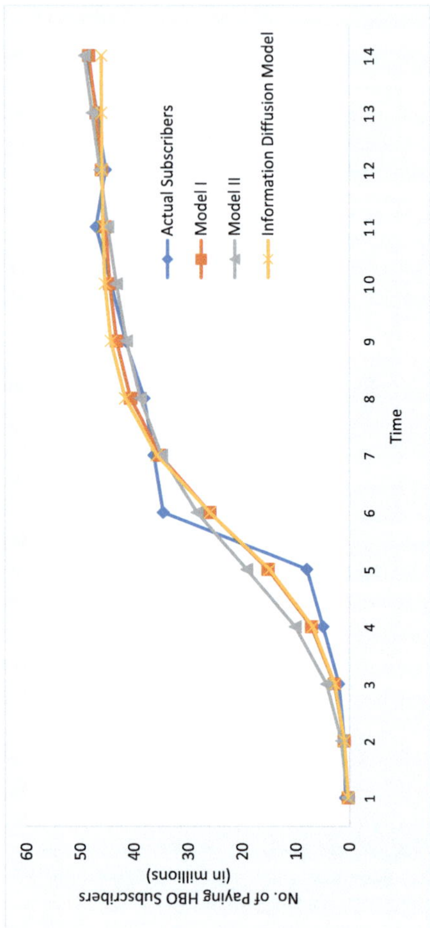

Fig. 11.5 Graphical analysis for HBO

testing various mathematical models, it provides a nuanced understanding of how different broadcasting approaches can enhance video visibility. The use of Distance-Based Analysis (DBA) to rank model performance offers a novel methodological approach in this context. Additionally, the study addresses the underexplored area of broadcasting phenomena in digital marketing, providing empirical evidence that fills a crucial gap. These contributions are poised to advance academic discourse and inform future research in digital content promotion.

11.5.2 Implications for Practice

This research provides valuable insights for professionals, particularly those working with YouTube content and digital marketing. By identifying effective broadcasting strategies, the study provides actionable insights that can be employed to optimize promotional efforts and enhance video view-counts. The use of ranked performance models enables content creators to select the most effective strategies tailored to their specific needs, thereby maximizing audience engagement and revenue. Additionally, OTT platforms considering the freemium model can utilize these insights to craft more effective marketing campaigns, improving user acquisition and retention rates. Ultimately, these practical applications can lead to more informed decision-making and strategic planning in digital content promotion.

11.5.3 Limitations and Future Research Directions

The current study only considers four major OTT platforms, namely Amazon Prime, Disney + , Hulu and HBO. There are many other OTT platforms that are popular in different regions of the world, and these platforms may have different characteristics and user behaviors. Thus, the researchers can base their study on those platforms.

Additionally, the proposed models do not take into account external factors that may impact subscriber growth, such as the COVID-19 pandemic or changes in market competition. Researchers can consider these factors in future for their study.

In addition, the research could explore the impact of different types of freemium models on subscriber growth, such as those that offer a limited free trial or those that offer a freemium version with no time or feature restrictions.

Finally, future research could analyze the impact of the freemium model on other aspects of an OTT platform's business model, such as revenue, customer retention, and overall user satisfaction.

11.6 Conclusion

The use of the freemium model in the OTT industry has become increasingly popular in recent years, with major players like Amazon Prime, Disney + , Hulu and HBO adopting this strategy to attract and retain subscribers. Through this research, we have analyzed the impact of the freemium model on subscriber growth for these four major OTT platforms, using two proposed models to predict the number of paying subscribers gained due to the freemium model. Our findings indicate that the freemium model is an effective strategy for attracting users and converting them into paying subscribers. However, it is important to note that there are limitations and drawbacks to this model, such as the potential for free users to consume a significant amount of resources without ever becoming paying subscribers. Overall, the freemium model has the potential to be a valuable tool for the OTT industry, but it is important to carefully consider its implementation and monitor its impact on subscriber growth [42, 43].

Acknowledgements "The work done in this paper has been supported by grants received by the first author from Institute of Eminence, DU, India as a part of Faculty Research Program via Ref no./IOE/2021/12/FRP and grants sanctioned to the second author from UGC NET via Ref. No. 220510230085."

References

1. Deubener, J., Velamuri, V.K., & Schneckenberg, D. (2016). *A typology of freemium business models for mobile applications.*
2. Kumar, V. (2014). Making freemium work. *Harvard Business Review, 92*(5), 27–29.
3. Liu, C. Z., Au, Y. A., & Choi, H. S. (2014). Effects of freemium strategy in the mobile app market: An empirical study of google play. *Journal of management information systems, 31*(3), 326–354.
4. Seufert, E. B. (2014). Freemium economics: Leveraging analytics and user segmentation to drive revenue. In *Freemium Economics: Leveraging Analytics and User Segmentation to Drive Revenue.* (pp. 1–233).
5. Semenzin, D., Meulendijks, E., Seele, W., Wagner, C. & Brinkkemper, S. (2012). Differentiation in freemium: where does the line lie?. In *Software Business: Third International Conference, ICSOB 2012, Cambridge, MA, USA, June 18–20, 2012. Proceedings 3* (pp. 291–296). Springer Berlin Heidelberg.
6. Hamari, J. (2015). Why do people buy virtual goods? Attitude toward virtual good purchases versus game enjoyment. *International Journal of Information Management, 35*(3), 299–308.
7. Nieborg, D., (2016). From premium to freemium: The political economy of the app. In *Social, casual and mobile games: The changing gaming landscape*, (pp. 225–240).
8. Iqbal, M. (2020). *App revenue statistics.* Business of Apps https://businessofapps.com/data/app-revenue
9. Gupta, K., Su, Y., Kunkel, T., & Funk, D. (2022). The super engagers of freemium gamified services: Using multimethod approach to examine why highly interactive consumers become paying consumers. *Internet Research, 32*(6), 1891–1909.

10. Rivelo, M. (2018). Can a freemium strategy coexist with the data center? *Forbes*. https://www.forbes.com/sites/forbestechcouncil/2018/01/10/cana-freemium-strategy-coexist-with-the-data-center
11. Wagner, T.M., Benlian, A., & Hess, T. (2013). The advertising effect of free—do free basic versions promote premium versions within the freemium business model of music services?. In *2013 46th Hawaii International Conference on System Sciences* (pp. 2928–2937). IEEE.
12. Subramanian, A., Seetharaman, A., Maddulety, K. (2021). From a free user to a premium subscriber: analysing the purchase decision of an OTT subscriber. *Research & Reviews: Journal of Engineering and Technology.*
13. Wagner, T. M., Benlian, A., & Hess, T. (2014). Converting freemium customers from free to premium—The role of the perceived premium fit in the case of music as a service. *Electronic Markets, 24*(4), 259–268.
14. Lambrecht, A., & Misra, K. (2017). Fee or free: When should firms charge for online content? *Management Science, 63*(4), 1150–1165.
15. Oestreicher-Singer, G., & Zalmanson, L. (2013). Content or community? A digital business strategy for content providers in the social age. *MIS quarterly*, (pp. 591–616).
16. Bapna, R., & Umyarov, A. (2015). Do your online friends make you pay? A randomized field experiment on peer influence in online social networks. *Management Science, 61*(8), 1902–1920.
17. Kurniawan, R., & Alversia, Y. (2021). Examining the antecedents of customer loyalty intentions in using freemium streaming service. In *Marketing and Smart Technologies: Proceedings of ICMarkTech 2020* (pp. 677–686). Springer Singapore.
18. Mäntymäki, M., & Islam, A. K. M. (2015). Gratifications from using freemium music streaming services: Differences between basic and premium users. In *Thirty Sixth International Conference on Information Systems, Fort Worth 2015.*
19. Ember S. (2017). *New York Times reports strong quarter on digital revenue growth.* The New York Times. https://www.nytimes.com/2017/07/27/business/new-york-times-company-2q-earnings
20. Shi, Z., Zhang, K., & Srinivasan, K. (2019). Freemium as an optimal strategy for market dominant firms. *Marketing Science, 38*(1), 150–169.
21. Koch, O. F., & Benlian, A. (2017). The effect of free sampling strategies on freemium conversion rates. *Electronic Markets, 27*(1), 67–76.
22. Southron, S. (2019). The CFO view of the freemium model. Forbes. https://www.forbes.com/sites/forbesfinancecouncil/2019/02/08/the-cfo-viewof-the-freemium-model.
23. Sack, D. (2013). When freemium beats premium. *The Wall Street Journal*. https://blogs.wsj.com/accelerators/2013/03/01/when-freemium-beats-premium
24. Over The Top Devices And Services Market Report, 2020–2027 [WWW Document], n.d. . Over The Top Devices And Services Market Report, 2020–2027. URL https://www.grandviewresearch.com/industry-analysis/over-the-top-ott-devices-services-market
25. Anand, A., Irshad, M. S., & Aggrawal, D. (2022). Application of intuitionistic fuzzy set TOPSIS in selecting most preferred OTT platform. *Nonlinear Studies, 29*(4), 1163–1181.
26. Anand, A., Irshad, M. S., & Dwivedi, Y. K. (2022). Modeling view count dynamics for YouTube videos: A multimodal perspective. *Kybernetes, 51*(10), 2964–2986.
27. Puthiyakath, H. H., & Goswami, M. P. (2021). Is over the top video platform the game changer over traditional TV channels in India? A niche analysis. *Asia Pacific Media Educator, 31*(1), 133–150.
28. Golwalkar, S., & Mishra, M. (2023). Rise of Indian regional content on video streaming platforms: A study of audience consumption and perception of Indian regional content on OTT platforms. In *AIP Conference Proceedings* (Vol. 2523, No. 1, p. 020107). AIP Publishing LLC.
29. Shim, D., Lee, C., & Oh, I. (2022). Analysis of OTT users' watching behavior for identifying a profitable niche: Latent class regression approach. *Journal of Theoretical and Applied Electronic Commerce Research, 17*(4), 1564–1580.
30. Lee, J., & Shin, D. H. (2017). Positive side effects of in-app reward advertising: Free items boost sales: A focus on sampling effects. *Journal of Advertising Research, 57*(3), 272–282.

31. Cheng, H. K., & Tang, Q. C. (2010). Free trial or no free trial: Optimal software product design with network effects. *European Journal of Operational Research, 205*(2), 437–447.
32. Holm, A. B., & Günzel-Jensen, F. (2017). Succeeding with freemium: Strategies for implementation. *Journal of Business Strategy, 38*(2), 16–24.
33. Fitzgerald, S. (2019). Over-the-top video services in India: Media imperialism after globalization. *Media Industries Journal, 6*(2), 00–00.
34. Kim, S., Baek, H., & Kim, D. H. (2021). OTT and live streaming services: Past, present, and future. *Telecommunications Policy, 45*(9), Article 102244.
35. Sundaravel, E., & Elangovan, N. (2020). Emergence and future of over-the-top (OTT) video services in India: An analytical research. *International Journal of Business, Management and Social Research, 8*(2), 489–499.
36. Aggrawal, N., Arora, A., Anand, A., & Dwivedi, Y. (2021). Early viewers or followers: A mathematical model for YouTube viewers' categorization. *Kybernetes, 50*(6), 1811–1836.
37. Anand, A., Babbar, G., & Aggrawal, N. (2024). Modelling dynamic nature of YouTube's digital ecosystem and its analysis using numerical methods. *International Journal of Systems Assurance Engineering and Management*. https://doi.org/10.1007/s13198-024-02436-2
38. Babbar, G., Anand, A., & Aggrawal, N. (2024). Modelling & analyzing view growth pattern of YouTube videos inculcating the impact of subscribers, word of mouth and recommendation systems. *International Journal of Mathematical, Engineering & Management Sciences, 9*(3), 435–450.
39. Bisht, M., Irshad, M.S., Aggarwal, N., & Anand, A., (2019). Understanding popularity dynamics for YouTube videos: an interpretive structural modelling based approach. In *2019 Amity International Conference on Artificial Intelligence (AICAI)* (pp. 588–592). IEEE.
40. Irshad, M. S., Anand, A., & Agarwal, M. (2020). Modeling active life span of Youtube videos based on changing viewership-rate. *Investigación Operacional, 41*(2), 249–262.
41. Aggrawal, N., Arora, A., Anand, A., & Irshad, M. S. (2018). View-count based modeling for YouTube videos and weighted criteria–based ranking. *Advanced mathematical techniques in engineering sciences*, (pp. 149–160).
42. Bapna, R., Ramaprasad, J., & Umyarov, A. (2016). Monetizing freemium communities: Does paying for premium increase social engagement? *Available at SSRN 2885681.*
43. Bass, F. M. (1969). A new product growth for model consumer durables. *Management Science, 15*(5), 215–227.

Chapter 12
Market Analysis of Cloud Service Providers in Singapore

Easwaramoorthy Rangaswamy, Vishal Bhasker, and Gemini V. Joy

Abstract The objective of this research was to analyse and understand the various factors which are driving the cloud adoption by the organisations for the Singapore market. Most of the Information technology vendors have changed their business model and evolved their product and services around cloud, hence it was interesting to analyse if the adoption is real and driven by the need stemming from the technology and business drivers. The current cloud ecosystems of the Singapore market were analysed to understand what the industry's expectations were from the cloud service providers. The data of the cloud users from Singapore was collected and analysed using quantitative research methods with deductive approach using random sampling method. It was further tested for the hypothesis developed around links between cloud adoption and the size of the organisation, various other factors, industry verticals and COVID as a situation. The results proved that adoption is dependent on the technical and business drivers and the actual challenges and benefits were captured which could help cloud vendors incorporate those views to refine their offerings. The results further proved that there is significant correlation between the size of organisation and the cloud adoption by incorporating the sample from the bigger enterprises and further expanding this research for the other Southeast Asian nations could get a bigger and better view of the cloud market in the region as most of the corporate treat ASEAN as one single integrated market for their business planning.

Keywords Cloud computing · Cloud service · Digitalisation · Cloud ecosystems · Market analysis · Information technology

---

E. Rangaswamy
Amity Global Institute, Singapore, Singapore
e-mail: erangaswamy@singapore.amity.edu

V. Bhasker
Oracle Corporation, Singapore, Singapore

G. V. Joy (✉)
VIT Business School, VIT University, Vellore, India
e-mail: geminivjoy@vit.ac.in

P. K. Kapur et al. (eds.), *Data-centric Approaches to Industrial Decisions*, Asset Analytics, https://doi.org/10.1007/978-981-96-7556-2_12

12.1 Introduction

In a simpler language, Cloud computing is the delivery of different IT (Information Technology) related services via the Internet as a medium. These services include tools and applications like data storage, servers, databases, networking, and software [1]. Mell and Grance [2] stated that National Institute of Standards and Technology US (NSIT) define cloud computing as "A model for enabling an everywhere, convenient, on-demand network access to a shared pool of configurable computing resources (e.g. networks, servers, storage, applications, and databases) that can be quickly provisioned and released with very less effort or service provider interaction". Over the last one decade, many business houses have also started to build their strategy around cloud computing and major IT vendors too have diversified their services portfolio to start offering cloud computing as their flagship offering.

12.1.1 Problem Statement and Background of the Study

Cloud computing is the most favourable topic in today's academic research and most of the IT organisations like HP, IBM, Microsoft, Oracle. VMware etc. have forayed into the cloud business and we have companies like AWS and Google being the leader in this space. Even e-commerce giants like Alibaba and telecom majors like Huawei have also branched their business into the space of cloud computing. Cloud computing is primarily driven by three main services, namely Infrastructure as a Service (IaaS), Platform as a Service (PaaS) and Software as a Service (SaaS). Based on a recent International Data Corporation (IDC) report the worldwide spending on cloud services and products will double in the next 5 years and it will be an almost 500 billion business by 2023. IDC report identifies that SaaS will be leading the next wave of cloud growth followed by IaaS for the public cloud market [3].

In the current scenario, Cloud is driving the information technology business for all the IT service providers. Almost all of the IT vendors have restructured their business to offer cloud services in some form, be it IaaS, PaaS or SaaS. However, the pertinent question is how mature the market towards adoption is and what are the pertinent challenges or benefits from the eyes of consumers. Idea for this research was to document the factors which may help cloud service providers refine their product and marketing strategy to align as per the industry need and consumer's psyche. Since Singapore is a developed economy, view of the consumer from a mature market matter to the Service providers to propel the next level of growth in this market.

12.1.2 Research Questions

Despite so much advancement in the field of cloud computing, still the biggest problem on why and how organisations shall adapt to cloud technology and is the adaption being driven by any specific factors like business need or industry types. What factors are being considered by cloud providers to penetrate deep into the market and convince business to move from the traditional IT setup to the cloud way. The current situation has led the researcher to seek answers for the questions below.

- What are the benefits and challenges of moving corporate data into cloud for Singapore market?
- What are the various factors driving decisions while choosing a cloud service provider?
- How do industry verticals impact cloud adoption in Singapore?
- What is the impact of COVID-19 on cloud adoption in Singapore?

Most of the academic research in the area of cloud computing is about the technology aspect of the services and the challenges it may pose for certain segments and markets, but none has been dedicated to the Singapore market. The knowledge about the factors that could affect the decision for cloud adoption and the experience of early movers into this space will help cloud service providers to refine their product and market strategies. In this research, the current cloud ecosystems of the Singapore market were analysed and it was also assessed as to how the cloud market and industry approach towards adoption is evolving and the Industry's expectations form the cloud service providers. This was an untouched field of conducting a study pertaining to the Singapore market.

12.1.3 Research Aim and Objectives

The area of research and focus was "Market Analysis of Cloud Service providers in Singapore" with aim to understand the difference in market position of various cloud service providers in Singapore and also to evaluate the factors driving their market positions in Singapore. Despite the predications by various agencies on cloud growth, there is a need to delve deeper to study and understand the perspective of Cloud providers and consumers who are using cloud technology. There is also a need to study the challenges business would face if they moved their critical operations to cloud. The situation stated above has given the objective of this research.

- To evaluate the benefits and challenges of moving corporate data into cloud for Singapore Market.
- To explain various factors driving decision while choosing a cloud service provider.
- To analyse whether Industry verticals help push cloud adoption.
- To evaluate Impact of COVID 19 on Cloud Adaption.

- To provide recommendations on best practices for cloud adoption for the Singapore market.

These objectives at a country level can help the cloud service providers assess the factors which drive cloud adaptation and also make them understand the way customers rate the benefits and challenges of cloud experience on various parameters. The study conducted based on these objectives can be used by cloud service providers to further refine their market penetration strategies. This study can be used as a best practices document towards cloud adoption by Singapore market and can be emulated by other regional neighbouring countries who always look up-to Singapore for the benchmarks.

12.2 Literature Review

The related literatures are required because it provides the conceptual background of the study, and it provides ample inputs to guide through the proposed research related to benefit of cloud migration and the factors driving cloud adoption.

12.2.1 Benefits and Challenges of Moving Corporate Data into Cloud for Singapore Market

In order to understand the growth of cloud business, it was interesting to analyse whether it is difficult for the corporate to move their business into cloud and would consumer view it as a benefit or the challenge on various parameters. Ayob [4] conducted a study to demonstrate that Cloud computing does have its benefits in terms of cost saving, scalability and computation power. The study was conducted to demonstrate the benefits compared to traditional on-premises computing. The research question whether cloud migration is beneficial or challenging and does it deliver what it promises on a technology scale was tried to be addressed to an extent in this study. The study goes into the depth of each service that can be offered on the cloud and lists how organisations can derive benefits from them as compared to the traditional IT set up. It summarised the benefits of Cloud. It was a detailed study conducted but that took references from various sources and available secondary data but didn't capture any primary data and the study appeared an accumulation of various available data without actually studying the pulse of the actual cloud users.

Bloom and Pierri [5] did research on whether cloud computing is helping smaller firms compete with the already established giants. It was analysed that cloud is helping these companies reach market faster and are they able to scale their operation as per the market need. These smaller companies can cut down the fixed cost of hiring the IT staff, servers and software and can satisfy the larger computing need by just going to cloud. It acts as a major advantage to the smaller and newer firms.

It was an interesting observation that the early adopters of cloud were the smaller companies in US from the urban and educated areas. This study was completely focused on companies in the US but didn't touch the view and data of the companies across the globe. Another research done by Forbes Technology Council (2017) listed few challenges that organisations would face while moving their business into cloud. Many successful CTOs (Chief Technology Officers) and CIOs (Chief Information Officers) who form the technology council wrote the paper based on their experience and industry trends. This paper captured the view only from the perspective of challenges that Business would face. None of the study so far has been done from the perspective of Singapore market and the consumers view on both the challenges and benefits of moving their data into the cloud. In this context, the first hypothesis of the study was tested as below,

- There is no significant correlation between the cloud adoption and the size of the organisation.

12.2.2 Various Factors Driving Decision While Choosing a Cloud Service Provider

The important aspect of cloud adoption is in understanding the factors which drive a decision of moving business to cloud. Smith [6] of Texas University did a study on whether an IT professionals age impact cloud adoption. This research is somewhat relevant to this research subject regarding the overall factors which affect the adoption of cloud computing. While this research was centred on finding answers to the pertinent questions if it was technology or the business factors that drive the cloud adoption, the study by Marcus had linked to the age of IT executives to the cloud adoption by the US based organisations. Responses of the IT executives based out of the US based companies were processed to establish a co-relation between the age and the decision to adopt cloud for the organisation. It was summarised and while data was skewed towards the younger respondents which appears that they are keen towards moving to cloud as compared to the older respondent who are more exposed to the traditional form of an on-premises IT setup, still more study is required to understand whether correlation changes with certain age bracket.

Another study by Foreman and Anthony [7] of University of Phoenix was done for the purpose to get an agreement from cybersecurity professionals regarding data security and privacy in cloud computing and possible mitigating solution for that. The paper does highlight security as one of the biggest consideration and the factors towards the corporation's cloud journey. These studies doesn't cover the overall factors driving the decision of cloud adoption by the organisations. Another research was carried out on role of security towards the cloud adoption. Khandelwal and Saini [8] researched on security challenges of cloud computing. Their analysis was to find an answer if technology factors are the critical drivers for cloud adoption. In conclusion the adoption shall be carefully considered only after evaluating the security attributes being provided by cloud Service providers. Many studies were conducted to evaluate cloud adoption on the technology factors and few have also being done to access if this can be done as per the type of industry. In this context, the second hypothesis of the study was tested as below,

- There is no significant correlation between cloud adoption and the technology and business factors.

12.2.3 Industry Verticals Help Push Cloud Adoption

Many scholars tried looking for an answer if an industry can be classified as the leader in cloud adoption, albeit indirectly without putting a direct reference to this question. There was also a research that was done on the use of cloud technology by financial institutions in Singapore [9]. It was observed that Cloud computing is beneficial to the financial institution in terms of helping them to adapt and respond to new business, regulatory compliances and market environments. It was also evaluated that the Board of Directors at any financial institutes have the responsibility to implement the governance framework for the cloud adaption within their firms. It was concluded that the cloud services would continue to evolve and so does the regulations related to the cloud for the Singapore market and the banks see cloud as a tool to reduce cost and increase efficiency. However this study didn't give any clarity on the position of various cloud providers in the Singapore market and the factors which are driving their market position. Baker, Fratto and Reiners [10] did somewhat similar study related to "Banking on the cloud'. This study analysed if the banking industry is leading the cloud adoption. This study was related to how a data concerned sector like Banking would adapt to cloud and on the flip side cloud technology may become a single point of failure for a banking system. This study highlighted that no financial regulatory agency has any direct supervising authority or control over cloud services providers and this issue shall be taken care as the cloud may be the most notable risk to financial market stability by becoming a single point of failure In addition this study didn't share any data on any other industry's readiness for cloud adaption and doesn't project a complete view on what must be done to enhance the adaption in the banking sector. There is no credible data on cloud adoption available for other industry verticals like manufacturing, production or government institutions or any other. In this context, the third hypothesis of the study was tested as below,

- There is no significant relationship between cloud adoption and type of industry.

12.2.4 Impact of COVID 19 on Cloud Adaption

While it has been observed that various factors like the age of IT executive, technology, security and industry segment impacting the cloud journey of an organisation, none has been devoted to the impact of COVID since this has been a very recent phenomenon. Another research by Navratil, Ubik and Peciva [11] was done on the subject "Cloud computing an alternative or the future". This was evaluated on the scale of education, healthcare, and culture domain. It was summarised one of the key benefits is that there is no requirement of a huge initial investment to start to use IT services. This study answered that Cloud is not merely a technology shift but also a mind shifts as well. According to Larry Downes [12] acknowledged that COVID has caused an economic turbulence in the world, and this has led to a situation that technological disruptions will transfer every business at a faster pace, and this will affect all kind of industry be if retail, entertainment, media or any to name. The interesting observation is that while this change due to COVID is temporary, however the shift in consumer/user behaviour is going to be permanent. This indeed suggest that COVID is a trigger for technology business like cloud computing.

12.2.5 Research Gap

In Summary, none of the research have been done to figure out if the size of the organisation matters for cloud adoption and none too has analysed the experience of actual cloud users in terms of the benefits and challenges they have witnessed during transition and after transition to cloud in Singapore. Moreover COVID -19 pandemic is a recent event hence the effect of this particular factor on adoption too hasn't been analysed. While a substantial amount of research has been done on cloud computing, but none has been devoted to the Singapore market in particular.

12.2.6 Conceptual Framework

A conceptual framework has been derived which depicts the flow of information and the results that were synthesised to arrive at the desired outcome of this study. It started with analysing the experience of cloud users about cloud adoption. Various factors for cloud adoption were analysed. An assessment was done about the primary reasons for choosing a cloud vendor based on their unique position and selling propositions. It was also helpful in understanding the user's experience of migrating to the traditional IT workload on the cloud. Simultaneously the assessment was done if organisations moved to cloud more during the COVID times over few months and this research would be helpful in generating the best practices approach for cloud adoption for the Singapore market (Fig. 12.1).

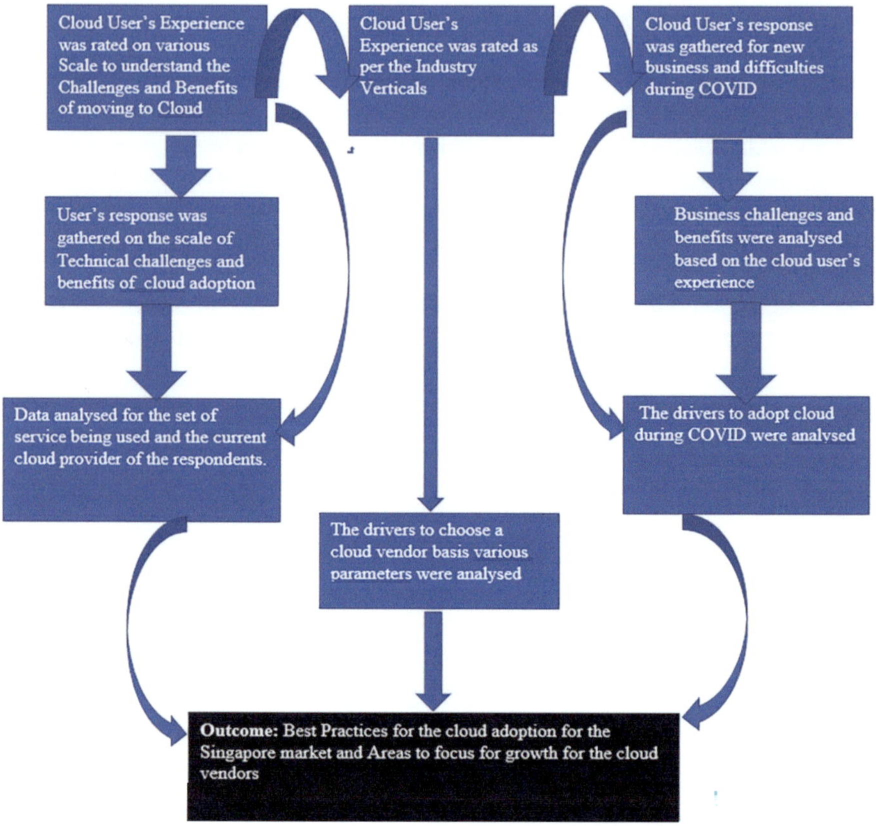

Fig. 12.1 Conceptual framework

12.3 Methods and Methodology

The purpose of this research was to come out with a best practice document for cloud adoption for the Singapore based customers and to further guide the cloud service providers in terms of upgrading their offerings to make the user experience much better. The realism research philosophy was adapted for this research exercise. It's a known fact that realism philosophy is based on scientific method to develop the knowledge and this research approach depends upon the fact the reality is independent of the human mind with an approach that what was once hypothetical becomes observable. The research method adopted for this study was quantitative research with a deductive approach. The variable related to the consumer's experience with cloud, their observation about certain parameters and other variables were quantified to come out with findings and recommendations. Hypotheses were tested against the processed quantified data.

In an attempt to establish the research process, survey strategy was adopted for this research work. A set of questions which were linked to the research objectives was sent to the set of respondents via online form method. Research choice adopted was mono method as the analysis was done based on the quantitative data and single data collection methods was used and associated quantitative data analysis methods and procedures would be applied. This study was based on the primary data being collected over a certain period. Cross sectional analysis method was adopted to conduct this study as data was collected over a definite period of time for collecting data from various respondent.

12.3.1 Population/Sample Selection

Next part was to source the data for the companies using cloud in Singapore market and to establish the population size to be selected for the survey. The Accounting and Corporate Regulatory Authority (ACRA) of Singapore has the data of the companies operating in Singapore and the information was sourced from them. The primary target for this analysis were the SME (Small and medium enterprise) of Singapore who are the consumers and early adapter of cloud as a preferred IT technology. The data shows that there are about 219,000 SME companies operating out of Singapore [13]. The actual user of cloud in some form are about 750 plus companies with an employee base of about 40 or more. In order to capture the more relevant experience of cloud users the employee strength of 40 or more was considered for this study. Based on the available data, the population size for this study was about 750 companies. The final sample size of around 100 enterprises was targeted for this survey as sample size with 10 percent of the population size is being considered good for the research work. Knowing that not all would respond, hence the sample size was kept little higher than the 10 percent of the population size. A simple random sampling method was adopted and about 100 enterprises were targeted for this research work. For this research, the primary data was collected using the online web-based questionnaire.

12.3.2 Validity and Reliability

Validity and reliability concepts were being used to test the quality of research being conducted. Cronbach's Alpha was applied to check the reliability whether questionnaire was stable or not. If the results would show a high level of correlation coefficient, which would range from 0.74 to 0.95, then it means that it was reliable to use [14]. The Cronbach's Alpha test conducted showed that the reliability of instrument is from 0.782 to 0.858 which is good. Validity would require a very well-designed questionnaire of the highest standard so as not to be misconstrued. In this study, the questions were discussed with 2 subject matter experts working as cloud solution architects with cloud service providers and 2 actual users of cloud technology who have been cloud users for the past couple of years to ensure the content validity of the questionnaire. For construct validity, questionnaire was discussed with experts on statistic and incorporated their inputs to improve the research instrument.

12.4 Discussion

The data received from 89 completed online surveys were transferred into SPSS i.e. a statistical tool for the further analysis of the data. The respondents were all from Singapore since the study is specific to the Singapore market and out of that a total of 67.4 percent were males and females were 32.6 percent of the total respondents also highlighting a fact that majority of the cloud user and decision makers are males. Further, for the age group frequency, it's seen that from the total respondents a total of 58.4 percent is in the age bracket of 20–40 years and 33.7 percent were being from the age bracket of 41–50 years and very less of about 7.9 percent constituted the age bracket of 51 years and above. This further signifies that the majority of the cloud users are a younger population who have adopted well to this technology. The study's age of respondents aligns with Smith [6] of Texas University who did conclude in his research that age of the IT executive does matter in the cloud adoption and most of the young executive adopts very well compared to the older executives. Analysis shown in Table 12.1 is about advantages of cloud computing. On a scale of 1–5 where "5" denotes "strongly agree" and "1" being "strongly disagree", the advantages of using the cloud computing in their organisations was rated by the respondents and maximum respondents rated "easy access to data" and "easier access to hardware and software" as the biggest advantages as on cloud, infrastructure can easily scale up and down depending upon the business requirement. Another advantage is having "reduced capital investment" on the cloud. The reduced capital investment stands out as one of the prime advantages of moving to cloud by the small and medium enterprises. Analysis in Table 12.2 shows that on a scale of 1–5 where 1 stands for 'very difficult" to 5 being "easy", the cloud users were asked to rate their experience of using cloud in their organisation. It was rated on various parameters to elicit a response in terms of what factors posed hindrance in adoption and what were the factors which worked in favours. The biggest challenge with a mean value of 2.55 that was faced by users was the first time "migration of their on-premises workload and application to cloud" followed by their concern around data security and billing clarity. However, the majority of the respondents experienced that Cloud technology was superior and business users faced no issues after the on –premises workload was successfully migrated to cloud. Parameter related to "Ease of use of cloud technology" didn't pose too much challenge to the cloud users with a mean value of 4.48.

As shown in Table 12.3, Respondents were further asked to state the reasons for using cloud on various scales and the majority of them chose cost saving as a primary driver followed by the flexibility to pay as per use the second preferred

Table 12.1 Advantage of using cloud

Details	Advantage of cloud—easy access to HW& SW	Adv. of cloud—better data & safety control	Adv. of cloud—quality of aervice	Adv. of cloud—better access to online services	Adv. of cloud easy access to data	Adv. of cloud—reduced capital investment	Adv. of cloud—reduced IT staff	Adv. of cloud—easy data recovery
N	89	89	89	89	89	89	89	89
Mean	4.35	3.93	4.12	4.14	4.43	4.32	3.88	4.10
Median	4.00	4.00	4.00	4.00	5.00	5.00	4.00	4.00

Table 12.2 Global warming related variables

Details	N	Minimum	Maximum	Mean	Std. deviation
Cloud Experience—Ease of use	89	3	5	4.48	0.559
Cloud experience—Apps Migration to Cloud	89	1	4	2.55	1.008
Cloud Experience—Billing Clarity	89	1	5	3.55	0.948
Cloud Experience—Data Security	89	1	5	3.25	0.793
Cloud Experience—Business Continuity during Migration	89	3	5	3.84	0.678
Cloud Experience—Service Performance after Cloud Migration	89	3	5	4.36	0.514
Cloud Experience—Business User Experience	89	4	5	4.57	0.499
Cloud Experience—RPO & RTO objectives are met	89	4	5	4.7	0.464

Table 12.3 Objective of using Cloud

Details	Objective cloud-cost saving	Objective cloud-ease of use	Objective cloud-no train IT staff	Objective cloud-scalability	Objective cloud-pay per use
Mean	0.72	0.59	0.48	0.55	0.61
Std. Deviation	0.45	0.495	0.503	0.501	0.492
N valid	89	89	89	89	89

option. Whereas the option that organisations don't need to train their IT staff on the changing technology was the least preferred reason for going to cloud.

As depicted in Table 12.4, it was further analysed about the drivers for opting for new cloud solution during COVID and the maximum response was garnered for the factor related to "Service level agreement" that the cloud companies would provide to the users /organisations to ensure that the services level guarantees are provided and to ensure the business continuity. Cost still remains the third best option for the cloud beside Service Level Agreement (SLA) and Business continuity.

12.4.1 Correlation

Table 12.5 shows that the Pearson correlation test was conducted between the new cloud services availed by Organisations during COVID and reasons for selecting

Table 12.4 Driver for Cloud during COVID

Details	N	Minimum	Maximum	Mean	Std. Deviation
Driver for cloud in Covid - Cost control	89	0	1	0.55	0.501
Driver for cloud in Covid – Business Continuity	89	0	1	0.67	0.475
Driver for cloud in Covid SLA	89	0	1	0.72	0.45
Driver for cloud in Covid - Lack of IT Staff	89	0	1	0.39	0.492
Driver for cloud in Covid - Risk Mitigation	89	0	1	0.54	0.502
Valid N (list wise)	89				

cloud vendor on the parameters of pricing and on the brand value and market position. A bivariate Pearson's correlation coefficient (r) was calculated. The bivariate correlation between these variables was positive and good with p ¡ 0.05 (Table 12.7) for both the factors of price and Brand value/Market position of cloud vendors' vs new services in COVID. The Pearson correlation (r) test results show that before computing value of "r", the assumptions of normality and linearity were evaluated, and those were in order and were found to be normal.

Hypothesis 1 The first null hypothesis which states that there is no co-relationship between cloud adoption and size of the organisation was tested using "Shapiro–Wilk" and "Kolmogorov–Smirnov" test (Table 12.6). The test was conducted between the parameter using "Size of employee" as a factor list and various advantages of using cloud by the organisations in the dependent list. The "sig" value is less than 0.05 for the test using "Shapiro–Wilk" and "Kolmogorov-" test. All the parameters of advantage of using cloud have "sig" value less than 0.05 hence the null hypothesis is rejected.

The alterative hypothesis is that there exists a correlation between cloud adoption and the size of the organisations. Various secondary data also validated the alternate hypothesis, and this stands true while the survey was conducted for the small and medium organisations of Singapore. Bloom and Pierri [5] conducted research for the US Market on cloud adoption and did conclude that smaller companies adopt cloud faster as the need to have a huge capital and to train their IT staff goes away if opting for the cloud. The data analysed for the Singapore market also validate the same finding hence the alternate hypothesis is that "there exists a correlation between the cloud adoption and the size of the organisation".

Hypothesis 2 To test second null hypothesis that there is no correlation between the cloud adoption and the technology and business factors, the one sample T test was conducted on the parameters of "cloud experience of the respondents on the technology and business factors". The test analysis of the all the technology and

Table 12.5 Pearson correlation for new cloud service in COVID

Correlations				
Details		Parameters for selecting vendor-Pricing	Parameters for selecting vendor-Market position/Brand	New cloud service in Covid
Parameters for selecting vendor-Pricing	Pearson Correlation	1	.161*	.230**
	Sig. (2-tailed)		0.034	0.002
	N	89	89	89
Parameters for selecting vendor-Market position/Brand	Pearson Correlation	.161*	1	.298**
	Sig. (2-tailed)	0.034		0
	N	89	89	89
New cloud service in Covid	Pearson Correlation	.230**	.298**	1
	Sig. (2-tailed)	0.002	0	
	N	89	89	89

*Correlation is significant at the 0.05 level (2-tailed)
**Correlation is significant at the 0.01 level (2-tailed)

business factors shows that the "sample t" test is statistically significant as Sig = 0 < 0.05 and Test Statistic, t value > 1.96 for all the factors as per the Table 12.7, The Shapiro–Wilk statistic and checking the pictures of the histogram confirmed that each variable were distributed normally. Hence the null hypothesis is rejected, and the alternative hypothesis is established that there exists a correlation between the cloud adoption and the technology and business factors.

Hypothesis 3 In order to test the third Null hypothesis that there exists no correlation between cloud adoption and the type of Industry, the one sample T test was conducted on the scale "challenges of using cloud as per industry that each respondent represents". The test analysis on all the parameters as per the respondent industry showed that the T test is statistically significant as Sig = 0 < 0.05 and Test Statistic, t value > 1.96 for all the factors as per the Table 12.8. Hence the null hypothesis is rejected, and the alternative hypothesis proves that there exists a correlation between the cloud adoption and the type of the industry. The hypotheses were tested, and a further analysis was done on the advantages of moving data to cloud and also the challenges that users experienced while moving data to cloud.

Table 12.6 Shapiro wilk statistics –hypothesis 1

Details	No of employee	Kolmogorov-Smirnov[a]			Shapiro–Wilk		
		Statistic	Df	Sig	Statistic	Df	Sig
Advantage of cloud—Easy access to HW & SW to HW & SW	100–200	0.388	24	0	0.685	19	0
	Greater than 200	0.286	65	0	0.759	50	0
Adv. of cloud—quality of service	100–200	0.43	24	0	0.591	19	0
	Greater than 200	0.314	65	0	0.771	50	0
Adv. of cloud—better access to online services	100–200	0.275	24	0.001	0.803	19	0.001
	Greater than 200	0.292	65	0	0.779	50	0
Adv. of cloud—easy access to data	100–200	0.241	24	0.005	0.814	19	0.002
	Greater than 200	0.401	65	0	0.616	50	0
Adv. of cloud—reduced capital investment	100–200	0.265	24	0.001	0.788	19	0.001
	Greater than 200	0.348	65	0	0.726	50	0
Adv. of cloud—easy data recovery	100–200	0.365	24	0	0.727	19	0
	Greater than 200	0.247	65	0	0.813	50	0

[a]Lilliefors significance correction

12.5 Conclusions and Recommendations

This research was conducted to understand the cloud adoption being done by the small and medium enterprises in the Singapore market and to correlate the factors which drives the adoption. The impact of COVID on the cloud market was also analysed through the data generated from the survey.

12.5.1 Conclusion

12.5.1.1 Benefits and Challenges of Cloud

Few of the observations based on the responses were that the key concerns of the organisations in adopting to cloud are related to, "being resistant to change" and the concern of integrating on the on-premises applications and infrastructure with

Table 12.7 One sample T test for hypothesis -2

One-sample test						
Details	Test value = 0					
	t	df	Sig. (2-tailed)	Mean difference	95% confidence interval of the difference	
					Lower	Upper
Cloud Experience—Ease of use	66.088	88	0	4.485	4.35	4.62
Cloud experience—Apps Migration to Cloud	21.023	88	0	2.551	2.31	2.79
Cloud Experience—Billing Clarity	31.122	88	0	3.551	3.32	3.78
Cloud Experience—Data Security	33.986	88	0	3.246	3.06	3.44
Cloud Experience—Business Continuity during Migration	47.063	88	0	3.841	3.68	4
Cloud Experience—Service Performance after Cloud Migration	70.544	88	0	4.362	4.24	4.49
Cloud Experience—Business User Experience	75.94	88	0	4.565	4.45	4.69
Cloud Experience—RPO & RTO objectives are met	84.153	88	0	4.696	4.58	4.81

the cloud. Support from the management is also being viewed as one of the key concerns by the organisations in adopting to cloud. The cloud helps companies to reduce the hours it takes to build IT infrastructure, accelerate their IT projects that are important for the business growth or cost control while a normal physical server could take many weeks to procure and provision, A server in cloud takes minutes to set up and sale this helping companies to respond faster to the changing need of business. The main advantages of moving to cloud was related to easy access to cloud and easy access to hardware and software to give the flexibility to scale as per the business requirement. Another observation from the data was that 84 percent of the organisations opted for the public cloud whereas only 13 percent opted for the private cloud. It clearly signifies that the majority of the organisations trusted the public cloud thus reinforcing the consumer's belief in the power of cloud. Wenhao et al. [15] aligned to the findings highlighting that the consumer may start analysing the standard for cloud service selection, thus enabling the effective and reliable selection of desired cloud services.

Table 12.8 One sample test for hypothesis -3

One-sample test						
Details	Test value = 0					
	t	df	Sig. (2-tailed)	Mean difference	95% confidence interval of the difference	
					Lower	Upper
Cloud Challenge—Not Dependable	37.488	88	0	3.841	3.64	4.05
Cloud Challenges—Lack of Motive	33.697	88	0	3.159	2.97	3.35
Cloud Challenges—Resistance to change	30.426	88	0	2.652	2.48	2.83
Cloud Challenges—Data Residency issues	24.086	88	0	2.246	2.06	2.43
Cloud Challenges—integration with on-premises systems	30.518	88	0	2.696	2.52	2.87
Cloud Challenges—Data security and Privacy	22.493	88	0	2.159	1.97	2.35
Cloud Challenges—Lack of trust on cloud	52.157	88	0	3.87	3.72	4.02
Cloud Challenges—cost of upgrading own network	47.271	88	0	3.957	3.79	4.12
Cloud Challenges—Lack of Training	54.563	88	0	3.841	3.7	3.98
Cloud Challenges—existing apps migration complexity	33.858	88	0	3.072	2.89	3.25
Cloud Challenges—recurring cost	69.81	88	0	4.42	4.29	4.55
Cloud Challenges—Initial migration is expensive	32.483	88	0	2.957	2.77	3.14
Cloud Challenges—No Management Support	22.845	88	0	2.478	2.26	2.69
Cloud Challenges—Legal, IT and Practical challenges	31.439	88	0	2.826	2.65	3.01

Out of the total respondents, 42 percent of them have stated to be using AWS for their cloud requirement, followed by 17 percent of them on Microsoft Azure with Google cloud being at 13 percent of the sample size used for the survey. Remaining cloud vendors are within 10 percent of the total spot. This analysis for Singapore relates to the secondary data done by [16] also states that AWS is the clear cloud leader, followed by Azure and Google Cloud and remaining vendors comes after them. Neha et al. [17] does agree and argue that cloud computing has changed the way Micro, Small, and Medium Enterprises provide and use computing services. The vendors and other stakeholders need to ensure that they follow the market requirements.

12.5.1.2 Factors Driving Cloud Decision

The null hypothesis that there exist no correlation between the size of the organisation and the cloud adoption was rejected after the relevant tests were conducted and this hypothesis was also supported by the secondary data that was available for the US market that there exist a correlation between the cloud adoption and the size of the organisations The smaller company decides faster to go for cloud as it takes way their requirement to invest high capital to set up their IT operations and train and hire the IT staff whereas it takes longer for an bigger enterprise to embark on their cloud journey due to the complexities involved in moving the already existing on-premises IT operations on to the cloud and also to get the management approvals to move data on to cloud. The survey also confirmed that in 44 percent of the cases Management takes the decision to move to cloud and in 29 percent of the cases the management still takes the decision with support from IT, hence it is evident that cloud decisions are seldom taken at an IT layer and its always a strategic management decision to move to cloud.

survey also confirmed that in 44 percent of the cases Management takes the decision to move to cloud and in 29 percent of the cases the management still takes the decision with support from IT, hence it is evident that cloud decisions are seldom taken at an IT layer and its always a strategic management decision to move to cloud.

12.5.1.3 Industry Vertical Push Cloud Adoption

The third hypothesis proves that there exists a correlation between cloud adoption and the type of industry. Data substantiate that few of the industry which are leading the adoption in Singapore are the financial sector and the Manufacturing /production sector. The government has also started to open up for the cloud adoption and "data residency issue" i.e. data to stay in the country does play an important role in the decision taken by the government departments to go for the cloud.

Further analysis of data during COVID also established that organisations had challenges in maintaining their IT infrastructure as there was a constraint of sending IT staff onsite. About 58 percent of the respondent reported that either it was very difficult, or they somehow managed their IT department during COVID, and these factors triggered decision to opt for offloading incremental workload on the cloud as it ensured the business continuity and ensured the Service level Agreement provided by the cloud service providers. Cost reduction in going to cloud also played an important role for organisation opting for new cloud solutions during COVID. The total cost of ownership for going to cloud appeared low to the companies as compared to having to maintain their own infrastructure.

12.5.1.4 COVID Impact on Cloud Adoption

The data substantiate that most of the organisations had issues in managing their IT infrastructure COVID as the movement were restricted and IT staff couldn't respond timely to urgent systems related issues thus triggering the Cloud demand as most of the organisations which struggled during this time had opted for some form of new cloud solution and with most of them selecting the public cloud to rely on their IT operations. Further analysis showed that primary drivers for opting for the cloud service providers during COVID were related to the Service level guarantees offered by the cloud vendors to ensure the business continuity of the organisations. This proved to be the biggest driver to opt for more and new cloud solutions during COVID. Another driver was pricing offered by the cloud vendors and their market and brand positions. The existing market position of cloud vendors played a part in getting them more business since the credibility of successfully managing multiple customer is higher and creates trustworthiness. Fuzan et al. [18] agreed and highlighted that the guaranteed compensation and pricing for cloud services does playing in the successful managing of the customer. They argued that the impact of participants' risk attitudes is different for existing and new service cases. Such situation would have the signalling effect of compensation when users do not know the service quality. A similar study by Hussain et al. [19] on cloud services also argued that consumers might have taken risk attitudinal behaviour when choosing the cloud vendors.

12.5.1.5 Recommendations

Cloud technology has proven to be a successful shift form the way organisations used to run their IT operations earlier. In order to successfully adopt to the cloud, organisations must pay key attention to the initial migration of their data and application on the cloud as that's the factor which is the biggest hindrance and fear of organisations from adapting to cloud. The vendors must provide a smooth transition and must address the concern related to business continuity during transition. Cloud vendors must also pay attention to their pricing model and must show case that the total cost of ownership of cloud is much cheaper compared to maintaining the on-premises infrastructure. Since data analysis substantiate that the cost and flexibility to pay per use is one of the biggest factors in driving a decision related to cloud adoption, hence the cloud service providers must be able to demonstrate the lower total cost of ownership to the customers. The cloud vendors must also focus in bringing few extra customers even at a lower price to consolidate their position in the market as this business is driven more by trust as more customers will buy cloud solution from a vendor who already has big names as reference. Since Security remains a greater concern for the cloud adoption, the solution providers must ensure that the best of the security features are built in their cloud offerings, and they must focus on building the local data centre. The local data centres in the country of residence will address the concern and legal requirement of the cloud users which is related to "data residing only in the country". This will help trigger business growth for the cloud vendors who haven't yet invested in terms of opening the data centres in the country of operations.

12.5.2 Research Contributions

This research is centred on the cloud market of Singapore and predominantly around the small and medium enterprises of Singapore. Reportlinker [20] stated that the cloud market in Southeast Asia and Singapore prominently is expected to grow at a Compound annual growth rate (CAGR) of over 6 percent during the period 2019–2025. Since the market will be growing at an impressive 6 percent CAGR it becomes imperative for the cloud vendor to understand the actual challenges being experienced by the Cloud users and what can be done to address them. This research will guide them in terms of what improvement cloud vendors need to bring into their product offerings to make it attractive for the market and further how to be prepared to address the issues surrounding the data security and residency and particularly the initial issue of migrating the application and data onto the cloud.

12.5.3 Future Research and Development

This research can be further exploited in terms of taking this to a larger set of cloud users from bigger enterprise as well. So far, the data and sample size were limited due to the circumstances related to COVID which had further limited the ability to interview few long term and early users of the cloud technology in order to incorporate their view for a detailed and deeper analysis in terms of the benefits and challenge of the cloud. It would have further given an insight into an aspect of how cloud technology has evolved over last few years and the way cloud vendors too have evolved their products, pricing and strategy over a period of time. By incorporating the sample from the bigger enterprises would give a holistic view of the Singapore cloud market and further this research can be expanded for the ASEAN (Association of the South East Asian Nations) economy to get a bigger view of the cloud market in the region as most of the corporate treat ASEAN as one single integrated market for their business planning, hence it will give cloud vendors a better view in terms of refining their market strategy for growth in the cloud business.

Author Contributions Each named author has substantially contributed to conducting the underlying research and drafting this manuscript. All authors read and approved the final manuscript.

Funding The authors did not receive support from any organisation for the submitted work.

Data Availability The authors confirms that all data generated or analysed during this study are included in this published article.

Declarations

Conflict of Interest The authors have no competing interests to declare that are relevant to the content of this article.

Consent to Participate Informed consent was obtained from all individual participants included in the study.

Human or Animal Rights This article does not contain any studies with human or animal subjects performed by any of the authors.

References

1. Frankenfield, J. (2020). Cloud computing. *Investopedia.* https://www.investopedia.com/terms/c/cloud-computing.asp, (Accessed 28 Dec 2022)
2. Mell, P., & Grance, T. (2011). The NIST definition of cloud computing.
3. Edward, J. (2020). Online available at https://kinsta.com/blog/cloud-market-share/, (Accessed 27 Dec 2022).
4. Sether, A. (2016) Cloud Computing Benefits (2016). Available at SSRN: https://ssrn.com/abstract=2781593 (Accessed 28 Dec 2022).
5. Bloom, N., & Pierri, N. (2018). Research: Cloud computing is helping smaller, newer firms compete. Harvard Business Review. https://hbr.org/2018/08/research-cloud-computing-is-helping-smaller-newer-firms-compete, (Accessed 26 Dec 2022)
6. Smith, M. L. (2020). Does the age of an IT executive impact adoption levels of cloud computing services? *International Journal of Business and Social Science, 11*(3), 1–8. (Accessed 3 Dec 2022)
7. Foreman, A. (2019). *Security limitations and mitigation of cloud computing*. ProQuest Dissertations Publishing, online available at https://search.proquest.com/pqdtglobal/docview/2300203638/6A3145F0342E49F5PQ/18?accountid=14565, (Accessed 27 Dec 2022)
8. Khandelwal, M., & Saini, H. (2019). Review on security challenges of cloud computing (October 2, 2019). In *International Conference on Advancements in Computing & Management (ICACM-2019)*, Available at SSRN: https://ssrn.com/abstract=3463271 (Accessed 28 Dec 2022).
9. Peihani, M. (2018). Financial regulation and disruptive technologies: The case of cloud computing in Singapore. *Singapore Journal of Legal Studies*, (pp. 77–99), Online Available from: https://papers.ssrn.com/sol3/results.cfm, (Accessed 2 Dec 2022)
10. Baker, C. M., Fratto, D., & Reiners, L. (2020) Banking on the Cloud (2020). *21 Transactions: The Tennessee Journal of Business Law 381*, Duke Law School Public Law & Legal Theory Series No. 2020-52, Available at SSRN: https://ssrn.com/abstract=3647392, (Accessed 27 Dec 2022).
11. Navratil, J., & Ubik, S., & Pečiva, P. (2018). Cloud computing alternative or the future (September 6, 2018). In *2018 ENTRENOVA Conference Proceedings*. Available at SSRN: https://ssrn.com/abstract=3282685 (Accessed 28 Dec 2022)
12. Larry, D. (2020). Online available at, https://hbr.org/2020/10/a-measured-approach-to-regulating-fast-changing-tech (Accessed 28 Dec 2022)

13. Synergix technologies. (2022). Online available at https://www.synergixtech.com/news-event/business-blog/business-challenges-smes-2019/, (Accessed 05 Dec 2022).
14. Grogan, S., Conner, M., Norman, P., & Porter, I. (2000). Validation of a questionnaire measuring patient satisfaction with general practitioner services. *Quality in Health Care, 9*, 210–215.
15. Wenhao, B., Junwen, M., Xudong, Z., Weixiang, W., & An, Z. (2022). Cloud service selection based on weighted KD tree nearest neighbor search. *Applied Soft Computing., 131*, Article 109780. https://doi.org/10.1016/j.asoc.2022.109780
16. Magoulas, T., & Swoyer, S. (2020). Cloud computing market trends and vendor positioning 2020. *BARC Research.* https://barc-research.com/cloud-computing-market-trends-2020/, (Accessed on 23 Nov 2022)
17. Neha, T., Avtar, S., & Sangal, A. L. (2022). Cloud services selection: A systematic review and future research directions. *Computer Science Review, 46*, Article 100514. https://doi.org/10.1016/j.cosrev.2022.100514
18. Fuzan, C., Aijun, L., Harris, W., & Minqiang, L. (2022). Compensation and pricing strategies in cloud service SLAs: Considering participants' risk attitudes and consumer quality perception. *Electronic Commerce Research and Applications, 56*, Article 101215. https://doi.org/10.1016/j.elerap.2022.101215
19. Hussain, W., Raza, M. R., Ahmed Jan, M., Merigo, J. M., & Gao, H. (2022). Cloud risk management with OWA- LSTM predictive intelligence and fuzzy linguistic decision making. *IEEE Transaction Fuzzy System* (pp. 1–1).
20. Reportlinker. (2020). Online available at https://www.prnewswire.com/news-releases/the-data-center-market-in-southeast-asia-is-expected-to-grow-at-a-cagr-of-over-6-during-the-period-20192025-301036630.html, (Accessed 28 Nov 2022)

Chapter 13
CMMI-Based Conceptual Project/Program Management & Measurement (PMM) Framework for Sustainable Lean Implementation & Mining Operations

Praveen Kumar Harkawat

Abstract As per EY, in 2023, rising hopes around ESG, Sustainability, and Climate change, will top the agenda for mining leaders. Global disruption will also put new pressure on enterprises to discover opportunities to reimagine business models and accelerate system & process innovation and transformation to ensure sustainable operations. While progressing their transformation into purpose-led, low-carbon, agile organizations; mining companies should implement new process frameworks Lean combined with CMMI to improve efficiency, operations & productivity. This research proposes a new lean measurement framework supported by CMMI's Project Management (PM) processes, which can give the best results to organizations. The CMMI (V1.3)-based PM framework for Lean implementation in Mining, will help in the effective planning & execution of lean initiative at the organizational level. The proposed Lean Project/Program Management & Measurement (PMM) framework is being built on CMMI's selected Process Area's practices/sub-practices & work products and provides rating/status of lean practices implementation in an organization. It will provide the AS-IS status of lean practices highlighting the strength & weaknesses, so the implementation team can plan for further focused improvements. It will help companies in implementing lean in a structured way based on a standard & proven project/program management and measurement framework ensuring sustainability, transparency, governance & compliance, and continuous quantifiable improvements.

Keywords Lean · Mining · Capability maturity model -integrated (CMMI) · ESG · Sustainability · Project management (PM)

P. K. Harkawat (✉)
Quality Group, L&T Technology Services, Vadodara, Gujarat, India
e-mail: praveen.harkawat@ltts.com

P. K. Kapur et al. (eds.), *Data-centric Approaches to Industrial Decisions*, Asset Analytics, https://doi.org/10.1007/978-981-96-7556-2_13

13.1 Introduction

India is a global player in the production of many minerals. It has a leadership position and capability in the mining industry, and the country generates a wide variety of minerals [1]. Many Indian mining companies are growing at a very fast pace and contributing to the overall growth of the economy. The sector is very significant, and to maintain future growth, it should embrace innovative systems.

The mining industry in India is relatively less mature when it comes to sustainability, methodical planning, and business procedures, and there is less use of traditional quality standards, systems, processes, and frameworks. Maintaining an efficient and successful business is critical to the enterprise's long-term viability and growth. The industry is dealing with problems like cyclical pricing, falling profitability, and varying demand, along with issues of sustainability, net zero, and stringent regulatory compliance. Indian companies need to keep an eye on the fast-changing global standards and should implement best practices followed by global mining giants.

In the current business climate, the industry should look for new chances by modernizing, utilizing cutting-edge technology, and implementing frameworks for process improvement. Due to the high capital requirements of the Indian mining sector, it is important to focus on reducing the waste of important assets, which can be effectively handled using a lean framework. Many organizations have benefited greatly from lean in terms of increasing the utilization of equipment, labor, and other crucial resources. CMMI is another well-established process framework used by IT/Technology companies across the globe and has helped in improving quality, delivery & customer satisfaction.

Through this paper/research, we intend to find the answers to some of the key questions as mentioned below:

- What are the key challenges in the mining industry?
- How do companies try to address those challenges in a systematic way?
- Are companies using any improvement frameworks like Lean, Six Sigma, Balanced Score Card, ISO, etc.?
- What are the key challenges of using frameworks like Lean?
- What is the status of the usage of lean in the mining industry?
- Are there any measurement frameworks that can be used to measure the effectiveness of lean implementation in the mining industry?
- Can CMMI's Project Management Process Areas be used for Lean for Sustainable Lean Implementation & Mining Operations?

In the author's view, CMMI's Project Management processes/practices are being used extensively by organizations for better project planning, monitoring, and control. A combined lean and CMMI Project/Program Management and measurement (PMM) framework will be of great help to the mining industry.

In this paper, we propose a new framework for the mining industry that will include 3 key process areas of CMMI V1.3: project planning, monitoring and control, and integrated project management. The same practices can also be used for lean program management in larger mining organizations with a complex structure involving multiple departments and stakeholders at all management levels.

13.2 Literature Survey

13.2.1 Challenges of the Mining Industry

In addition to a more unstable geopolitical environment, the top concerns for leaders in the mining and metals industries in 2023 will be rising expectations for ESG, climate change, sustainability, and operating licenses. The effects of the global disruption will also put more strain on labor, productivity, and expenses, which will compel companies to explore opportunities to accelerate innovation and reimagine their business models. The question is: How can mining and metals companies move ahead with their transformation into purpose-led, low-carbon, digitally enabled organizations while navigating urgent difficulties like growing costs and supply chain disruption? As next year approaches, the mining and metals sector is responding by utilizing new frameworks and technologies to make increasingly significant changes to corporate and operating structures [2].

The mining segment is evolving and taking on a new role in the economy. Mining businesses must exactly measure their effects in order to establish their positive sides. This would facilitate investors to make more learned decisions about capital allocation and offer them an improved hold of the performance of the industry. Utilizing technology, processes, and systems is essential to raising business productivity and expanding advancements in health and safety. The industry may increase efficiency and set itself up for future expansion by leveraging new technical and process/systems instruments. In order to supply vital minerals for life, the mining sector must be sustainable. Now is the perfect time for mining firms to implement new systems/frameworks and optimize their processes for long-term impact.

One very important and suitable tool for transformation in mining companies is Lean implementation. Lean has been implemented by mining and allied industries and has given good benefits to the organization. It can also help mining companies ensure sustainable operations.

13.2.2 Lean and Its Usage in the Mining Industry

In the existing competitive macroeconomic environment in which the firm is operating, key assets and the way they are handled are becoming gradually more significant. Lean techniques intend to eliminate waste from systems. As part of the Toyota Production System (TPS), Taiichi Ohno, Toyota's Chief Engineer, established the prime seven wastes (Muda). Overprocessing, Defects, Overproduction, Transportation, Inventory, Motion, and Waiting are the seven wastes. The implementation of the Toyota Production System in European countries throughout the 1990s saw the introduction of the eighth waste of underutilized talent, human capital, or employee "skills" [3, 4].

Lean is all about getting rid of unnecessary or redundant processes, enhancing workflow, and adding more value to what businesses do. In mining also it can be used in all the functions & operations to reduce waste and improve safety & productivity. Rather than developing products and services based on marketing projections, a lean enterprise fulfills consumer requests through innovative ways. As a result, the cycle time to market for new goods and services is shortened significantly by cutting down on development times. Lean manufacturing capabilities are the outcome of the use of more adaptable, automated, computer-controlled production equipment [2].

Lean is being used by many businesses to enhance systems, quality, customer happiness, and procedures. There are possible risks and uncertainties associated with the implementation programs which has to be addressed through a structured program management system. Lean initiatives and programs involve several departments and stakeholders in their conception, implementation, and oversight. Coordination, communication, governance, and management are among the issues. One of the possible causes of the low success rate of lean adoption is the absence of organized project/program management. So, there is a need for a new and innovative lean project/program management framework.

In terms of practice, using lean principles can help businesses lower physical waste and increase the amount of value extracted from ore, but lean can offer observable advantages across all facets and procedures of mining operations. Mining businesses must come up with new strategies due to the industry's ongoing unpredictability and volatility brought on by shifting demand, cyclical pricing, and a downward trend in profitability. It will aid in process simplification and efficiency improvement in order to maintain competitiveness and be responsive to a complex and dynamic market [3, 5].

Lean in mining has been used for a few years now. The mining industry has benefited from it in the past through cost savings, increased productivity and quality, and improved safety. A lean system established upon company environment and requirements, has been used by the mining and metal industries. It is assisting businesses in reaping certain benefits, but more may be done by pursuing additional advancements to fulfill the organizations' and businesses' ever-changing needs. However, there are numerous obstacles to implementing lean in the mining sector that must be overcome for the good of the sector [6, 7].

Implementing lean in mining requires systematic project/program management. The key objective and success can be achieved by controlling several factors, such as risk, cost, procurement, and safety, all of which must be balanced to finish a lean project on time.

13.2.3 Challenges of Lean Implementation in Mining and Possible Solutions

Due to the lack of ability to manage key factors, lean implementation in mines is more difficult than in most industrial plants. In controlled factories, it is simpler to use methods like 5S and the visual workspace to increase cleanliness and order. Though it demands methods for the successful implementation and sustainability of lean projects, operational excellence is still attainable. Businesses may make sure they are making the greatest profit by using their time and resources as efficiently as possible by using Lean Six Sigma [3].

Mines can significantly enhance operational efficiency in areas with the highest losses. For instance, standardizing material regions that can be replenished regularly can save waiting times and material waste. This could improve the number of ore face cycles per day and cut expenses per meter excavated by about some percentage. Value stream mapping can be used to identify crucial components that can decrease lead times and enhance cycle times, increasing the throughput of mines. The environment gains when process flow improvements lead to lower operating footprints and less energy wasted, which aids in sustainable operations [3, 4].

Additional potential exists with new mining-related technologies, such as autonomous operations which will enable a reduction in the number of personnel needed inside mines. Instead of creating additional waste and unused potential, embracing the latest technologies related to lean can increase the payback that can be gained [3].

Any process improvement methodology's application is rife with unknowns and uncertainties. In mining, lean and other frameworks aren't used very often, and the industry lacks conceptual models that are coherent enough to direct the creation of new systems. To improve outcomes in sustainable mining operations, innovative frameworks and processes are required. Lean may produce greater results if it is used in conjunction with other frameworks for improvement for monitoring and tracking purposes since it will give management and lean practitioners more visibility & insight [6, 8].

To fulfill the need for innovative systems & frameworks and to achieve better results in sustainable mining operations, lean can be combined with other improvement frameworks for monitoring & tracking purposes. The new process improvement framework starts with a set of planning activities that must be managed systematically and systematically.

This paper presents a CMMI V1.3-based planning, monitoring, measurement, and tracking framework for lean implementation in the mining industry. The proposed framework will include the planning, monitoring, and control (project management) related practices of CMMI as a base and add relevant practices/processes to make it more robust & practical. The new framework's process/practices will be mapped to the equivalent project management practices, which will cover all the phases/stages of the lean project/program, starting with initiation/planning to execution & closure.

13.3 Research Gap and Necessity of New Framework

Many businesses are implementing lean initiatives and programs to increase productivity, customer satisfaction, processes, quality, and sustainability. There are numerous problems and restrictions with the implementation of programs. Lean project/program planning, implementation, and monitoring involve numerous stakeholders and departments. The lack of structured, systematic monitoring during the implementation of the lean initiative may be one of the reasons why the success rate of lean implementation is not very high. To the best of the author's knowledge, there is no management and measurement framework for lean implementation in mining that is based on the PM process area & practices of the CMMI V1.3.

Therefore, it is essential to suggest a framework that can be utilized to ensure that lean projects and programs at the organizational level are implemented successfully and practises are maintained for a longer length of time. Many IT/technology organizations have implemented and exploited the PM process area of CMMI. Organizations managing large & complex IT projects/programs utilize PM techniques as part of the CMMI model. To promote better monitoring during the execution of projects and programs, organizations have implemented a structured project management framework.

13.4 Objectives of the Study

This paper's goal is to suggest a fresh management framework or paradigm for the application of lean. The information at hand implies that Lean and other frameworks are only sometimes used, and that there aren't enough clear conceptual models in the market to direct the adoption of new systems. Innovative structures and processes are needed in industry. Therefore, it is necessary to promote lean and other frameworks that would aid in producing better outcomes in sustainable mining operations.

If lean is combined with other improvement frameworks for monitoring & tracking purposes, it will provide better visibility to management & practitioners and may give superior results & quantifiable improvements. There are improvement frameworks like CMMI, PCMM, PMP-PMI, etc., which can be used to build a Lean PMM system for structured & systematic planning & monitoring of lean practices in the organization.

The paper suggests the application of CMMI's project management practices with lean implementation in mining for planning, monitoring & tracking of lean improvement initiatives /programs across the departments and organization.

13.5 Methodology

The researcher decided to rely on secondary data from the lean implementation in several industries in order to construct the new framework. The researcher also looked into how IT organizations use the existing project management frameworks and processes. Many organizations utilize CMMI V1.3 to measure organizational maturity and capabilities and enhance processes. The researcher made the choice to create a new framework for managing lean implementation in mining using the practices and sub-practices of the CMMI's Project Management process area. A new suggested Framework for Mining organizations was investigated, and all Project Management (PM) process areas' practices were then mapped to it. A lean project or program must adhere to the processes, practices, and sub-practices outlined in the new framework.

13.6 Discussion/Results: CMMI-Based Lean PMM Framework for Effective Monitoring of Lean Implementation in Mining

The proposed framework will reuse and combine the following practices/processes of CMMI-DEV V1.3:

- Project Planning (PP): It's a CMMI ML2 process area and includes practices/sub-practices related to project scoping, estimation, project initiation, kick-off, resourcing, planning, etc.
- Project Monitoring & Control (PMC): It's a CMMI ML2 process area and includes practices/sub-practices related to project monitoring at various stages/phases of execution and controlling the deviations as needed
- Integrated Project Management (IPM): It's a CMMI ML3 process area and includes practices/sub-practices related to the integration of all planning & monitoring and control-related processes.

The above-mentioned proven processes/practices have been used by IT/Technology companies for project/program management for almost 3 decades [6] and have given optimal results, as they provide a structured framework for handling complex projects/programs.

By making use of the CMMI Process areas for project/program management, researcher is recommending a conceptual Lean Project/Program Management & Measurement (PMM) Framework, which will consist of Lean Project/Program Planning (LPP), Lean Project/Program Monitoring & Control (LPMC) and Lean Integrated Project/Program Management (LIPM). The details are provided in the following section.

- Lean Project/Program Planning (LPP): This process area can be used to describe lean project/program activities and will include creation and updation of plans and related activities. This will also include estimating/scoping of lean project/program planning based on some key considerations/ factors. The scope definition and estimation will help the team in setting up a high-level plan based on the scope of the lean project/program requirements. Then, there will be practices related to the calculation of the efforts needed for the lean project/program's high-level activities/tasks. Once scope and estimation are available, the team will identify the lean project/program's stages/phases. Then there will be practices related to the assessment of the lean project/program's cost for the tasks on the basis of efforts.

After initial planning, scoping, and estimation, practices will be related to setting up a detailed plan for sustained management of the lean project/program. This will include setting up and maintaining the lean project/program's financial plan and calendar. Next, practices will be related to finding and analyzing the lean project/program's risks and issues. For a process improvement program like lean, there will be lots of data/report generation, so practices for managing the lean project/program's data are required.

Further, there will be practices for planning of resources/assets required to undertake & execute lean project/program activities. It will be supported by practices related to planning for competence, training & abilities required for lean project/program activities. All these practices will help in setting up & sustaining a comprehensive/master plan for a lean project/program. Having defined, initial planning activities, there will be practices related to re-evaluation and reconciliation of the plan as per commitments from relevant stakeholders. The re-evaluation and reconciliation need to be re-checked and re-examined, so there will be practices related to changed commitments. All this will result in changed accountability for various tasks that were planned for the lean project/program.

- Lean Project/Program Monitoring & Control (LPMC): For a process improvement plan, there will be need to define practices to monitor and control the key tasks and activities included as a part of Lean Project Planning (LPP) practice area. The new proposed lean project/program Monitoring & Control (LPMC) practice area will include practices/sub-practices related to the actual implementation & advancement of lean project/program. As a part of this practice, the planned activities will be supervised & then compared to overall lean implementation plan.

This will be supported by practice to ensure supervision of available existing/factual values of lean project/program planning factors versus lean project plan. To strengthen the tracking of lean project/program plan through regular a well-established monitoring mechanism, there will be practices for regular monitoring. As an add-on practice, there will be tracking of risks through a robust risk monitoring system. For data, there will be a practice to track management of project data as planned. For a lean project/program, it's very important to track participant's contribution/support as planned along with regular assessment/checking of the advancement and issues, there will be practices related to data and issues monitoring at a predefined & chosen stages/time. The outcome of every phase/stage will be monitored through a specific practice. After monitoring of outcomes, if any corrections are needed, it will be done through a specific practice. It will ensure that require corrections are tracked to conclusion & closed. During all this tracking, its suggested to take further actions if results of lean project/program vary substantially wrt to the plan. There will be another practice to gather & analyze the concerns and decide corrections as necessary to address the concerns. Lean team has to take necessary corrections for the identified concerns related to implementation and deal with corrective actions related to project/program and ensure closure & conclusion.

- Lean Integrated Project/Program Management (LIPM): Any lean project/program need to be managed in a holistic and integrated manner to ensure that all the required activities are being planned and monitored in a collaborative way. To ensure proper coordination and collaboration, integrated lean project/program management practice area will be defined including related sub-practices.

The first practice is related to the execution of a project/program using a defined lean process that is customized from the organization's set of standard lean processes. The next practice will be related to setting up the lean project's/initiative's defined processes/systems to ensure uniformity across the organization. Then, in the model, we will add a very important practice related to the utilization/employment of the organizational lean process assets and data planning of the lean activities. To provide guidance for an integrated approach, there will be a practice to ensure that the team combines the lean project/program plan and the other related plans and manages the project/program using the plan. As a part of the same group of practices, there will be another important practice related to sharing the documented experiences to the organizational lean process assets repository. It will ensure that all the learnings, best practices and knowledge is being shared and stored at a central place and will be available for future references for the planning of similar lean projects/programs. There will be another practice related to the establishment of teams for lean project/program implementation, coordination, and collaboration with relevant stakeholders. It will ensure better management and will help in the participation of the related participants in the lean project/program. All the above key practices will ensure active participation from relevant interested parties to discover, collaborate, and track vital dependencies during lean implementation and resolve issues related to project/program execution with related participants.

Below is the pictorial representation of the Conceptual Lean Project/Program Management & Measurement (PMM), which includes lean project management related specific goals (SG) and specific practices (SP). All SPs are being supported by generic goals (GG) and generic practices (GP) (Fig. 13.1).

The below mapping shows the available Processes, Practices, and sub-practices of CMMI and the proposed Lean Project/Program Management & Measurement (PMM) framework. Each specific practice of the new framework can be rated based

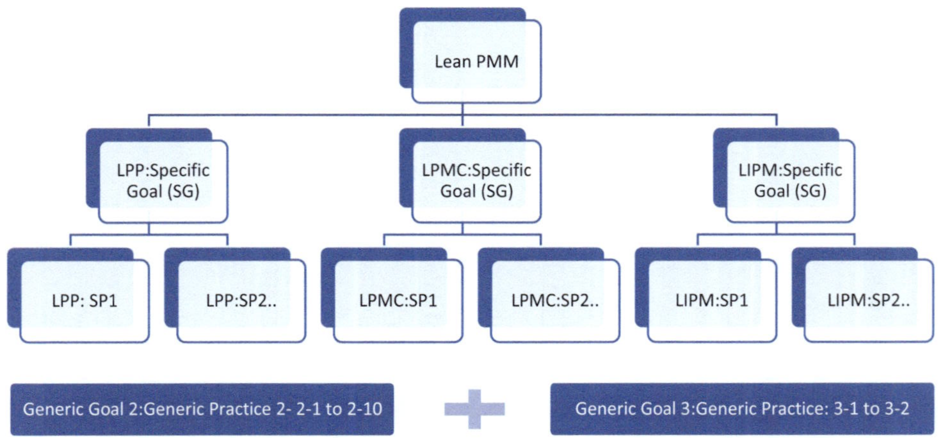

Fig. 13.1 Pictorial view of lean project/program management (PMM) model/framework. (*Source* Self-learning and Research)

on the level of implementation of lean in a particular function, department, or company. The rating given to each practice will be presented cumulatively to give a high-level view of the organization's lean maturity/capability rating (Table 13.1).

We are proposing an integrated lean project/program management framework, including all the required elements of planning, monitoring, and control. The framework will have 3 key components, where the integrated system will be commination of 2 key practice groups (planning and monitoring & control) as shown in the following architectural block diagram (Fig. 13.2):

The proposed framework will help in planning the implementation of lean projects/programs in a phased manner, starting with Lean Project/Program initiation, kick-off and planning. After planning, next phase will be monitoring of the planned activities using pre-defined set of metrics and performance parameters/reports. These reports/dashboards will help management in getting a high level of the status of lean project/program's activities and support lean team is better decision-making. The required, corrective actions to control any deviations wrt KPIs can be taken by the team during project/program evaluation/execution as shown below (Fig. 13.3):

The Enterprise-wide Lean PMM System will include measurement processes, governance, tools, monitoring systems, risks & issues database, best practices, lesson learned and other related components/modules as shown in Fig. 13.4:

New PMM framework can be used by lean practitioners and management for planning, monitoring & tracking lean improvement initiatives in mining organizations. The reports/dashboards can be designed & used to show the regular status/progress of lean implementation. It can work as a decision support system for executives and managers.

13.6.1 Reports/Dashboards for Project/Program Monitoring & Tracking

The following set of Reports/Dashboards can be designed/prepared/published:

- Lean Projects/Program Under Implementation related Metrics/Reports showing following key metrics:
  - o % or Days of Delay in-

    Milestones
    Deliverables
    Stage-Gate
    Phase/Stage
    Key Event

  - o Number of Open Risks including:

    High Values Risks (with High-Risk Priority Number)
    Low-Value Risks (with Low-Risk Priority Number)

Table 13.1 Mapping of the available processes, practices/sub-practices of CMMI, and the proposed Lean PPM

CMMI PM PA's Practices & Processes [6]	Proposed Lean PMM Framework's Specific Goals (SG) /Practices (SP) (Re-produced as per CMMI Processes, Practices and Sub-practices) Format followed is, < Lean PA Name. SP. Serial Number > : < Practice Detail >
Planning 1: SG	LPP1 SG: Estimates/Scope of LEAN project/program planning considerations/ factors are Set up and maintained
1.1	LPP.SP 1.1: Set up a high-level plan to understand about the scope of the LEAN project/program
1.2	LPP.SP 1.2: Set up and maintain efforts needed for the LEAN project/ program's high level activities/tasks
1.3	LPP.SP 1.3: Identify the LEAN project/program's stages/phases
1.4	LPP.SP 1.4: Assess the LEAN project/program's cost for the tasks on the basis of efforts
Planning 2: SG	LPP2 SG: A detailed plan is Set up and sustained for management of the LEAN project/program
2.1	LPP.SP 2.1: Set up and sustain LEAN project/program's financial plan & calendar
2.2	LPP.SP 2.2: Find and analyze LEAN project/program's risks
2.3	LPP.SP 2.3: Plan for managing the LEAN project/program's data
2.4	LPP.SP 2.4: Plan for resources/assets to undertake & execute LEAN project/ program activities
2.5	LPP.SP 2.5: Prepare outline for competence, training & abilities required for LEAN project/program activities
2.6	LPP.SP 2.6: Prepare outline for participation with teams for LEAN
2.7	LPP.SP 2.7: Set up & sustain comprehensive/master plan of LEAN project/ program
Planning 3: SG	LPP3 SG: Lean Plan is re-evaluated and reconciles as per commitments from relevant stakeholders
3.1	LPP.SP 3.1: Re-Check and Re-Examine LEAN project/program to understand project/program commitments
3.2	LPP.SP 3.2: Re-arrange LEAN project/program plan as needed
3.3	LPP.SP 3.3: Get assurance from teams accountable for various tasks
Monitoring & Control 1: SG	LPMC 1 SG: Actual implementation & advancement of LEAN project/ program are supervised & compared to LEAN implementation plan
1.1	LPMC.SP 1.1: Supervise available existing/factual values of LEAN project/ program planning factors versus LEAN project plan
1.2	LPMC.SP 1.2: Track LEAN project/program plan through regular a well-established monitoring mechanism
1.3	LPMC.SP 1.3: Track risks through risk monitoring system
1.4	LPMC.SP 1.4: Track management of project data as planned
1.5	LPMC.SP 1.5: Track participant's contribution/support as planned

(continued)

Table 13.1 (continued)

CMMI PM PA's Practices & Processes [6]	Proposed Lean PMM Framework's Specific Goals (SG) /Practices (SP) (Re-produced as per CMMI Processes, Practices and Sub-practices) Format followed is, < Lean PA Name. SP. Serial Number > : < Practice Detail >
1.6	LPMC.SP 1.6: Regularly assess/check the advancement and issues as needed/planned
1.7	LPMC.SP 1.7: Examine/Assess planned outcomes at predefined & chosen stages/time
Monitoring & Control 2:	LPMC 2 SG: Required corrections as needed are tracked to conclusion & closed as & when results vary substantially
2.1	LPMC.SP 2.1: Gather & analyze the concerns and decide corrections as necessary to address the concerns
2.2	LPMC.SP 2.2: Make necessary corrections for the identified concerns related to LEAN
2.3	LPMC.SP 2.3: Deal with corrective actions related to LEAN project/program to closure & conclude
Integrated Project Management 1: SG	LIPM 1 SG: The LEAN project/program is conducted/executed using a defined LEAN process that is customized from the organization's set of standard LEAN processes
1.1	LIPM.SP 1.1: Set up & sustain the LEAN project's/initiative's defined processes/systems
1.2	LIPM.SP 1.2: Utilize/Employ the organizational LEAN process assets and data planning the LEAN activities
1.3	LIPM.SP 1.3: Incorporate/Combine the LEAN project/program plan and the other plans
1.4	LIPM.SP 1.4: Manage the LEAN project/program using the plan
1.5	LIPM.SP 1.5: Share the documented experiences to the organizational LEAN process assets repository
1.6	LIPM.SP 1.6: Establish Teams for Lean Project/Program Implementation
Integrated Project Management 2: SG	LIPM 2 SG: Coordination and collaboration of the LEAN project/program with relevant stakeholders is conducted
2.1	LIPM.SP 2.1: Manage the participation of the related participants in the LEAN project/program
2.2	LIPM.SP 2.2: Participate with relevant interested party to discover, collaborate, and track vital dependencies in LEAN implementation
2.3	LIPM.SP 2.3: Resolve issues related to LEAN implementation with related participants

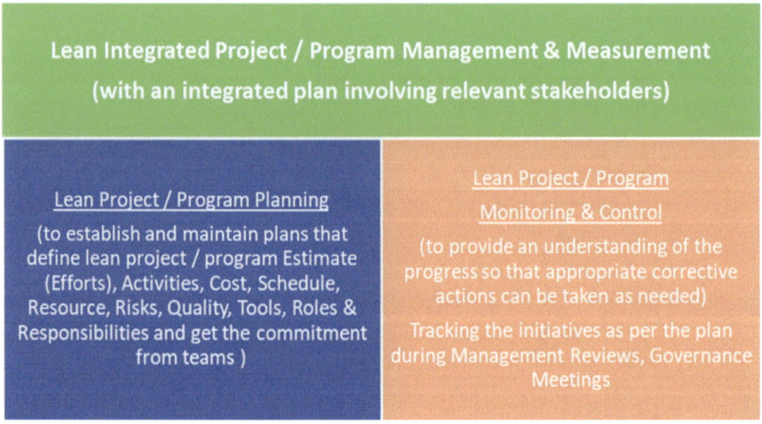

Fig. 13.2 CMMI-based PMM framework for lean project/program/initiative monitoring & tracking. (*Source* Self-learning and Research)

Risks Materialized or Occurred
Number of days the risk is open
Department-wise/function-wise number of risks

- o % Waste Reduction in:

  Materials
  Key Assets
  Resources

- o % Increase in Value (Value addition in $ or Rupees terms)
- o Number of Lessons Learned & Best Practices
- o Internal Stakeholder's Satisfaction Index
- o Stakeholder's Engagement Index
- o Customer Satisfaction Index
- o Key Attention Areas/Details
- o Resource Utilization
- o Success Rate of Lean projects/program: Department/Function-wise
- o Training Metrics:

  Number of Person Days of Lean Training per Employee/Year
  Lean Training Effectiveness %

- Number of Lean Projects/Programs Under Implementation at the Department Level/Unit Level

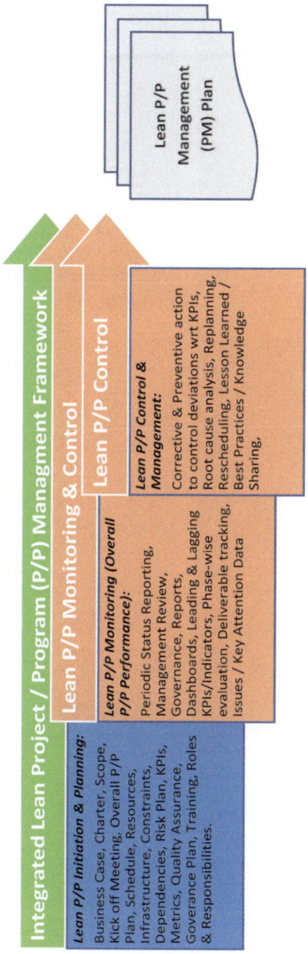

Fig. 13.3 High-level view of CMMI-based lean PM plan for lean initiative monitoring & tracking. (*Source* Self-learning and Research)

Fig. 13.4 High-level/enterprise-wide lean PMM system. (*Source* Self-learning and Research)

13.6.2 Reports/Dashboards for Lean Mining Operation's Monitoring & Tracking

The proposed framework can also include the following Lean related Metrics (indicative list) for tracking/ensuring sustainable mining operations:

- % Compliance with Lean processes/practices in the operations department
- % Improvement in Cycle Time of key mining operations
- % Improvement in Productivity at mine site/plant
- % Improvement in Equipment Availability
- % Improvements in Mean Time Between Failure (MTBF) of major equipment
- % Reduction in Mean Time to Repair (MTTR) of major equipment
- % Reduction in cost per 1000 tons of mineral extracted
- % Reduction in accidents per 1000 tons of mineral extracted
- % Reduction in Fuel Usage in Heavy Machines
- % Reduction in Empty Travel Distance of Machines/Vehicles
- % Reduction in Waiting Time
- % Improvement in Machine Uptime
- % Improvement in Waste Recycling
- % Improvement in employees involved in Lean improvement initiatives

To complement the above set of Metrics, organization can publish % Rating (Earned Value) of the project management process areas. The % rating will provide the status of various practices followed for lean implementation. The identified measures/numbers need to be included in the selected project management process & practices and should be monitored/tracked by lean practitioners and management.

A typical Rating/Capability/Maturity report will be presented in a tabular and graphical form as below. The blue and green colored cell indicates a very high and high level of project management compliance respectively. The red and yellow status indicates low level of compliance and team need to focus & plan for further improvement to increase the score of related process areas & practices of lean implementation (Fig. 13.5).

The above dashboard can be scaled up to meet large-sized lean initiatives/program information & monitoring needs, starting from the department to multiple mines to the organization.

Management and lean practitioners should use the reports, dashboards, outputs, and databases very sparingly. When monitoring and tracking, the emphasis should be on the improvement of the processes and systems rather than on the people or groups. While using the reports and dashboards focus should be on KPIs and progress at the different levels of management/hierarchy. The framework should be customized as per the size of the implementation/team and the type of organization/company.

Mines A	Lean Process Area's Rating (%)		
	LPP	LPMC	LIPM
SG 1 % Complete	70%	100%	83%
SG 2 % Complete		86%	100%
SG 3 % Complete		67%	0%
GG 2 % Complete	0%	60%	60%
GG 3 % Complete	0%	0%	0%
Specific Practices	13	10	9
Generic Practices	12	12	12
Current Date - Current Status	23%	62%	49%
Previous Date - Previous Status	0%	10%	5%
Previous Date - Previous Status	0%	5%	0%

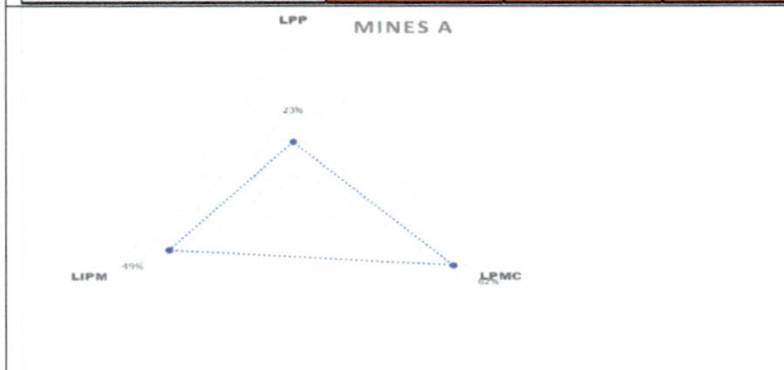

Color	Earned Val
red	0-14
yellow	15-64
green	65-89
blue	90-100

Earned Value= % Rating of Process Area of Model (indicative capability / maturity of practices followed)

Legend / Abbreviations:

SG: Specific Goals
GG: Generic Goals
SP: Specific Practices
GP: Generic Practices
LPP: Lean Project / Program Planning
LPMC: Lean Project / Program Monitoring & Control
LIPM: Lean Integrated Project / Program Management

Note: All data / info is indicative and used for representational purpose

Fig. 13.5 Typical/representative dashboard showing process areas-wise rating (%) for mine-A showing capability/maturity of all 3 project/program management process of the proposed lean PMM framework. (*Source* Self-learning and Research)

13.7 Implications

The proposed framework, if used effectively, shall provide an overall view of the AS-IS status of the lean project/program's implementation in mining organizations. It would help organizations in better understanding of the status of lean implementation & related practices in quantitatively. It would provide an overall view of lean implementation to the management at the level of the organization. The best practices incorporated in the proposed framework will strengthen the existing measurement systems used by the mining organizations. The proposed framework can be presented as a strong evidence of monitoring & tracking during the ISO and other certification audits in the organization.

13.8 Conclusion

The proposed framework is built using the CMMI's project management processes/practices, starting from project planning, monitoring & control, and including integrated project management. It covers to complete range of activities for project management and measurement. All these processes/practices can help in end-to-end project/program planning, monitoring & control involving multiple stakeholders and departments. The framework also proposed required reports/dashboard for management reviews & governance. The regular management governance will help the organizations in ensuring success of lean project/programs.

13.9 Limitations and Future Research Scope

The proposed conceptual framework/model is based on a pre-defined set of project management practice areas of CMMI V1.3. It's being designed based on certain concepts and assumptions about the usage and implementation in the mining and allied industry. It must be validated by lean practitioners, experts and industry professionals in a real-time environment in a typical organizational setup with multiple departments and stakeholders. Further re-calibration, changes, and re-designing will be required as per the inputs after review & validation by mining companies and lean practitioners. However, in its current form, it can be piloted and will be of great help to the industry.

In the future, more practices from Six Sigma (DMAIC), CMMI (Risk Management and Measurement process areas), PCMM (training, knowledge management, and competency-related process areas), Balanced Scorecard (BSC), and PMP-PMI's (select knowledge areas) can be added to make it more robust and comprehensive. New process areas/practices of the latest CMMI V2.0 can also be added to make the framework broader and more inclusive. Such inclusions will help the lean practitioners and management in better decision-making.

References

1. FICCI's report on sector overview, https://ficci.in/sector-details.asp?sectorid=81
2. Paul Mitchell, Top 10 business risks and opportunities for mining and metals in 2023, EY Report, (https://www.ey.com/en_gl/mining-metals/risks-opportunities)
3. Waste Extraction: Lean Process Improvement within the Mining Industry, (https://www.amsaxum.com/blog/process-improvement-within-the-mining-industry-2/)
4. Nawras Skhmot, The 8 Wastes of Lean, (https://theleanway.net/The-8-Wastes-of-Lean)
5. CARLY LEONIDA, C-Leaning up the mining industry, https://theintelligentminer.com/2020/07/08/c-leaning-up-the-mining-industry/
6. CMMI: https://www.wibas.com/cmmi/project-management-cmmi-dev
7. Harkawat, P. K. (2022). Goal question metrics (GQM) based measurement framework/approach for lean implementation in mining. *IJRAR22D2404 International Journal of Research and Analytical Reviews (IJRAR).* www.ijrar.org
8. Harkawat, P. K. (2022). KPI-based performance measurement framework/approach for lean implementation in mining industry. *International Journal of Innovative Science and Research Technology*. https://doi.org/10.5281/zenodo.7467453

Chapter 14
Applying Hybrid MCDM Techniques for Optimization of Productivity in Industry—A Case Study of Material Handling System

Vishal Akarker, Mahesh Dhawalikar, and Milind Sakhardande

Abstract Multi-Criteria Decision-Making is an area of study covering a multitude of subject fields ranging from daily life, business, engineering etc. The decision-making process is a complex process involving comprehensive reasoning when taking a decision involving multiple factors that must be considered & multiple choices to the original decision, from which a sensible choice must be made. Multi-Criteria Decision-Making is one of the techniques designed for solving such problems. This paper presents a case study for the optimization of productivity in the industry. A case study of the material handling system is selected for this paper. The material handling system is a complex process & accounts for a significant percentage of operating costs in the manufacturing industry due to high labor requirements and associated labor costs. It is also a significant contributor to high operating cycle time, high damage rate, crowded operating conditions, and obstruction to material flow. The proposed study combines AHP, TOPSIS & Cost Benefit Analysis to obtain the optimal solution among various alternatives. During the study, a particular facility's material handling processes were analyzed and the most significant factors for inefficiency were found. Ten different alternatives were proposed from which one system was selected using the Hybrid AHP-TOPSIS method and Cost- Benefit Analysis. This paper's results will be a valuable aid to decision-makers in evaluating the complex problems in the industry to obtain the optimal solution.

Keywords MCDM · AHP · TOPSIS · COST–BENEFIT ANALYSIS

V. Akarker (✉) · M. Dhawalikar · M. Sakhardande
Department of Mechanical Engineering, Goa College of Engineering, Goa, India
e-mail: vakarker@gmail.com

M. Dhawalikar
e-mail: maheshd@gec.ac.in

M. Sakhardande
e-mail: milind@gec.ac.in

P. K. Kapur et al. (eds.), *Data-centric Approaches to Industrial Decisions*, Asset Analytics, https://doi.org/10.1007/978-981-96-7556-2_14

14.1 Introduction

In industry, Decision-makers face a myriad of challenges when it comes to optimizing operations at the workplace. To maintain and enhance organizational productivity and profitability, various strategies and new systems must be carefully integrated into day-to-day operations. This integration often involves navigating conflicting factors such as cost, cycle time, and safety [1]. Balancing these factors is crucial for the efficient operation of facilities and the overall success of the organization.

One powerful approach to tackling these complex decision-making scenarios is Multi-Criteria Decision-Making (MCDM). MCDM is a set of techniques that enables the decision-makers to evaluate multiple quantitative and qualitative criteria simultaneously, allowing for a comprehensive assessment of different alternatives [2]. By considering a wide range of factors, MCDM helps decision-makers arrive at the most reasonable and optimal solutions.

Several MCDM techniques have been developed and widely used in various industries to address complex problems involving multiple alternatives. These techniques include the Analytic Hierarchy Process (AHP), Technique for Order Preference by Similarity to Ideal Solution (TOPSIS), VIKOR, ELECTRE and Simple Additive Weighting (SAW) among others [3–5]. Each technique has its strengths and is suitable for different types of decision-making scenarios.

AHP, for example, is particularly useful for breaking down a complex problem into a hierarchy of simpler sub-problems, which can then be analyzed independently. This method facilitates a structured and systematic comparison of different alternatives based on various criteria. TOPSIS, on the other hand, helps in identifying the alternative that is closest to the ideal solution by considering the distance of each alternative from the ideal and the negative ideal solutions. VIKOR focuses on ranking and selecting from a set of alternatives, particularly when dealing with conflicting criteria. ELECTRE is effective for decision-making problems where the alternatives need to be ranked or sorted into categories, while SAW provides a straightforward approach to evaluating alternatives by summing the weighted performance ratings of each alternative.

The material handling system is one critical area where decision-makers frequently struggle to find the most appropriate and effective solution. This system encompasses various equipment, practices, and methodologies that need to be considered to achieve optimal outcomes. Material handling is a pivotal part of an organization's operations as it significantly contributes to the total operational costs and directly influences the productivity of day-to-day activities [6, 7]. Effective material handling is essential for a company's growth and profitability.

An ideal material handling system aims to achieve several key objectives: optimal manpower utilization with reduced labor costs, timely and efficient movement of materials, and the implementation of systems that can transport items safely without causing any material damage. Achieving these objectives enhances the productivity of the organization and contributes to its overall success.

Given the complexity and importance of material handling, the introduction and application of MCDM techniques in solving material handling problems is crucial. These techniques provide a structured approach to evaluate and compare different systems and equipment, considering various factors that influence the decision-making process. By using MCDM, decision-makers can systematically analyze alternatives and select the most suitable option that balances cost, efficiency, safety, and other relevant criteria.

The primary goal of this paper is to explore the application of MCDM techniques in addressing complex material handling system problems. The paper employs a hybrid model that integrates multiple MCDM methods to provide a comprehensive evaluation of the factors involved. This hybrid approach ensures a more robust and nuanced decision-making process, allowing for sensible and reasonable decisions that enhance the efficiency and effectiveness of material handling systems.

In conclusion, the application of MCDM techniques in industrial decision-making, particularly in the context of material handling, offers a powerful tool for optimizing operations. By systematically evaluating multiple criteria and alternatives, MCDM enables decision-makers to make informed choices that improve productivity, reduce costs, and enhance overall operational efficiency. This paper underscores the importance of these techniques and demonstrates their practical application in solving real-world material handling challenges.

14.2 Research Methodology

In this paper, a hybrid model combining AHP (Analytic Hierarchy Process), TOPSIS (Technique for Order Preference by Similarity to Ideal Solution), and Cost–Benefit Analysis is proposed for evaluating complex decision problems. This integrated approach leverages the strengths of each method to provide a comprehensive framework for decision-making. The model aims to define the objectives of the decision problem, identify various criteria that influence the decision-making process, and ultimately find the most beneficial alternative in terms of cost.

First, the AHP method is employed to determine the relative importance of different criteria. AHP is particularly well-suited for weighing criteria because it allows for structured pairwise comparisons [8, 9]. Decision-makers can express their judgments about the relative importance of one criterion over another, which are then used to calculate a set of weights representing the relative importance of each criterion. This step is critical as it lays the foundation for the subsequent analysis by establishing the priority of each criterion.

Step 1: Defining Objectives and Criteria Identification

The initial step involves defining the objectives of the decision problem clearly. Objectives guide the direction of the analysis and ensure that all relevant factors are considered. Once the objectives are set, various criteria affecting the decision-making process are identified. These criteria could include factors such as cost, time, efficiency, safety, and quality, among others.

Step 2: Application of AHP for Weighting Criteria

The AHP method is used to find the relative importance of each criterion through judgments and pairwise comparisons. Each criterion is compared with every other criterion, and decision-makers assign a relative importance score to these comparisons. These scores are then used to create a pairwise comparison matrix, from which the weights of each criterion are derived. This process ensures that the weights accurately reflect the decision-makers' priorities.

Step 3: Pairwise Comparison for Alternatives

Pairwise comparison for all alternatives with respect to each criteria is to be performed as done in the previous step. This involves evaluating how each alternative performs relative to others for a specific criterion. The results of these comparisons are tabulated, and matrix multiplication is performed with the weighted eigenvalues of each criterion to derive the overall scores for each alternative.

Step 4: Application of TOPSIS for Ranking Alternatives

TOPSIS is to be utilized for ranking various alternatives to find the optimal solution by the principle of choosing the solution which is shortest to the ideal solution and farthest from the negative ideal solution [10, 11]. First, the Normalized decision matrix is to be computed with the normalized values found using the square root of the sum of the criteria weights of each alternative.

Step 5: Computation of Weighted Standardized Matrix

The weighted standardized matrix is to be computed by multiplying attribute weights by each respective alternative value.

Step 6: Determination of Ideal and Negative Ideal Solutions

From the matrix table, find the Ideal solution & Negative Ideal solution. These solutions serve as benchmarks against which all alternatives are compared.

Step 7: Calculation of Separation Measures

Calculate the separation of alternatives from the Ideal & Negative Ideal solution

Step 8: Relative Closeness to the Ideal Solution

Determine the relative closeness to the ideal solution as alternative score values

Step 9: Cost–Benefit Analysis

After ranking the alternatives, Cost–Benefit Analysis is used to determine the most beneficial alternative in terms of cost. This involves tabulating the cost values for each alternative and normalizing these costs. Normalization ensures that the cost values are on a comparable scale. Finally, the cost–benefit ratio for each alternative is calculated by dividing the respective alternative score by the normalized cost value. This ratio helps identify the optimal solution by highlighting the alternative that offers the best balance between performance and cost.

By following this comprehensive methodology, decision-makers can systematically evaluate complex problems, consider multiple criteria, and select the most beneficial alternative. This integrated AHP-TOPSIS-Cost Benefit Analysis model provides a robust framework for making informed and rational decisions that enhance operational efficiency and organizational productivity.

14.3 Case Study

14.3.1 Data Collection & Analysis

EDI Ltd, a company located in Goa, operates a facility that is vital to its manufacturing & global sales operation. This facility is responsible for both supplying raw materials to the production area and storing finished goods once the products are made. Given the central role that material handling plays in these daily operations, it significantly influences the overall productivity of the plant. Therefore, understanding and optimizing the material handling system is crucial for enhancing operational efficiency. To address this, an in-depth study of the current facility was conducted to identify various factors that affect productivity and pinpoint inefficiencies and bottlenecks within the existing operation. The strategy and framework adopted for this assessment were based on the methodologies outlined by De Koster in his influential paper on warehouse optimization [12]. De Koster's framework provides a systematic approach to evaluating warehouse operations, identifying weaknesses, and proposing improvements.

The current material handling system at EDI Ltd's facility relies on a combination of equipment and manual labor. The primary equipment includes one pallet truck and one reach truck, which are used for the movement of materials within the facility. The existing system utilizes the reach truck for placing pallets into the racking system and the pallet truck for general material movement. Despite the presence of these mechanized tools, a significant portion of the material handling tasks is performed manually by workers. This includes loading and unloading materials at the dock, which is labor-intensive and time-consuming. One of the key shortcomings identified in the current system is the absence of lifting or leveling platforms. These platforms are essential for efficient loading and unloading operations, as they can adjust the height of materials to match the height of transportation vehicles or storage racks, thereby minimizing the physical effort required from workers and reducing the time taken for these tasks. Without such platforms, the facility relies on simple hand pallet trucks moved by workers, leading to substantial time wastage and inefficiencies. The manual nature of the material transportation process, performed with hand pallet trucks, means that the operating speed of the facility is heavily dependent on

Fig. 14.1 Pallet truck

the workers' physical capabilities and pace. This reliance on manual labor introduces variability and potential delays into the process, as the speed at which materials are moved can fluctuate based on the workers' fatigue levels, physical condition, and other human factors. As a result, the overall productivity of the facility is compromised, with considerable time being lost in manual handling tasks that could be automated or mechanized for greater efficiency.

Furthermore, the lack of specialized equipment for lifting and leveling not only slows down the material handling process but also poses ergonomic risks to the workers. Repeated manual lifting and moving of heavy materials can lead to injuries and health issues, which in turn can affect the workforce's availability and productivity. Addressing these ergonomic concerns is essential for maintaining a healthy and efficient working environment. In summary, the study of EDI Ltd's material handling system revealed several critical inefficiencies and bottlenecks. The reliance on a single pallet truck and reach truck, coupled with extensive manual labor, limits the facility's operational efficiency. The absence of lifting or leveling platforms exacerbates these issues, leading to wasted time and reduced productivity. By adopting the strategic framework proposed by De Koster, the facility can systematically identify and address these weaknesses, implementing solutions that enhance material handling efficiency, reduce reliance on manual labor, and ultimately improve the overall productivity of the plant's operations . The current material handling systems at the facility is shown in Figs. 14.1, 14.2, 14.3, 14.4 & 14.5.

14.3.2 Problem Statement

The current material handling system at EDI Ltd's facility is labor-intensive which significantly impacts the company's efficiency and operational costs. This outdated system is not only costing the company time and money but also compromising the safety of its workers. Several critical issues have been identified within the current

Fig. 14.2 Reach truck

Fig. 14.3 Manual loading/unloading

Fig. 14.4 Manual transportation

Fig. 14.5 Existing system

system, which necessitate an urgent review and overhaul. One of the primary problems is the excessive amount of time spent on loading and unloading operations. The facility lacks essential lifting and leveling platforms, meaning that workers must manually handle heavy loads using simple hand pallet trucks. This manual process is time-consuming and inefficient, leading to considerable delays in the material handling workflow. Additionally, the reliance on manual labor introduces variability in the speed of operations, as it depends heavily on the physical capabilities and conditions of the workers.

The labor-intensive nature of the current system also poses significant safety risks to the workers. Repeated manual lifting and moving of heavy materials increase the likelihood of workplace injuries and health issues, which can further disrupt operations and reduce productivity. Ensuring the safety of workers is paramount, and the current system falls short in providing a safe and ergonomically sound working environment. Moreover, the frequent handling of products by workers has resulted in an increased incidence of product damage. The manual nature of the process means that products are more likely to be mishandled, leading to damage that incurs additional costs for the company. Damaged products can result in lost revenue, increased waste, and a negative impact on the company's reputation for quality.

Another issue is the increase in idle time among workers, which adds an unnecessary cost burden to the company. The inefficiencies in the current system mean that workers often spend time waiting for tasks to be completed or for materials to be moved, rather than engaging in productive activities. This idle time represents lost potential and contributes to higher labor costs without corresponding increases in productivity.

The confined nature of the workplace area further exacerbates these issues. The movement of workers in a limited space leads to unnecessary obstructions of material flow and machinery, disrupting the overall efficiency of operations. These obstructions can cause delays and increase the complexity of coordinating material handling tasks, further reducing productivity.

To address these issues, the study involves proposing ten different alternatives to the existing system. These alternatives will be evaluated using a Hybrid Multi-Criteria Decision-Making (MCDM) technique, specifically the Analytic Hierarchy Process (AHP) combined with the Technique for Order Preference by Similarity to Ideal Solution (TOPSIS). This method allows for a comprehensive and systematic evaluation of each alternative based on multiple criteria, ensuring that the most efficient system is selected.

The current system will be compared with the alternative systems using four key criteria: manpower, time, skill, and damages. Manpower criteria will assess the number of workers required and their associated costs. Time criteria will evaluate the efficiency of operations and the potential for reducing delays. Skill criteria will consider the need for specialized labor and the ease of training. Damages criteria will examine the potential for reducing product damage and associated costs.

By systematically evaluating these alternatives, the study aims to identify the most efficient and cost-effective material handling system for EDI Ltd. This new system should enhance productivity, reduce costs, improve worker safety, and minimize product damage, thereby addressing the critical issues present in the current system.

14.3.3 Proposed Systems & Findings

The proposed material handling systems were developed based on the following key assumptions:

(i) Division of Material Handling Activities: The entire material handling process is divided into three distinct operations:

  (a) Loading/Unloading: This includes all activities related to loading materials onto transport vehicles or unloading them from transport vehicles.
  (b) Material Movement: This refers to the movement of materials within the facility, including transferring them from one location to another.
  (c) Storage/Retrieval Operations: This encompasses the storage of materials in designated areas and the subsequent retrieval of these materials when needed.

(ii) Focus on Semi-Finished Goods: It is observed that 90% of the time, the materials being handled are semi-finished goods in boxes. Therefore, the analysis and proposed systems primarily focus on the handling of these semi-finished goods boxes. This assumption simplifies the scenario by considering a more uniform type of material flow and ensuring the proposed systems are optimized for the most common type of material handled.
(iii) Time Measurement for Unitized Pallets: For the purpose of this analysis, the time required for handling operations is measured in terms of the time needed to fill or move a unitized pallet. This means that the time readings for various material handling activities are based on the handling of a single pallet load. This assumption provides a standard unit of measurement, allowing for consistent and comparable time readings across different operations.

By considering these assumptions, the proposed material handling systems can be evaluated and optimized to ensure efficient and effective handling of semi-finished goods, thereby improving overall operational efficiency.

The proposed systems are illustrated in Table 14.1:

The following time estimates are standard time values obtained from standard time calculations under ideal working conditions, vendor brochures and specification sheets available of the all products. These is used only as base data for judgment in the pairwise comparison table. Different attributes of the proposed individual systems are represented in the Table 14.1.

The cost of system implementation is given in Table 14.2.

The total number of workers employed in the current system was 11. The time taken to load pallets inside the truck was 450 s/pallet. The time for the movement of pallets from the staging area to storage was 120 s/pallet. % Reduction in manpower is

Table 14.1 Proposed systems

System	Activity	Equipment	Manpower	Time	Skill	Damages
Existing System	Loading/ Unloading	Manual	8 Nos	450 Sec	No	High
	Material Movement	Manual Hand Pallet Truck	3 Nos	120 Sec	No	High
	Storage/ Retrieval	Reach Truck	1 No	70 Sec	Yes	Low
System 1	Loading/ Unloading	Yard Ramp	4 Nos	215 Sec	No	Low
	Material Movement	Electric Pallet truck	2 Nos	70 Sec	Yes	Low
	Storage/ Retrieval	Reach Truck	1 No	70 Sec	Yes	Low
System 2	Loading/ Unloading	Dock leveler	4 Nos	195 Sec	Yes	Low
	Material Movement	Electric Pallet truck	2 Nos	70 Sec	Yes	Low
	Storage/ Retrieval	Reach Truck	1 No	70 Sec	Yes	Low
System 3	Loading/ Unloading	Lift Table	4 Nos	205 Sec	Yes	Low
	Material Movement	Manual Hand Pallet Truck	2 Nos	120 Sec	Yes	Low
	Storage/ Retrieval	Reach Truck	1 No	70 Sec	Yes	Low
System 4	Loading/ Unloading	Lift Table	4 Nos	205 Sec	Yes	Low
	Material Movement	Electric Pallet truck	2 Nos	70 Sec	Yes	Low
	Storage/ Retrieval	Reach Truck	1 No	70 Sec	Yes	Low
System 5	Loading/ Unloading	Manual	4 Nos	236 Sec	No	High
	Material Movement	Scissor Pallet Truck	2 Nos	120 Sec	Yes	High
	Storage/ Retrieval	Reach Truck	1 No	70 Sec	Yes	Low
System 6	Loading/ Unloading	Manual	4 Nos	236 Sec	No	High
	Material Movement	Scissor Pallet Truck (Electric)	2 Nos	70 Sec	Yes	High
	Storage/ Retrieval	Reach Truck	1 No	70 Sec	Yes	Low

(continued)

Table 14.1 (continued)

System	Activity	Equipment	Manpower	Time	Skill	Damages
System 7	Loading/ Unloading	Manual	6 Nos	210 Sec	No	High
	Material Movement	Flexible Roller Conveyer	2 Nos	145 Sec	Yes	Low
	Storage/ Retrieval	Reach Truck	1 No	70 Sec	Yes	Low
System 8	Loading/ Unloading	Manual	4 Nos	225 Sec	No	High
	Material Movement	Manual Stacker Machine	2 Nos	70 Sec	No	Low
	Storage/ Retrieval	Reach Truck	1 No	70 Sec	Yes	Low
System 9	Loading/ Unloading	Manual	4 Nos	225 Sec	No	High
	Material Movement	Semi-Electric Stacker Machine	2 Nos	120 Sec	Yes	Low
	Storage/ Retrieval	Reach Truck	1 No	70 Sec	Yes	Low
System 10	Loading/ Unloading	Manual	4 Nos	225 Sec	No	High
	Material Movement	Electric Stacker Machine	2 Nos	70 Sec	Yes	Low
	Storage/ Retrieval	Reach Truck	1 No	70 Sec	Yes	Low

Table 14.2 Total cost of systems

System	Purchase cost	Cost/Year	Labor Cost/Year	Maintenance cost (10% of purchase cost)	Total Cost/Year
System 1	10,50,000/-	210,000/-	9,07,200/-	105,000/-	12,22,200/-
System 2	11,00,000/-	220,000/-	9,07,200/-	1,11,000/-	12,38,200/-
System 3	3,92,000/-	78,400/-	9,07,200/-	39,200/-	10,24,800/-
System 4	8,50,000/-	170,000/-	9,07,200/-	85,000/-	11,62,200/-
System 5	98,000/-	19,600/-	14,25,600/-	9,800/-	14,55,000/-
System 6	250,000/-	50,000/-	14,25,600/-	25,000/-	15,00,600/-
System 7	2,60,000/-	52,000/-	11,66,400/-	26,000/-	12,44,400/-
System 8	1,80,000/-	36,000/-	9,07,200/-	18,000/-	9,61,200/-
System 9	4,40,000/-	88,000/-	9,07,000/-	44,000/-	10,39,000/-
System 10	10,00,000/-	200,000/-	9,07,200/-	1,00,000/-	12,07,200/-

Table 14.3 % Reduction in cost and time

System	% Reduction in costs	% Reduction in time
System 1	23.01	52.22
System 2	20.38	56.67
System 3	34.10	54.44
System 4	25.27	54.44
System 5	6.44	47.64
System 6	3.51	47.64
System 7	19.98	67.78
System 8	38.19	50
System 9	33.19	50
System 10	22.37	50

obtained by subtracting 1 from the ratio of the workers employed in the current system divided by the workers in the proposed system. % Reduction in costs is obtained by subtracting 1 from the ratio of the current labor costs divided by proposed labor costs.

% Reduction in cost & time is shown in Table 14.3.

14.3.4 Application of Methodology

In the process of selecting the most efficient material handling system, a Hybrid Analytic Hierarchy Process (AHP) and Technique for Order of Preference by Similarity to Ideal Solution (TOPSIS) method is employed.

Detailed Breakdown of methodology is illustrated below:

Analytic Hierarchy Process (AHP): This method is used to assign weights to various criteria based on their relative importance. AHP involves breaking down the decision-making problem into a hierarchy of more easily comprehended sub-problems, each of which can be analyzed independently. The pairwise comparison method is employed to determine how each criterion compares to others in terms of importance, leading to the calculation of a set of weights that reflect the relative significance of each criterion.

Technique for Order of Preference by Similarity to Ideal Solution (TOPSIS): This method ranks the various systems based on their distance from an ideal solution. TOPSIS evaluates each option by comparing its performance to the best and worst possible scenarios (ideal and anti-ideal solutions). The approach helps in identifying which system is closest to the ideal and farthest from the worst-case scenario, providing a clear ranking of the systems.

Combination of Methods: The use of both AHP and TOPSIS is strategic. AHP helps in assigning meaningful weights to the criteria, which reflects their importance in the decision-making process. TOPSIS uses these weights to evaluate and rank the systems based on their performance across the criteria. Combining these methods leverages the strengths of each while minimizing their individual limitations. AHP provides a structured way to quantify subjective judgments, while TOPSIS offers a quantitative ranking of alternatives.

Cost–Benefit Analysis: After ranking the systems using the hybrid AHP-TOPSIS approach, a Cost–Benefit Analysis (CBA) is performed. This analysis evaluates the financial implications of each system by comparing the costs involved with the benefits gained. The most efficient system is chosen based on this analysis, ensuring that the selected system offers the best value relative to its cost.

Exclusion of Manual Systems: Systems 3, 5, 8, and 9 are excluded from the analysis. These systems are manual and rely on human effort and skills. They are deemed less effective because they do not significantly reduce the manpower or activity time compared to existing systems. Additionally, these manual systems do not offer substantial improvements in reducing damages, making them less viable options for consideration.

Pairwise Comparison for Weight Determination: To determine the weights of each criterion, a pairwise comparison is conducted. This involves comparing each criterion directly with every other criterion to assess their relative importance. The results of these comparisons are used to calculate the final weights, which are then utilized in the TOPSIS method for ranking the systems.

The Hybrid AHP-TOPSIS method, combined with Cost–Benefit Analysis, provides a comprehensive approach to selecting the most efficient material handling system. The exclusion of manual systems and the systematic determination of criterion weights ensure that the final choice is both practical and effective.

Pairwise comparison made between the criteria to determine the weights of each criteria is illustrated below in Table 14.4.

Further Pairwise comparison was made between the systems with respect to each criteria. A pairwise comparison between the systems with respect to the Manpower criteria is shown in Table 14.5.

Similarly, Pairwise Comparison was made between the systems with respect to Time, Skill & Damage criteria and is shown in Tables 14.6, 14.7, and 14.8 respectively.

After obtaining weights derived for each alternative with respect to all individual criteria, the values were tabulated in a matrix table as shown in Table 14.9.

Table 14.4 Pairwise comparison table for criteria

	Manpower	Time required	Skill	Damages	Weights	Maximum Eigen value	Consistency ratio
Manpower	1/1	2/1	4/1	1/2	0.294344	4.05659	0.0188618
Time		1/1	3/1	1/2	0.19247		
Skill			1/1	1/5	0.0738113		
Damages				1/1	0.439317		

Table 14.5 Pairwise comparison table between systems wrt manpower criteria

	System 1	System 2	System 4	System 6	System 7	System 10	Weights	Maximum Eigen	Consistency ratio
Manpower									
System 1	1/1	1/1	1/1	4/1	1/2	3/1	0.225316	6.05456	0.0109118
System 2		1/1	1/1	5/1	1/3	4/1	0.225316		
System 4			1/1	7/1	3/1	5/1	0.225316		
System 6				1/1	1/5	1/2	0.0339878		
System 7					1/1	6/1	0.0647493		
System 10						1/1	0.225316		

Table 14.6 Pairwise comparison table between systems wrt time criteria

	System 1	System 2	System 4	System 6	System 7	System 10	Weights	Maximum Eigen value	Consistency ratio
Time									
System 1	1/1	1/5	1/3	6/1	1/8	3/1	0.106652	8.76084	0.0552169
System 2		1/1	3/1	5/1	1/4	6/1	0.220571		
System 4			1/1	1/7	1/5	1/4	0.0662739		
System 6				1/1	1/9	1/3	0.0776113		
System 7					1/1	9/1	0.448987		
System 10						1/1	0.0799049		

Table 14.7 Pairwise comparison table between systems wrt skill criteria

	System 1	System 2	System 4	System 6	System 7	System 10	Weights	Maximum Eigen value	Consistency ratio
Skill									
System 1	1/1	2/1	1/5	4/1	1/2	3/1	0.204945	6.4744	0.0948801
System 2		1/1	1/5	5/1	1/3	4/1	0.145083		
System 4			1/1	7/1	3/1	5/1	0.261587		
System 6				1/1	1/5	1/2	0.043057		
System 7					1/1	6/1	0.298436		
System 10						1/1	0.0468929		

Matrix Multiplication of the values in Table 14.9 was then done with the criteria weights derived in Table 14.4. Table 14.10 shows the final values after multiplication along with the square root of the sum of weights highlighted in bold letters. These square root values are later used for deriving normalized values for each system wrt all criteria i.e., Manpower, Time, Skill and Damages.

Table 14.11 represents the standardized values after dividing each column value by the square root value.

The values obtained in Table 14.11 are then multiplied with criteria weights using matrix multiplication. Table 14.12 represents the values obtained after matrix multiplication.

As per the TOPSIS methodology, the ideal & negative ideal solution values are to be obtained. The ideal solution represents the best performance values for all criteria together for any particular solution & negative ideal represents the worst performance

Table 14.8 Pairwise comparison table between systems wrt damage criteria

	System 1	System 2	System 4	System 6	System 7	System 10	Weights	Maximum Eigen value	Consistency ratio
Damages									
System 1	1/1	2/1	1/5	4/1	1/2	3/1	0.132566	6.39906	0.0798113
System 2		1/1	1/3	5/1	1/3	4/1	0.112053		
System 4			1/1	7/1	5/1	5/1	0.443977		
System 6				1/1	1/5	1/2	0.035041		
System 7					1/1	6/1	0.226568		
System 10						1/1	0.049794		

Table 14.9 Matrix table of weights derived for each alternative

	Manpower	Time	Skill	Damages
System 1	0.225316	0.106652	0.204945	0.132566
System 2	0.225316	0.220571	0.145083	0.112053
System 4	0.225316	0.0662739	0.261587	0.443977
System 6	0.0339878	0.0776113	0.043057	0.035041
System 7	0.0647493	0.448987	0.298436	0.226568
System 10	0.225316	0.0799049	0.0468929	0.049794

Table 14.10 Matrix multiplication table with square root of the sum of weights

	Manpower	Time	Skill	Damages
System 1	0.0663204	0.0205273	0.0151273	0.0582385
System 2	0.0663204	0.0424533	0.0107088	0.0492268
System 4	0.0663204	0.0127557	0.0193081	0.1950466
System 6	0.0100041	0.0149378	0.0031781	0.0153941
System 7	0.0190586	0.0864165	0.0220279	0.0995352
System 10	0.0663204	0.0153793	0.0034612	0.0218754
	0.134376	0.1015571	0.0349803	0.2334116

Table 14.11 Standardized table after dividing each column by square root values

	Manpower	Time	Skill	Damages
System 1	0.493544	0.202126	0.432450	0.249510
System 2	0.493544	0.418024	0.306137	0.210901
System 4	0.493544	0.125602	0.551969	0.835634
System 6	0.074449	0.147088	0.090854	0.065953
System 7	0.141830	0.850916	0.629724	0.426436
System 10	0.493544	0.151435	0.098948	0.093720

Table 14.12 Weighted standardized matrix by multiplying attribute weights to each system values

	Manpower	Time	Skill	Damages
System 1	0.1452716	0.0389032	0.0319197	0.1096139
System 2	0.1452716	0.0804571	0.0225963	0.0926525
System 4	0.1452716	0.0241746	0.0407416	0.3671082
System 6	0.0219135	0.0283101	0.006706	0.0289741
System 7	0.0417469	0.1637758	0.0464807	0.1873407
System 10	0.1452716	0.0291467	0.0073035	0.0411728

values for all criteria together for any particular solution [11]. Here it is to be noted that in the initial judgment for inputting values in a pairwise comparison table, Less manpower, Less time, Less Skill, and Less Damages were given higher numbers in the pairwise comparison table i.e., Lesser the requirement, higher beneficial value & hence higher preference values were given. The ideal & negative ideal solution is given below.

Ideal Solution: {0.1452, 0.1637, 0.04648, 0.3671}.

Negative Ideal Solution: {0.0219, 0.0241, 0.0067, 0.0289}.

After obtaining ideal & negative ideal solutions, the separation of alternative values from both solutions was found. The separation of alternative values from Ideal solution values is given in Table 14.13.

Similarly, the Separation of Alternative values from Negative Ideal Solution values is given in Table 14.14.

The relative closeness of alternative values to the ideal solution were found in Table 14.15. The alternative score values for each system are found in Table 14.15.

Ai^+ denotes the separation from the ideal solution for a particular system and Ai^- denotes the separation from the negative ideal solution for a particular system. Table 14.16 represents the normalized Cost values of each system tabulated in Table 14.2. These normalized values are to be used in cost–benefit analysis, to determine the most beneficial alternative with respect to cost as shown in Table 14.17.

System 4 has a maximum value of 4.6854, hence it is the most optimal solution.

Table 14.13 Separation of alternatives from ideal solution

	Manpower	Time	Skill	Damages	Ai^+
System 1	0.0000	0.0156	0.0002	0.0663	0.2865
System 2	0.0000	0.0069	0.0006	0.0753	0.2878
System 4	0.0000	0.0195	0.0000	0.0000	0.1397
System 6	0.0152	0.0184	0.0016	0.1143	0.3866
System 7	0.0107	0.0000	0.0000	0.0323	0.2074
System 10	0.0000	0.0181	0.0015	0.1062	0.3548

Table 14.14 Separation of alternatives from negative ideal solution

	Manpower	Time	Skill	Damages	Ai^-
System 1	0.0152	0.0002	0.0006	0.0065	0.1502
System 2	0.0152	0.0032	0.0003	0.0041	0.1506
System 4	0.0152	0.0000	0.0012	0.1143	0.3615
System 6	0.0000	0.0000	0.0000	0.0000	0.0041
System 7	0.0004	0.0195	0.0016	0.0251	0.2157
System 10	0.0152	0.0000	0.0000	0.0001	0.1241

Table 14.15 Relative closeness to the ideal solution

	System 1	System 2	System 4	System 6	System 7	System 10
Ai +	0.2865	0.2878	0.1397	0.3866	0.2074	0.3548
Ai-	0.1502	0.1506	0.3615	0.0041	0.2157	0.1241
Ai + + Ai-	0.4368	0.4385	0.5013	0.3908	0.4232	0.4789
Ai− / Ai + + Ai-	0.344	0.3436	0.7213	0.0106	0.5098	0.2591

Table 14.16 Normalized cost values of each system

System	Total cost	Normalized value
System 1	12,22,200/-	0.16135
System 2	12,38,200/-	0.16346
System 4	11,62,200/-	0.15342
System 6	15,00,600/-	0.19811
System 7	12,44,400/-	0.16428
System 10	12,07,200/-	0.15937

Table 14.17 Cost–benefit analysis representing alternative score values

System	Alternative score/normalized cost value	Ratio
System 1	0.3439/0.16135	2.1313
System 2	0.34356/0.16346	2.1017
System 4	0.7212/0.15342	**4.6854**
System 6	0.0105/0.19811	0.0532
System 7	0.5098/0.16428	3.0929
System 10	0.2590/0.15937	1.6202

14.4 Result & Discussion

For this study, an extensive analysis of the Material Handling system within the facility was conducted. The primary objective was to evaluate and optimize the efficiency and cost-effectiveness of the current system. To achieve this, ten different alternative systems were proposed, each incorporating various types of material handling equipment and strategies. These alternatives were meticulously analyzed using the Hybrid Analytic Hierarchy Process (AHP) combined with the Technique for Order Preference by Similarity to Ideal Solution (TOPSIS) method, alongside a comprehensive Cost–Benefit Analysis (CBA). Among the ten proposed systems, System 4 emerged as the optimal choice. This system included a combination of a lift table, an electric pallet truck, and a reach truck. The Hybrid AHP-TOPSIS method was instrumental in determining this outcome, as it enabled a multi-criteria decision-making process that considered both quantitative and qualitative factors. The Cost–Benefit Analysis further reinforced this decision, highlighting the financial advantages of System 4. System 4 achieved a maximum ratio of 4.6854, signifying its superior cost–benefit performance compared to the other alternatives analyzed. This high ratio indicates that System 4 is significantly more beneficial in terms of cost efficiency and operational effectiveness.

Conversely, System 6 exhibited the lowest ratio, indicating that it was the least beneficial alternative among those evaluated. Several key factors contributed to the optimality of System 4. Firstly, the system required only seven workers, which is a reduction of five workers compared to the existing system. This reduction in labor resulted in substantial cost savings. Specifically, the annual labor cost for System 4 was calculated to be Rs. 9,07,200/-, which is approximately Rs. 6,50,000/- less than the cost associated with the existing system. This significant decrease in labor costs directly contributes to the overall cost efficiency of System 4. Furthermore, System 4 demonstrated a remarkable improvement in operational efficiency. The time required for material handling operations was reduced from 450 to 205 s. This reduction in time translates to increased productivity and faster processing times, which are critical factors in enhancing the overall efficiency of the facility's material handling system. Importantly, the implementation of System 4 did not necessitate the addition of extra skilled laborers to the existing workforce. This aspect is crucial as it indicates that the transition to the new system can be achieved seamlessly without incurring additional costs or facing challenges related to recruiting and training new personnel.

In summary, the detailed analysis and evaluation of the ten proposed material handling systems, using the Hybrid AHP-TOPSIS method and Cost–Benefit Analysis, conclusively identified System 4 as the optimal choice. The combination of a lift table, an electric pallet truck, and a reach truck not only provided superior cost–benefit performance but also significantly reduced labor costs and operational time. These improvements were achieved without the need for additional skilled labor, making System 4 a highly efficient and cost-effective solution for the facility's material handling needs.

14.5 Conclusion

The case study discussed in this paper on the Material Handling System presents a comprehensive and detailed methodology. This methodology not only addresses the specific challenges within material handling but also demonstrates its versatility in tackling a diverse range of industrial problems and facilitating intricate decision-making processes across various sectors and industries.

In addressing industrial problems, it is crucial to integrate both quantitative and qualitative criteria into the analysis process. This holistic approach ensures a comprehensive evaluation that considers not only measurable factors but also nuanced qualitative aspects. One effective methodology that combines these dimensions is the integration of Analytic Hierarchy Process (AHP), Technique for Order of Preference by Similarity to Ideal Solution (TOPSIS), and Cost–Benefit Analysis (CBA).

AHP facilitates the structuring of complex decision-making by hierarchically organizing criteria and alternatives, allowing decision-makers to weigh their relative importance. This structured approach not only enhances transparency but also ensures that all relevant factors are considered in the decision-making process.

TOPSIS, on the other hand, complements AHP by providing a quantitative framework to determine the best alternative based on proximity to the ideal solution. By calculating the distance of each alternative from the ideal solution and the worst-case scenario, TOPSIS helps in ranking alternatives objectively, thereby aiding decision-makers in selecting the most optimal solution.

In conjunction with AHP and TOPSIS, CBA provides a rigorous method to assess the economic feasibility of alternatives. By quantifying costs and benefits associated with each alternative, CBA enables decision-makers to evaluate potential returns and risks, ensuring that financial considerations are factored into the decision-making process.

Moreover, the integration of these methodologies allows decision-makers to analyze the impact of each criterion on the final solution comprehensively. Through sensitivity analysis and scenario testing, decision-makers can optimize criteria and parameters, thereby refining their decisions and identifying opportunities for further improvement.

Ultimately, this integrated approach not only facilitates informed decision-making but also empowers decision-makers to identify and prioritize criteria essential for achieving better solutions. By balancing quantitative rigor with qualitative insights, organizations can navigate complex industrial problems effectively, ensuring sustainable outcomes aligned with strategic objectives and stakeholder expectations. Both quantitative & qualitative criteria can be included in the analysis work for any industrial problem.

References

1. Pedro, M. et al. (2012). First Look at MCDM: Choosing a Decision Method, Advances in Smart System Research. ISSN 2050-8662
2. Rekik, R., et al. (2016). Using multiple criteria decision making approaches to assess the quality of web sites. *International Journal of Computer Science and Information Security, 14*, 747–761.
3. Jato-Espino, et al. (2014). A review of the application of multi-criteria decision-making methods in construction. *Automation in Construction., 45*, 151–162. https://doi.org/10.1016/j.autcon.2014.05.013
4. Tapas Kumar, B. et al. (2019). MCDM technique application to the selection of an Indian Institute of Technology. *Operational Research in Engineering Sciences: Theory and Applications 2*(3), 65–76
5. Ho, W., et al. (2010). Multi-criteria decision-making approaches for supplier evaluation and selection: A literature review. *European Journal of Operational Research, 202*(1), 16–24.
6. Ghaleb, A. M., et al. (2020). Assessment and comparison of various MCDM approaches in the selection of manufacturing process. *Advances in Materials Science and Engineering*. https://doi.org/10.1155/2020/4039253

7. Pourjavad, E., et al. (2011). A MCDM approach for prioritizing production lines: A case study. *International Journal of Business and Management*. https://doi.org/10.5539/ijbm.v6n10p221
8. Fatma, E. et al. (2020). Multi-criteria decision-making methods and their applications—a literature review. In *Proceeding of the International Conference on Industrial Engineering and Operations Management*, Dubai, UAE, March 10–12, 2020.
9. Kumar, K., et al. (2017). Evaluation and selection of best priority sequencing rule in job shop scheduling using hybrid MCDM technique. *IOP Conference Series: Materials Science and Engineering, 197*, Article 012059. https://doi.org/10.1088/1757-899X/197/1/012059
10. Roszkowska, E. (2011). Multi-criteria decision making models by applying the TOPSIS method to crisp and interval data. *Multiple Criteria Decision Making, 6*, 200–230.
11. Marqués, A. I., et al. (2020). Ranking-based MCDM models in financial management applications: analysis and emerging challenges. *Progress in Artificial Intelligence*. https://doi.org/10.1007/s13748-020-00207-1
12. De Koster et al. (2008). Warehouse assessment in a single tour. *Warehousing in the Global Supply Chain: Advanced Models, Tools, and Applications for Storage Systems*. https://doi.org/10.1007/978-1-4471-2274-6_17.

Chapter 15
Image Caption Generation Evaluation Based on XceptionNet, VGGNet, and ResNet

Battula Srinivasa Reddy, Gagan Sharma, and Naived George Eapen

Abstract A caption generation for the image is called image captioning. Image captioning means giving the scenario of an image based on attributes presented in the image. Artificial Intelligence (AI) based technologies have been increasing rapidly for over a decade. Deep neural networks are one of the critical ingredients in AI. By using these neural networks, we are generating captions for the image. Convolutional Neural Network (CNN) is the first neural network that is used for the image caption generation and feature extraction of the image. Later, when AI-based research was increased, different types of neural networks replaced CNN and got better results. This paper aims to give the introduction of image captioning and compare the three best neural networks (XceptionNet, VGGNet, ResNet) for image captioning for an image with the BLEU scores.

Keywords Artificial intelligence · Deep learning · Convolutional neural network (CNN) · Xception · VGG-16 · ResNet neural networks

15.1 Introduction

Image caption generation is a complex process that involves using neural networks to analyze an image and produce a descriptive caption that accurately reflects its content. The process typically relies on a combination of Convolutional Neural Networks (CNNs) and Recurrent Neural Networks (RNNs) to achieve this goal. The CNN is primarily responsible for image extraction, where it applies techniques like max and average pooling to capture the essential features of the image. These features are then

B. S. Reddy (✉) · G. Sharma · N. G. Eapen
Department of Data Science, Christ (Deemed To Be University), Lavasa, Pune, India
e-mail: battula.srinivasa@msds.christuniversity.in

G. Sharma
e-mail: gagan.sharma@msds.christuniversity.in

N. G. Eapen
e-mail: naived.eapen@christuniversity.in

P. K. Kapur et al. (eds.), *Data-centric Approaches to Industrial Decisions*, Asset Analytics, https://doi.org/10.1007/978-981-96-7556-2_15

used as input for the caption generation process. While RNNs are commonly used for generating captions, there are other types of neural networks that can also work with these image features to produce captions [1–3]. In the early stages of development, many researchers explored the use of semantic embeddings for image captioning, which allowed for a more nuanced understanding of the relationship between images and their associated text [2]. Generative Adversarial Networks (GANs) have also been employed for image captioning, where the network consists of a generator and a discriminator that work together to minimize errors in the generated captions by reducing the gradient. One specific type of GAN, known as VanillaGAN, is considered a simpler architecture but can still be effective in this task. Within the broader context of image captioning, injecting and merging are two distinct methodologies used to integrate image features with textual data. These methodologies include architectures like Init-inject, Pre-inject, Par-inject, and merge, each with its unique approach to how image vectors and textual information are combined [4].

According to a study by Vinyls et al., the Pre-inject architecture has shown superior performance compared to other architectures in terms of generating accurate and relevant captions. Another study has compared LSTM-based and Encoder-Decoder-based image caption generators. In the Encoder-Decoder framework, the encoder processes the image to generate a representation, which is then passed to the decoder to produce the corresponding caption. Long Short-Term Memory (LSTM) networks are a key component of RNNs used in caption generation, as they are particularly effective at capturing long-term dependencies within the sequence of words that form the caption. A [5] LSTMs can determine which words are relevant based on the overall context of the caption. Natural Language Processing (NLP) techniques also play a crucial role in image captioning. Within NLP, there are two primary methods for summarizing text: abstractive and extractive. The abstractive approach generates summaries that may include words and phrases different from the original text, effectively rephrasing the content. In contrast, the extractive approach involves selecting and extracting key words or phrases directly from the input text to form a summary [6]. While generating captions, there is a possibility of encountering errors, often referred to as gradient descent issues. To address this, Gated Recurrent Units (GRUs) are sometimes used, as they help mitigate the vanishing gradient problem that can occur in RNNs. GRUs are similar to LSTMs but use an update gate and a reset gate to control the flow of information, ensuring that the most relevant information is passed on to the output [7]. In scenarios where the process is reversed—where a caption is provided as input to generate an image—NLP techniques are again crucial [8]. However, a significant advancement in this area came in 2021 when OpenAI introduced a multimodal network called CLIP (Contrastive Language-Image Pre-training). CLIP is capable of working in both directions, generating captions from images and images from captions, without requiring direct optimization for each specific task [9]. It also possesses zero-shot learning capabilities, similar to those found in GPT-2 and GPT-3, allowing it to recognize and generate content with high accuracy, even for images or concepts it has not been explicitly trained on [10].

15.2 Background

15.2.1 Image Searching

15.2.1.1 Keyword-Based Image Query

Keyword-based image search means giving a keyword as a search to get the images based on that word. For this, there are different mechanisms that work to give the related images according to the given keyword. While searching the images there is a chance to get the ambiguity of images of keyword search [11]. Before the rise of deep learning most used content-based image retrieval to get the images [12]. For sparse feature vector dimensionality reduction using latent semantic indexing and using similarity matching functions can exploit the correlation of keywords [13]. The latent semantic index uses the SVD for dimensionality reduction and to measure the similarity uses the cosine angle between the query and document vectors.

15.2.1.2 Reverse Image Search

Using a keyword-based search for images is a keyword-based image query and using the image as a search to get the related images is called reverse image search [14, 15]. Compact composite descriptors are very well known for getting the low-level features of the image and there are different kinds of descriptors like color coherence descriptors, Discrete Wavelet Transform Spectral Hashing, etc., [16]. When Yiltan Bitirim et al conducted research on google for reverse image search with various categories they got 52% to 57% of accuracy [17]. Even companies like Google are not up to the accuracy so still there is a lack of reverse image search. There are different kinds of platforms that are using this technique company like online shopping providers, Google and Amazon, etc., [18]. At the first stage, there is only one technique to reverse an image. We have to select the image from the gallery but nowadays different platforms provide the lens which can take any picture randomly on the spot.

15.2.1.3 Large-Scale Image Retrieval

Large-scale image retrieval is a process of getting more similar images as given input of image [19]. In the early stage using the descriptors based on local features, a bag of words (A bag of visual words is a vector of occurrence counts of a vocabulary of local image features in computer vision), compressed fisher vectors, and hamming embedding [20, 21]. But later mostly using convolutional based CNN's [22]. While large-scale image retrieval it is very important to get a large level of semantic images. Semantic hierarchy is very important here because our aim is to get more images related. To exploit hierarchical image comparison is very important, for comparison feature indexing, and hierarchical comparison is better to get the more semantic hierarchical images [23].

15.2.2 Image Captioning

15.2.2.1 Generating Keywords Based on Image

Generating keywords based on an image is a challenging task due to the complex nature of images. When features are extracted from an image, the process often involves analyzing every element within the image, from the most prominent objects to the finer details. This comprehensive extraction can make it difficult to distinguish which elements are truly central to the image's meaning and should therefore be emphasized in the keyword generation process. The difficulty lies in focusing on the primary subject of the image, as the extraction process does not inherently prioritize one aspect of the image over another. Identifying and generating keywords that accurately represent the main focus or theme of the image requires advanced techniques to filter out less relevant details and highlight the most significant elements. This selective process is crucial because the goal is to produce keywords that effectively convey the core content of the image, rather than an exhaustive list of all its features.

15.2.2.2 Generating Descriptions Based on Image

Description generation based on the image is as similar to image captioning of image. For image captioning, it is important to get the features of the image to know the attributes in based. Then only based on it will give a description for it. For feature extraction there are different kinds of neural networks like VGG-16, Xception, ResNet50, InceptionV3, VGG-19, and AlexNet etc., Will give descriptions based on those features of the image and techniques to give the description by using the LSTM, RNN, GRU and decoder methods. Other than the English language it is difficult to get the description of an image but by using attention-based DenseNet-BiLSTM and using this method with China language based description for image and got good accuracy [24].

15.2.2.3 Generating Summaries Based on Image

Generating descriptions from images is a process closely related to image captioning, where the goal is to produce a textual summary that encapsulates the content and context of an image. Both tasks share a fundamental reliance on accurately extracting and interpreting the features of the image. To generate a meaningful description, it's essential first to identify the various attributes and elements within the image. This process of feature extraction is typically carried out using sophisticated neural networks designed to capture different levels of detail and structure within an image. Various neural network architectures are employed for this purpose, each with unique strengths. Models such as VGG-16, Xception, ResNet50, InceptionV3, VGG-19, and AlexNet are commonly used for their ability to extract detailed and hierarchical features from images. These models are trained to recognize patterns, shapes, and objects within an image, and they break down the visual information into a form that can be processed by other components of the system responsible for generating the textual description. Once the features of an image are extracted, these are then used as inputs to networks that specialize in generating text. Techniques such as Long Short-Term Memory (LSTM) networks, Recurrent Neural Networks (RNNs), Gated Recurrent Units (GRUs), and encoder-decoder frameworks are commonly employed to translate the extracted visual features into coherent and contextually appropriate descriptions. These methods work by processing the visual data in a sequence that allows the network to generate text that captures the essence of what is depicted in the image. While generating image descriptions in English is relatively well-developed, extending this capability to other languages presents additional challenges. Language-specific nuances, cultural contexts, and the structural differences between languages can complicate the description generation process. However, advancements in techniques like attention-based DenseNet-BiLSTM have shown promise in overcoming some of these challenges. For example, when applied to generate image descriptions in Chinese, these methods have demonstrated good accuracy, suggesting that with the right techniques, it is possible to effectively extend image description generation beyond English and into other languages. This represents a significant step forward in making image-based content more accessible across different linguistic and cultural contexts [25].

15.2.3 Object Recognition

Every image contains a variety of attributes, which could include people, animals, places, or various objects. To accurately interpret and understand an image, it is crucial to identify and analyze these attributes during the feature extraction process. This process is fundamental in the field of computer vision, where object recognition plays a central role. Object detection, which is a key aspect of object recognition, has evolved significantly over time. Initially, traditional methods were employed for this task, relying on handcrafted features and algorithms that were often limited in their effectiveness and adaptability. However, with the advent of neural networks and the subsequent rise of deep learning, more advanced methods have been developed, offering significantly improved accuracy and performance. In object recognition, the process typically involves three major steps. The first step is informative region selection, where the algorithm identifies the most relevant areas of the image that are likely to contain the objects of interest. This step is critical as it helps to focus computational resources on the most pertinent parts of the image, increasing the efficiency and accuracy of the recognition process [26]. The second step is feature extraction, where the selected regions are analyzed to identify specific characteristics and patterns that define the objects within those regions. This could involve recognizing shapes, textures, colors, and other visual features that distinguish one object from another. The third and final step is classification, where the extracted features are used to categorize the objects into predefined classes or labels. This step is where the system determines what each object is based on the features that have been extracted and analyzed. One of the powerful methods used for object detection in modern computer vision is Regions with Convolutional Neural Networks (R-CNN). R-CNN revolutionized object detection by introducing a way to combine region proposals with deep learning features, allowing for more precise and reliable identification of objects within an image. By segmenting an image into regions and then applying a CNN to each region, R-CNN effectively integrates the strengths of deep learning with traditional object detection techniques, leading to a more robust and accurate recognition process [27].

15.3 Methodology

The proposed model is designed to process an input image, denoted as I, and is trained to maximize the probability of generating a sequence of words, represented as p (S|I). Here, S refers to the sequence of words generated by the model, where each individual word, St, is selected from a vocabulary that has been constructed from the training dataset. The model operates by first feeding the input image I into a deep vision Convolutional Neural Network (CNN). This CNN is responsible for analyzing the image and identifying key objects and features, which are then encoded into a format that can be further processed. Once the image has been encoded by the CNN, these encodings are passed to a Long Short-Term Memory (LSTM) network, which is specifically designed for language generation. The role of the LSTM in this context is to take the encoded image data and use it to generate a coherent and meaningful sequence of words that describe the image. The LSTM works by sequentially predicting each word in the caption, drawing from the vocabulary learned during training, and forming a phrase or sentence that accurately reflects the content of the image. This model can be compared to a language translation model that uses LSTM networks, where the objective is to maximize the probability of p (T|S), with T representing the translation of an input text S. In a typical language translation model, an encoder LSTM processes the input text and converts it into a fixed-length vector, which is then used by a decoder LSTM to generate the translated text. However, in the image captioning approach, the CNN serves as the encoder, taking the place of the encoder LSTM. Instead of encoding an input text, the CNN encodes the input image into a fixed-length vector, which is then used by the LSTM to generate the descriptive text. This approach effectively combines the strengths of deep learning for both visual and textual data, enabling the model to translate visual information into language in a way that is analogous to translating text from one language to another (Fig. 15.1).

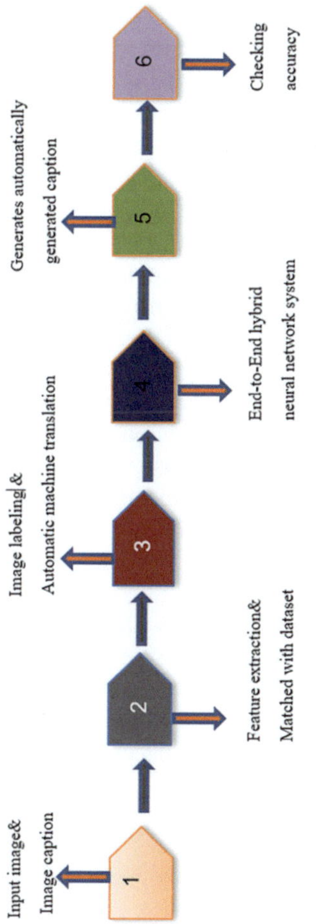

Fig. 15.1 Flow diagram of image captioning

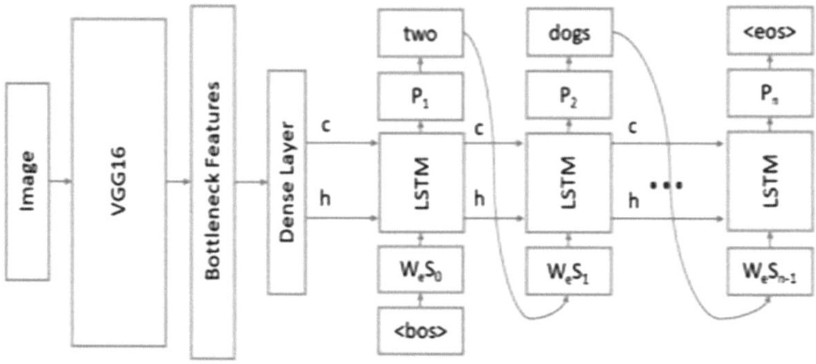

Fig. 15.2 VGG + LSTM

We use a VGG-16 pre-trained model for the job of classifying images. The next section discusses the models' specifics. The VGG-16 is followed by an LSTM network. Language creation is accomplished using the LSTM network. Since a current token depends on the preceding tokens for a sentence to make sense, LSTM networks are different from typical Neural Networks in that they take this into consideration.

The model that existed before the project has two distinct input streams, one for picture characteristics and the other for captions from the preprocessed input. To represent the picture features in a new dimension, the image features are transferred via a thick, completely linked layer. The embedded layer is passed after the input captions. Then, these two input streams are combined and sent into an LSTM layer. The caption embedding serves as the LSTM's input while the image serves as its starting state (Fig. 15.2).

And finally checks the generated caption with the BLEU score for accuracy.

Similarly, we use the Xception Net and ResNet with replacement to VGG-16, and run with the same LSTM model by checking accuracy with BLEU Score. For the 3 neural networks Xception, VGG-16, and ResNet are used for image captioning. Finally, we produce the results of any two images with these three models and their accuracy.

15.3.1 ResNet

ResNet, or Residual Network, is a pioneering deep learning architecture introduced by Kaiming He and colleagues in 2015. It addresses the degradation problem in deep neural networks by using residual learning. This involves adding shortcut connections that bypass one or more layers, effectively allowing the network to learn residual functions. This design enables the construction of much deeper networks without the issues of vanishing gradients or degraded performance. ResNet architectures, such as ResNet-50 and ResNet-101, have significantly improved the accuracy of image recognition tasks and remain influential in the development of modern deep learning models (Fig. 15.3).

As deep learning has evolved over the years, the number of layers in neural networks has increased, leading to better performance in object classification and recognition tasks. However, as more layers are added, training the model and achieving the desired accuracy becomes increasingly challenging. This complexity arises because deeper networks are harder to optimize, and issues such as vanishing gradients can impede effective training. To address this problem, the ResNet architecture was introduced. In their influential paper "Deep Residual Learning for Image Recognition" published in 2015, Shaoqing Ren, Kaiming He, Jian Sun, and Xiangyu Zhan presented ResNet as a solution. ResNet employs a novel approach called residual learning, which helps mitigate the degradation problem often encountered with very deep networks. Let's consider a scenario where we use a single layer on a particular dataset and observe high loss. To reduce this loss, additional layers are added. However, after adding nine layers, if the loss starts increasing, and even with thirty layers, the loss remains similar to the nine-layer model, it indicates a degradation problem. The degradation problem refers to the phenomenon where adding more layers to a neural network results in higher training error, contrary to the expectation that deeper networks should perform better. ResNet addresses this by introducing shortcut connections that skip one or more layers. These shortcuts create residual blocks, allowing the network to learn residual functions with reference to the layer inputs, rather than learning unreferenced functions. This technique simplifies the training process for very deep networks, making it possible to build models with hundreds or even thousands of layers without suffering from the degradation problem. As a result, ResNet has become a foundational architecture in the field of deep learning, enabling significant advancements in image recognition and other computer vision tasks [28].

In the context of residual networks (ResNet), the function h(x) is defined as:

$$h(x) = g(x) + x$$

where g(x) represents the function learned by the network layers, and xxx denotes the input which is added through a skip connection. The term g(x) can be considered a residual function, meaning it represents the change or difference that needs to be learned from the input.

The primary goal is to make h(x) an identity function, ideally making h(x)≈xh(x) \approx xh(x)≈x. To achieve this, g(x) should ideally approach zero, meaning the residual function would learn minimal or no changes. This simplifies the learning process, as the network can learn the identity mapping more easily if the layers only need to adjust the residual function g(x) to zero. Thus, by structuring the network to learn g(x), we facilitate the learning of identity mappings, thereby addressing the degradation problem observed in very deep networks.

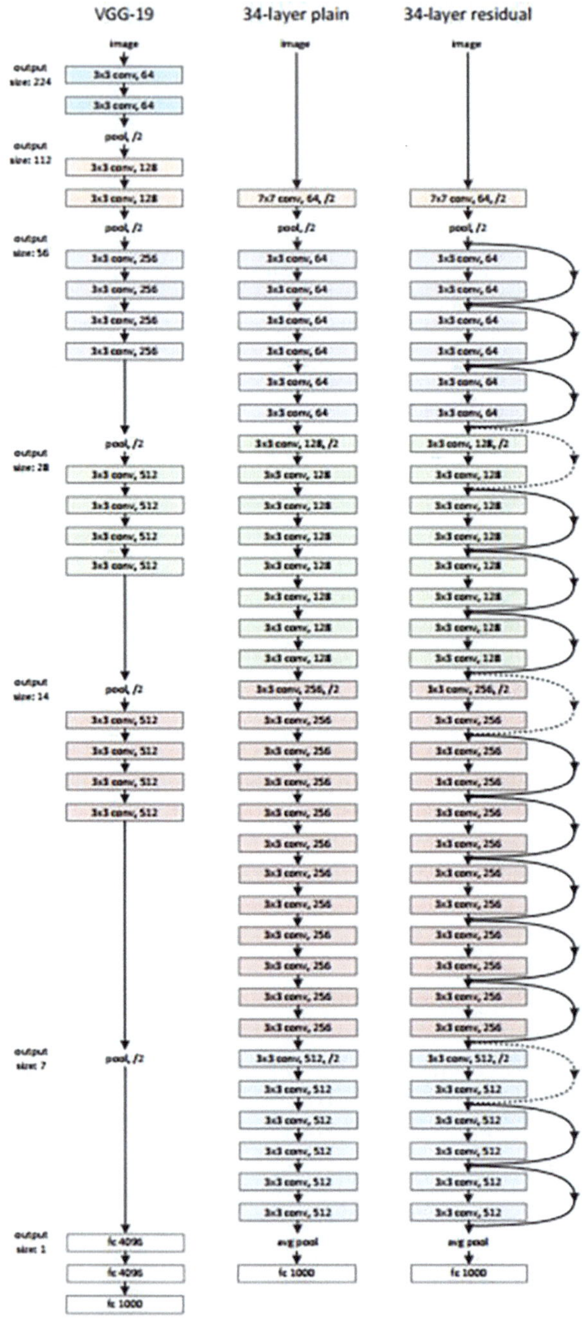

Fig. 15.3 ResNet architecture

15.3.2 VGGNet

VGGNet, developed by the Visual Geometry Group at the University of Oxford, is a convolutional neural network (CNN) architecture that has significantly influenced the field of computer vision. Introduced by Karen Simonyan and Andrew Zisserman in their 2014 paper "Very Deep Convolutional Networks for Large-Scale Image Recognition," VGGNet's design philosophy emphasizes depth, simplicity, and uniformity, which has proven effective in various image recognition tasks. The VGGNet architecture is characterized by its use of very small (3×3) convolution filters, which are applied throughout the entire network. This choice of filter size is smaller than those used in previous networks, allowing the model to capture fine-grained details in images. Despite their small size, these filters are highly effective because they are applied in multiple layers, enabling the network to learn complex features. One of the most notable aspects of VGGNet is its uniform architecture. The network consists of multiple layers of 3×3 convolutional filters, followed by a ReLU (Rectified Linear Unit) activation function, and then max-pooling layers. This pattern is repeated throughout the network, with each convolutional layer stack being followed by a max-pooling layer. The max-pooling layers reduce the spatial dimensions of the feature maps while retaining important spatial information. This regularity and simplicity in the architecture help streamline the network design and implementation. VGGNet's architecture comes in several variants, denoted by the number of layers, such as VGG-11, VGG-13, VGG-16, and VGG-19, where the numbers indicate the total layers in the network. Among these, VGG-16 and VGG-19 are the most popular due to their balance between depth and performance.

VGG-16, for example, has 13 convolutional layers and 3 fully connected layers, resulting in a total of 16 weight layers. A typical VGG-16 architecture begins with two convolutional layers with 64 filters each, followed by a max-pooling layer. This pattern is repeated with increasing filter depths of 128, 256, and 512 filters in subsequent layers. After the convolutional layers, the network includes three fully connected layers: the first two with 4096 neurons each, and the final one with 1000 neurons, corresponding to the 1000 classes in the ImageNet dataset. The output layer uses a softmax activation function to produce probability distributions over the classes.

Despite its impressive performance, VGGNet has some drawbacks. The network's depth and use of small filters lead to a large number of parameters, which increases computational cost and memory usage. Training VGGNet requires substantial computational resources and time. However, the network's design simplicity, effectiveness, and the quality of the learned features have made it a popular choice for various transfer learning tasks, where pre-trained VGGNet models are fine-tuned for specific applications.

VGGNet's impact extends beyond its initial implementation, as its principles have influenced many subsequent architectures, promoting the idea that deeper networks can capture more complex patterns. Its design has been foundational in developing

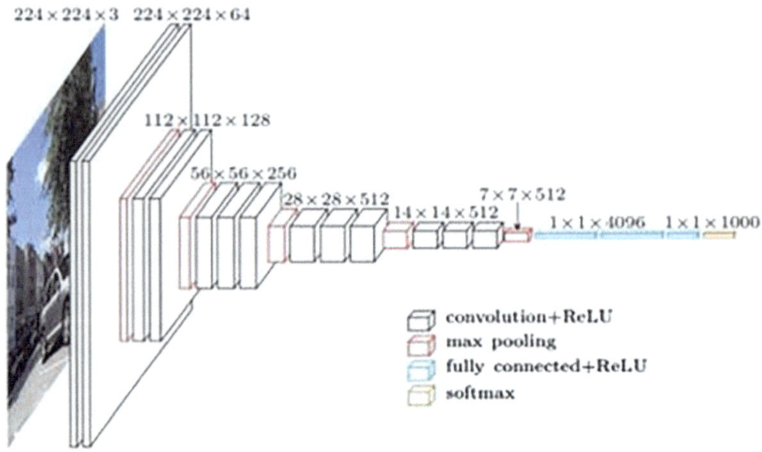

Fig. 15.4 VGGNet architecture

more advanced models, including those integrating skip connections and other innovations to address the limitations of deep networks. In summary, VGGNet is a landmark architecture in the evolution of CNNs, known for its depth, uniformity, and effective use of small convolutional filters. It has set a benchmark for subsequent neural network designs and continues to be a valuable tool in the field of computer vision (Fig. 15.4).

This is general VGGNet architecture.

- **Input:** It takes input of an image 224*224
- **Convolutional Layers:** It leverages the minimal receptive field 3*3 which is small but captures the left/right and up/down. And stride is 1 which means no of pixels shifting in the input matrix.
- **Hidden layers:** Here every layer uses the ReLU. Local Response Normalization (LRN) is often not used with VGG as it increases memory use and training time. Furthermore, it doesn't increase overall accuracy.
- **Fully Connected Layers:** Three fully connected layers make up the VGGNet. The first two levels each have
- 4096 channels, while the third layer has 1000 channels with one channel for each class.

VGG-16 and VGG-19 are two models of VGGNet. VGG-16 means it uses the 16 layers deep neural network. Similarly VGG-19 uses the 19 deep neural networks.

15.3.3 Xception

Xception is a neural network architecture that utilizes depthwise separable convolutions, a technique that significantly reduces the computational cost of deep learning models while maintaining high performance (Fig. 15.5). This architecture was introduced by François Chollet, the creator of Keras, and is also referred to as the extreme inception module. Xception's architecture consists of Depthwise Separable Convolution blocks combined with Maxpooling layers, all interconnected through shortcut connections similar to those found in ResNet architectures. What sets Xception apart is its unique order of operations: it applies a Pointwise Convolution before a Depthwise Convolution, reversing the typical sequence seen in other models. This approach allows for more efficient processing and has proven to enhance the model's accuracy across various image recognition tasks. The architecture's design is innovative in its ability to capture intricate patterns while maintaining computational efficiency, as shown in Figure 15.6:

This model is very hard to train but when compared to inception it has good improvements.

15.4 Results

We are going to write our results in Tables 15.1 and 15.2. Recently, but mostly for retrieval framework evaluation, Flickr datasets have been used. An important exception is a case where they combined retrieval and production, which results in the best performance on the available Flickr datasets. Train and test each model with Flickr8k dataset for image captioning. This dataset has 8,000 images and per image has 5 captions. After training and testing with this dataset giving the own image with a human-generated caption for each model and then comparing the human caption with the machine-generated with BLEU scores.

As shown in Table 15.1 these are the model-generated captions and human-generated captions. Based on these captions by comparing with the original given and model-generated captions with BLEU scores as shown in Table 15.2. The Bilingual Evaluation Understudy Score, commonly known as BLEU, is a statistical measure used to compare a generated sentence to a reference sentence in tasks such as machine translation or image captioning. BLEU scores range from 0.0 to 1.0, where 1.0 indicates a perfect match. This evaluation metric works by matching N-grams, which are sequences of words of length N, between the produced sentence and the reference sentence. For instance, a 1-gram score considers individual word matches, while a 2-gram score looks at pairs of words. BLEU allows you to assign different weights to N-gram matches; for example, assigning a weight of 1 for 1-grams and 0 for higher-order N-grams calculates the BLEU score solely based on single-word matches. For BLEU-4, which is a more comprehensive evaluation, the scores for 1-gram, 2-gram, 3-gram, and 4-gram matches are equally weighted at 25% or 0.25 each. The weight

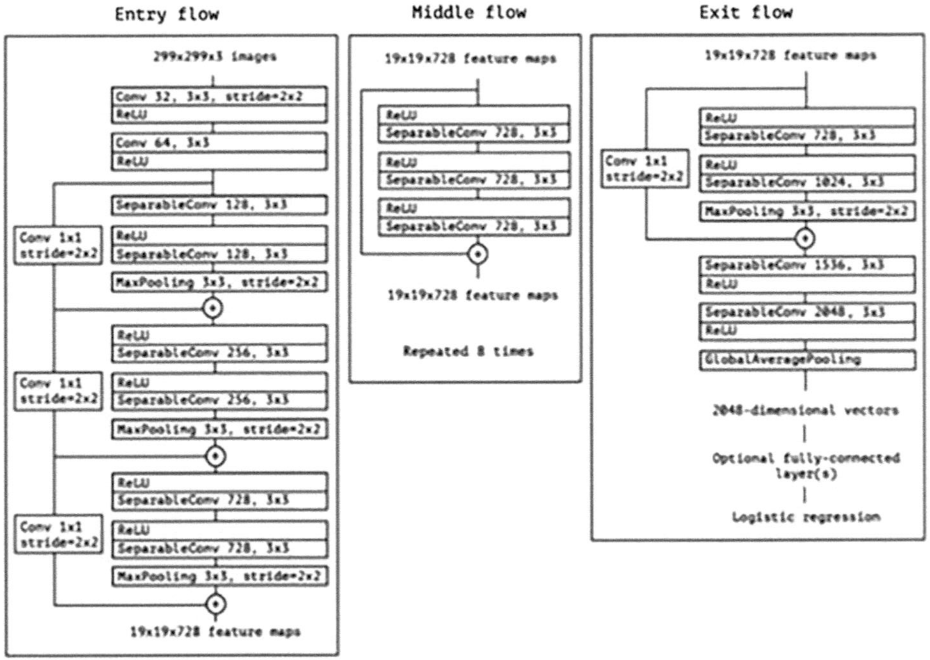

Fig. 15.5 XceptionNet architecture

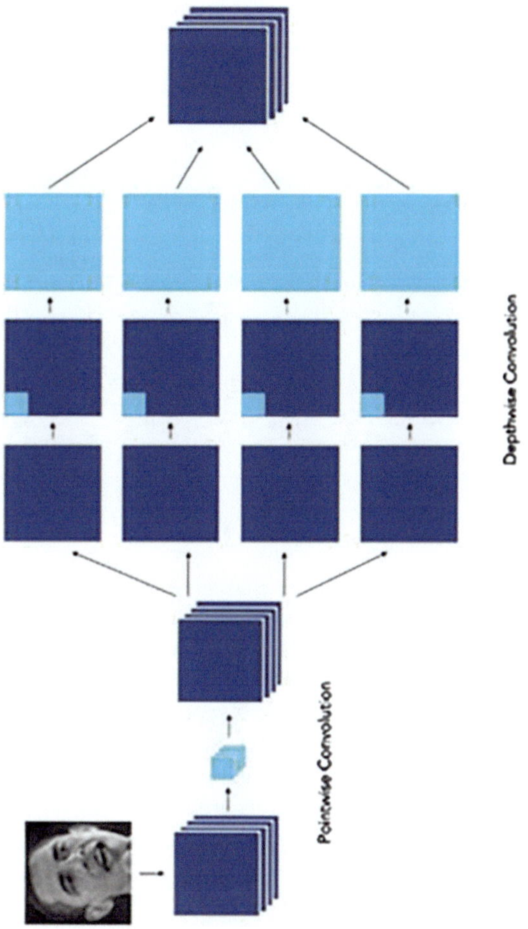

Fig. 15.6 Depthwise convolution

Table 15.1 Resulted captions of given image of each model

Xception Net	Four people are standing	A dog is running	Two people are moving
ResNet	Four people are discussing	A dog is sitting on grass	Two children playing football
VGGNet	Four people in a row	Dog is playing on grass	A boy in a white shirt with football
Human	Four people are discussing at the table	A dog is sitting in the green grass	Two people are playing with a football in the ground

Table 15.2 BLEU scores of each model

	BLEU-1	BLEU-2	BLEU-3	BLEU-4
Xception Net + LSTM	0.64	0.61	0.43	0.56
ResNet + LST M	**0.69**	**0.72**	0.59	**0.68**
VGGNet + LST M	0.53	0.47	**0.66**	0.57

configuration (1,0,0,0) is used for both cumulative and individual 1-gram BLEU scores, focusing solely on single-word matches. When considering 2-gram weights, both 1-gram and 2-gram scores are weighted at 50% each, while for 3-gram weights, the first three N-gram scores are evenly distributed at 33% each. In evaluating the performance of different neural networks like XceptionNet, ResNet, and VGGNet combined with an LSTM model, BLEU scores have shown that ResNet tends to generate more accurate captions, consistently achieving better BLEU scores across various evaluations. This suggests that ResNet's architecture may be more effective in capturing and translating the visual features into meaningful and contextually appropriate sentences.

15.5 Observations

Image caption generation is a difficult task to process the image. The important task is to give the appropriate caption based on contents present in the image. There is a rapid development in recent years for artificial intelligence. After using artificial intelligence, it became simpler than previous methods used. Observing the whole criteria of image captioning there are three different kinds of approaches to generate the image [29]. As said earlier, feature extraction is the main thing for extracting the features from images [30]. Most of the researchers used CNN for it. CNN is a neural network which has three layers : convolutional layer, pooling layer and fully connected layer. There are different kinds of neural network architectures which are also used for feature extraction like LeNet, ResNet, AlexNet, MobileNetV1, GoogleNet, ZfNet, Depth based CNN, Wide ResNet, VGG, PolyNet, Inception V2, Inception V3, V4 and Inception-ResNet, DenseNet, Pyramidal Net, Xception, Attention Based CNNs and Feature-Map based CNNs etc, Each neural network had their own architecture and different from others [29]. After using the architectures for image feature extractions they have to be decoded to detect the objects presented in the image and based on those images have to give the caption. For decoding there are different neural networks to process like RNN, LSTM, GRU, Attention based mechanism and GAN etc., Most of the time using MSCOCO, Flickr8K and Flickr30K datasets [31].

15.6 Change Over a Period of Time

The process of image captioning early done by using retrieval based methods [32]. Retrieval process gives a caption for the query image by a pre-specified sentence pool of one or a set of sentences. The caption generated is a pre- existing or formed sentence used by retrieved from image. Another type is template based and it is a commonly used method [33]. In template-based methods, generating captions through a procedure that is syntactically and semantically limited [34]. The initial step in this process involves detecting a specified set of visual concepts. Once these concepts are identified, they are assembled using sentence templates, specific language grammar rules, or combinatorial optimization techniques. To enhance efficiency, speed, and reliability, these methods are increasingly being refined and executed using deep neural networks [35]. In deep neural networks, there are various frameworks to consider: retrieval and template-based methods augmented by neural networks, multimodal learning, encoder-decoder architectures, attention-based mechanisms, compositional architectures, and novelty processes. Each of these frameworks contributes uniquely to the advancement of neural networks, enhancing their ability to process and understand complex data [36].

15.7 Applications

We have explored various image caption-generating techniques and their evaluation methods, highlighting their applications in NLP-based tools, aiding visually impaired individuals, and enhancing social media content. These techniques are also integral to everyday technologies like self-driving cars and Google image searches [37]. Despite their progress, ongoing research aims to make these methods more reliable and applicable to real-world challenges [38, 39]. For example, law enforcement could use image captioning to document vehicles involved in traffic violations or to capture details such as a prohibited vehicle's registration number, improving accuracy and efficiency in such scenarios. As this technology evolves, it has the potential to be even more impactful across diverse fields, making it indispensable for solving complex real-world problems.

15.8 Conclusion

Focused on the crucial application of automatic image captioning to benefit picture search engines and aid individuals who are blind or visually impaired, we leveraged advanced neural networks such as Xception, VGG-16, and ResNet in conjunction with the LSTM network. These models were used to generate descriptive captions for images in the Flickr8k dataset. Each neural network's architecture and functioning were meticulously examined, and their results evaluated using BLEU scores. The findings clearly demonstrated that the ResNet neural network outperformed the other models, except for two cases, exhibiting its superior capabilities in generating accurate and contextually relevant image captions. Additionally, ResNet's performance highlights its robustness in understanding and interpreting complex image features. As part of future work, we aim to explore additional models with improved accuracy, such as Transformer-based approaches, to further enhance the automatic image captioning process. This research represents a significant contribution to advancing the field of image captioning and opens up possibilities for enhancing accessibility and search functionality for visually impaired individuals and image-based applications.

References

1. A region-based image caption generator with refined descriptions. *Neurocomputing, 272*, 416–424.
2. Tanti, M., Gatt, A., & Camilleri, K. P. (2018). Where to put the image in an image caption generator. *Nature Language Engineering, 24*(3), 467–489.
3. Sharma, G., Kalena, P., Malde, N., Nair, A., & Parkar, S. (2018) Visual image caption generator using deep learning. *SSRN Electronic Journal.*
4. Sulabh K., & Samir Kumar, B. (2020) Comparative evaluation of CNN architectures for image caption generation. International Journal of Advanced Computer Science and Applications (IJACSA), 11(12).
5. Hossain, M. Z., Sohel, F., Shiratuddin, M. F., Laga, H., & Bennamoun, M. (2021). Text to image synthesis for improved image captioning. *IEEE Access, 9*, 64918–64928.
6. Jugran S., Kumar, A., Tyagi, B. S., & Anand, V. (2021) Extractive Automatic Text Summarization using SpaCy in Python & NLP. In Presented at the 2021 International Conference on Advance Computing and Innovative Technologies in Engineering (ICACITE), Greater Noida, India, Mar. 2021.
7. Vinyals and Alexander Toshev, "Show and Tell: A Neural Image Caption Generator".

8. Merchant, K. & Pande, Y. (2018). NLP based latent semantic analysis for legal text summarization. In *Presented at the 2018 International Conference on Advances in Computing, Communications, and Informatics (ICACCI)*, Bangalore.
9. You, Q., Jin, H., Wang, Z., Fang, C., Luo, J. (2016) Image captioning with semantic attention. In *Proceedings of the IEEE conference on computer vision and pattern recognition* (pp. 4651–4659).
10. Bhatia, S. (2021). A comparative study of opinion summarization techniques. *IEEE Transactions on Computational Social Systems, 8*(1), 110–117.
11. Yao, Y., Sun, Z., Shen, F., Liu, L., Wang, L., Zhu, F., Ding, L., Wu, G., Shao, L. (2019). *Dynamically visual disambiguation of keyword-based image search.* arXiv preprint arXiv: 1905.10955
12. Ji Wan Institute of Computing Technology, CAS, Beijing, China et al., Deep Learning for Content-Based Image Retrieval. In *ACM Conferences.* (Accessed 21 Sept 2022).
13. Rahman, M., Desai, B., & Bhattacharya, P. (2006). Visual keyword-based image retrieval using latent semantic indexing, correlation-enhanced similarity matching and query expansion in inverted index. In *Presented at the 2006 10th International Database Engineering and Applications Symposium (IDEAS'06)*, Delhi, India.
14. Agrawal, P., & Verma, J. S. (2013) *A survey of linear and non-linear filters for noise reduction January 2013(IJARCSMS).*
15. Gaillard, M., & Egyed-Zsigmond, E. (2017) *Large scale reverse image search - A method comparison for almost identical image retrieval.* Accessed: Sep. 21, 2022. [Online].
16. Chutel, P. M., & Sakhare, A. (2014) Evaluation of compact composite descriptor based reverse image search. In *Presented at the 2014 International Conference on Communications and Signal Processing (ICCSP)*, Melmaruvathur, India, Apr. 2014.
17. Bitirim, Y., Bitirim, S., Celik Ertugrul, D., & Toygar, O. (2020) An evaluation of reverse image search performance of Google. In *Presented at the 2020 IEEE 44th Annual Computers, Software, and Applications Conference (COMPSAC)*, Madrid, Spain.
18. Reverse image search for scientific data within and beyond the visible spectrum. (2018). *Expert System Application, 109*, 35–48.
19. Noh, H., Araujo, A., Sim, J., Weyand, T., & Han, B. (B.). *Large-scale image retrieval with attentive deep local features.*
20. Zheng, L., Wang, S., & Tian, Q. (2014). Coupled binary embedding for large-scale image retrieval. *IEEE Transaction Image Processing, 23*(8), 3368–3380.
21. Perronnin, F., Liu, Y., Sanchez, J., & Poirier, H. (2010) Large-scale image retrieval with compressed Fisher vectors. In *Presented at the 2010 IEEE Conference on Computer Vision and Pattern Recognition (CVPR)*, San Francisco, CA, USA.
22. Radenović, F., Iscen, A., Tolias, G., Avrithis, Y., & Chum, O. (2018) *Revisiting Oxford and Paris: large-scale image retrieval benchmarking.* arXiv.1803.11285.
23. Deng, J., Berg, A. C., & Fei-Fei, L. (2011). Hierarchical semantic indexing for large scale image retrieval. In *Presented at the 2011 IEEE Conference on Computer Vision and Pattern Recognition (CVPR)*, Colorado Springs, CO, USA.
24. Lu, H., Yang, R., Deng, Z., Zhang, Y., Gao, G., & Lan, R. (2021). Chinese image captioning via fuzzy attention- based DenseNet-BiLSTM. *ACM Transactions on Multimedia Computing, Communications, and Applications, 17*(1s), 1–18.
25. Extractive summarization of documents with images based on multi-modal RNN. (2019). *Future Generation Computer System 99*, 186–196.
26. Zou, Z., Chen, K., Shi, Z., Guo, Y., & Ye, J. (2023). *Object detection in 20 years: A survey.*
27. Zhao, Z.-Q., Zheng, P., Xu, S.-T., & Wu, X. (2018). *Object detection with deep learning: A review.*
28. Galatolo, F.A., Cimino, M.G., & Vaglini, G. (2021). Generating images from caption and vice versa via CLIP-Guided Generative Latent Space Search. ArXiv, abs/2102.01645.
29. Kiliç, V. (2021). Deep gated recurrent unit for smartphone-based image captioning". *Sakarya University Journal of Computer and Information Sciences, 4*(2), 204–215.

30. Ren, M., Kiros, R., Zemel, R. (2015) Exploring models and data for image question answering. *Advances in Neural Information Processing Systems.*
31. Lebret, R., Pinheiro, P. O., Collobert, R. (2014) *Simple image description generator via a linear phrase-based approach.* arXiv preprint arXiv:1412.8419
32. Kiran, A., et al. (2022). Reverse image search using deep unsupervised generative learning and deep convolutional neural network. *NATO Advance Science Institute Series E: Applied Science, 12*(10), 4943.
33. Poongodi, M., Hamdi, M., & Wang, H. *Image and audio caps: automated captioning of background sounds and images using deep learning.*
34. Xu, K., Ba, J., Kiros, R., Cho, K., Courville, A., Salakhudinov, R., Zemel, R., Bengio, Y. (2015) Show, attend and tell: Neural image caption generation with visual attention. In *International conference on machine learning*, (pp. 2048–2057). PMLR.
35. Mathur, P., Gill, A., Yadav, A., Mishra, A., & Bansode, N. K. (2017) *Camera2Caption: A real-time image caption generator*. In *Presented at the 2017 International Conference on Computational Intelligence in Data Science (ICCIDS).*
36. Merdivan, E. et al. (2019). Image-based text classification using 2D convolutional neural networks. In *Presented at the 2019 IEEE SmartWorld, Ubiquitous Intelligence & Computing, Advanced & Trusted Computing, Scalable Computing & Communications, Cloud & Big Data Computing, Internet of People and Smart City Innovation (SmartWorld/SCALCOM/UIC/ATC/CBDCom/IOP/SCI)*, Leicester, United Kingdom.
37. Panicker, M. J., et al. (2021). Image Caption Generator. *International Journal of Innovative Technology and Exploring Engineering, 10*(3), 87–92.
38. Hinz, T., Heinrich, S., & Wermter, S. (2020). Semantic object accuracy for generative text-to-image synthesis. *IEEE transactions on pattern analysis and machine intelligence., 44*(3), 1552–1565.
39. Singh, A., Anekar, S., Shenoy, R., & Patil, S. (2022). Text to image using deep learning. *International Journal of Engineering and Technical Research.*

Chapter 16
Exploring the Potential of Conversational AI for Supporting Children with Autism

Ishani Kathuria, Kamad Saxena, and Archana Singh

Abstract Autism or Autism Spectrum Disorder is the world's third most common developmental disability. In India, every 1 in 100 children below ten years of age has autism. With this significant rise in the disorder, better tools are required to diagnose it as well as treat it. The improvements in Artificial Intelligence make it an increasingly powerful tool in today's date. Conversational Artificial Intelligence is a subfield of Artificial Intelligence which essentially include virtual agents that users can talk to. Using the power of Conversational AI, we have proposed a virtual agent that can be used by children with Autism to learn English, develop their social skills and discover how to cope with anxiety. The proposed virtual chatbot is based on Natural Language Processing and other Deep Learning techniques. This application aims to improve and make treatment easier for children suffering from Autism.

Keywords Artificial intelligence · Natural language processing · Autism

16.1 Introduction

Autism Spectrum Disorder (ASD) is a developmental disorder in which people find it hard to interpret what others are feeling and is caused due to differences in the brain [1]. People diagnosed with ASD have a known difference, such as genetic conditions. There are still various other causes that are not yet identified. Due to this disorder, people get very anxious in social conditions, find it hard to explain their emotions and also find it hard to make friends [2]. Diagnosing ASD is defined into different levels that indicate the impact of autism on a person's ability to interact and

I. Kathuria (✉) · K. Saxena · A. Singh
Amity School of Engineering and Technology, Amity University, Noida, Uttar Pradesh, India
e-mail: ishani@kathuria.net

K. Saxena
e-mail: kamad.11.saxena@gmail.com

A. Singh
e-mail: asingh27@amity.edu

P. K. Kapur et al. (eds.), *Data-centric Approaches to Industrial Decisions*, Asset Analytics, https://doi.org/10.1007/978-981-96-7556-2_16

Fig. 16.1 The three levels of autism spectrum disorder (ASD)

communication and self-care. There are 3 levels—1, 2 and 3 of ASD as shown in Fig. 16.1.

Artificial Intelligence (AI) is understood as the ability to replicate human intelligence, but from a deeper perspective, it provides better functionality than a human brain and can perform some tasks better and more efficiently than a human brain. Its applications are evident in every sector and especially in the healthcare sector. With advancements presented by AI, diagnosing any sort of disorder is also made easier and achievable. Also, various new drug discovery, treatment methods and procedures are introduced which are taking the global healthcare sector to a completely new level [3]. The supercharged ability to diagnose diseases using AI has also been used to develop models to diagnose ASD [4].

Natural language processing can be understood as analyzing the human language and making the machine understand how humans speak and verbally interact with each other. It also adds up to the capabilities of Artificial Intelligence. One of the main applications of NLP is chatbots. Chatbots are trained AI models that can interact with humans in the human way possible [5]. Chatbots are now used in industries like customer care, help centers, education models, etc. In the field of healthcare, chatbots can take the place of doctors that can assist their customers at any given time and can handle multiple patient queries at the same time [6]. It also adds that personalized interaction experience to the people and does not make them feel like they are not interacting with a real doctor. Recently, chatbot's ability is also recognized in providing help to people dealing with mental disabilities or disorders. It can be used by such people to interact with someone always ready to listen to them, make them feel comfortable, respond to them without any judgment and caters to their specific needs and provide a personalized friendly experience.

16.2 Literature Review

There has been various research on developing different methods to help children and adults with autism learn how to socialize better, manage their emotions, and learn more about themselves. We have done a comprehensive review of the latest methods developed using Artificial Intelligence methods. In 2017, [7] developed a web application which uses Active Machine Learning (AML) to teach object recognition to children with ASD. The results show that AML was more effective than passive learning for 4 out of the 5 children they tested. An AI virtual character called Andy was designed and tested on children aged 4–14 years old who have Autism Spectrum Conditions (ASC) in a blended way with a human practitioner to engage and teach the children in [8].

Robotic assisted therapy with a personalized model was proposed in [9] to help children with ASD. They tested their therapy techniques on a dataset of video, audio, and physiological data of 35 children from Asia and Europe which achieved 60% agreement with human experts. Another robot assisted therapy method was developed with capabilities to extract facial expressions, gaze features, etc. thus helping in classification of therapist-specified behavioral classes [10].

Rouhi et al. [11] proposed an educational game using a Machine Learning (ML) system to help children with ASD learn how to express their emotions. The game, Emotify, is able to predict four emotions (happy, sad, angry, and neutral) from speech with an accuracy of 72% using a Random Forest Classifier. Begel et al. [12] designed a video calling application to help people with ASD understand what emotion their facial features will be interpreted as by neurotypical people so that they can learn how to socialize better. Lissa, a conversational bot was developed to have informal conversations with teenagers who have ASD and give real time feedback about their nonverbal cues [13]. Using conversational bots, specifically voice based conversational bots can help children with ASD explore their personal interests and promote self-care skills as shown in [14].

Emotion recognition through a Deep Learning architecture of Convolutional Neural Network was proposed in [15] to help ASD patients recognize the emotions represented by facial expression. Abdel Hameed et al. [16] showcases an AI-enabled Internet of Things system to monitor emotions of children with ASD through their heartrate using Support Vector Machine and Random Forest Classifiers.

All the reviewed literature showcases the various possibilities of supporting ASD patients using AI technologies. The reviewed research articles are summarized in Table 16.1.

Table 16.1 Literature review

Reference	Overview
Radwan AM et al. [7]	A web and touch-based application for teaching object recognition was created using Active Machine Learning (AML)
Porayska-Pomsta K et al. [8]	Proposed Andy, a virtual agent, to teach and engage with children aged 4–14 years old with ASD
Rudovic O et al. [9]	A personalized approach to robotic assisted therapy for children with ASD that achieved 60% agreement with human practitioners
Cai H et al. [10]	AI based robotic assisted therapy to analyze behavior of ASD patients
Rouhi A et al. [11]	A game developed with ML techniques to teach children with ASD how to correctly identify and express emotions
Begel A et al. [12]	A video calling application was proposed to help people with ASD identify how their facial features will be interpreted by neurotypical people
Ali MR et al. [13]	Virtual agent called Lissa to converse with teenagers diagnosed with ASD and give real time feedback about their behavior
Cha I et al. [14]	Engage children with ASD using a voice based conversational bot and help them explore their personal interests
Pavez R et al. [15]	Proposed a CNN framework to recognize emotions from facial features of user and give feedback about how the user's behavior will be interpreted as by a neurotypical person
Abdel Hameed M et al. [16]	Monitor emotions of children with ASD through their heart rate using an IOT system and inform guardian in case of emergency

16.3 Proposed Methodology

In this section, the proposed methodology of our chatbot is discussed including the emotion recognition models for speech and text emotion recognition. Our proposed model is a conversational virtual agent that will be able to teach socialization skills as well as basic English grammar to children diagnosed with autism spectrum disorder (ASD), aged between 5 and 10 years old. The agent will give feedback to the children in real time by conversing with them. As humans, during a conversation, we don't just focus on the words said by the other person but the tone behind those words also tells a lot about the emotional state of the person. Hence, our methodology takes text as well as speech in account while classifying the emotion.

The model will be trained using a large corpus of emotional speech and text data, as well as text data specific to language learning. For text emotion recognition, the sentences go through a pipeline in which NLP techniques like segmentation, tokenization, lemmatization, stop words removal etc. After going through the NLP pipeline, a transformer model will be trained passing the preprocessed sentences along with their emotion label as the input data to the model for training. On the other side, with speech emotion recognition, the model will be trained by extracting features from audio data in the form of Mel spectrograms. A Convolutional Neural Network will be used for the initial layers followed by transformer layers for better

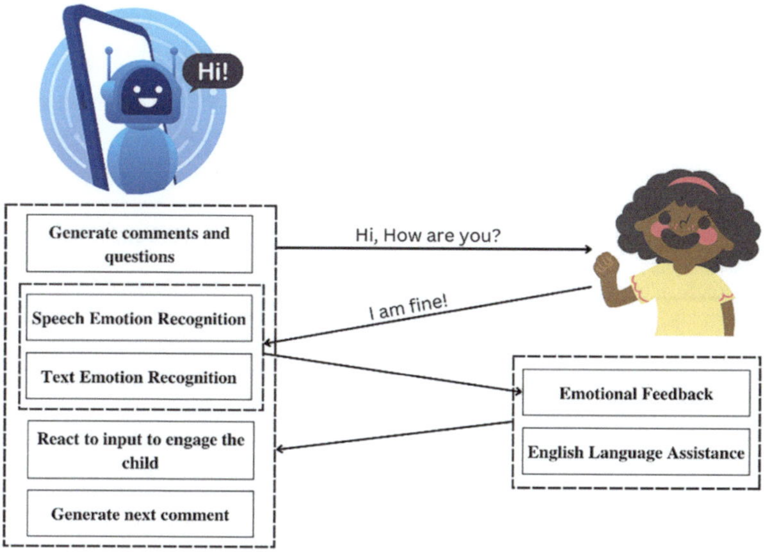

Fig. 16.2 Proposed methodology

feature extraction and classification. The AI virtual agent will then respond to the child's input by generating responses that are appropriate for their emotional state and language learning needs, with a focus on teaching social skills and English language skills.

Figure 16.2 demonstrates the working of the model in identifying the emotion of the autistic person as well as responding accordingly in order to continue the conversation and ultimately improving communication skills of the person.

16.4 Conclusion and Future Work

Autism Spectrum Disorder affects millions of people every year, a large part of which being children. Artificial Intelligence and Natural Language Processing technologies offer autistic people a one-of-a-kind opportunity to engage in safe, comfortable, and effective social interactions. As seen in previous literature, personalized and adaptive therapy is a significant advantage of virtual agents like chatbots and robotic therapies, and the advancement of these technologies is expected to improve outcomes for people with autism. While more research is needed to determine the long-term efficacy of these treatments, current evidence indicates that AI-powered interventions have the potential to significantly improve the lives of people with autism and their families. Keeping this in mind, we have proposed a Deep Learning framework for children with ASD to learn how to socialize, improve their language skills and be able to deal with excessive stress and anxiety in a safe way. The proposed model can be developed as a speech conversational virtual agent that will be able to have conversations and give real time feedback to ASD patients [17, 18].

References

1. Lord, C., Brugha, T. S., Charman, T., et al. (2020). Autism spectrum disorder. *Nature Reviews Disease Primers, 6*, 5. https://doi.org/10.1038/s41572-019-0138-4
2. Chandler, S., Howlin, P., Simonoff, E., et al. (2016). Comparison of parental estimate of developmental age with measured IQ in children with neurodevelopmental disorders. *Child Care Health Development, 42*, 486–493. https://doi.org/10.1111/cch.12346
3. Kathuria, I., Bhatia, M., Grover, A. (2022). Applications of deep learning in healthcare: A systematic analysis. In *Proceedings of International Conference on Recent Innovations in Computing*, 1st ed. Springer
4. Joudar, S. S., Albahri, A. S., & Hamid, R. A. (2022). Triage and priority-based healthcare diagnosis using artificial intelligence for autism spectrum disorder and gene contribution: A systematic review. *Computers in Biology and Medicine, 146*, Article 105553. https://doi.org/10.1016/j.compbiomed.2022.105553
5. Caldarini, G., Jaf, S., & McGarry, K. (2022). A literature survey of recent advances in chatbots. *Information, 13*, 41. https://doi.org/10.3390/info13010041
6. Kathuria, I., Saxena, K., Singh, A., Bhatia, M. (2023). SelfMed: Artificial intelligence enabled telemedicine chatbot. In *Recent Advances in Computational and Optimization Engineering*. Springer
7. Radwan, A. M., Birkan, B., Hania, F., & Cataltepe, Z. (2017). Active machine learning framework for teaching object recognition skills to children with autism. *International Journal of Development Disabilities, 63*, 158–169. https://doi.org/10.1080/20473869.2016.1190543
8. Porayska-Pomsta, K., Alcorn, A. M., Avramides, K., et al. (2018). Blending human and artificial intelligence to support autistic children's social communication skills. *ACM Transactions on Computer-Human Interaction, 25*, 1–35. https://doi.org/10.1145/3271484
9. Rudovic, O., Lee, J., Dai, M., et al. (2018). Personalized machine learning for robot perception of affect and engagement in autism therapy. *Science Robotics, 3*, eaao6760. https://doi.org/10.1126/scirobotics.aao6760
10. Cai, H., Fang, Y., Ju, Z., et al. (2019). Sensing-enhanced therapy system for assessing children with autism spectrum disorders: A feasibility study. *IEEE Sensors Journals, 19*, 1508–1518. https://doi.org/10.1109/JSEN.2018.2877662
11. Rouhi, A., Spitale, M., Catania, F., et al. (2019) Emotify: emotional game for children with autism spectrum disorder based-on machine learning. In *Proceedings of the 24th International Conference on Intelligent User Interfaces: Companion. Association for Computing Machinery*, (pp. 31–32). New York, NY, USA.
12. Begel, A., Tang, J., Andrist, S., et al. (2020). Lessons learned in designing AI for autistic adults. In *Proceedings of the 22nd International ACM SIGACCESS Conference on Computers and Accessibility. Association for Computing Machinery*, (pp. 1–6). New York, NY, USA
13. Ali, M. R., Razavi, S. Z., Langevin, R., et al. (2020). A virtual conversational agent for teens with autism spectrum disorder: experimental results and design lessons. In *Proceedings of the 20th ACM International Conference on Intelligent Virtual Agents*. ACM, Virtual Event Scotland UK, (pp. 1–8)
14. Cha, I., Kim, S.-I., Hong, H., et al. (2021). Exploring the use of a voice-based conversational agent to empower adolescents with autism spectrum disorder. In *Proceedings of the 2021 CHI Conference on Human Factors in Computing Systems*, (pp. 1–15). Association for Computing Machinery, New York, NY, USA.
15. Pavez, R., Diaz, J., Arango-Lopez, J., et al. (2021). Emo-mirror: A proposal to support emotion recognition in children with autism spectrum disorders. *Neural Computer Application*. https://doi.org/10.1007/s00521-021-06592-5

16. Abdel Hameed, M., Hassaballah, M., Hosney, M. E., & Alqahtani, A. (2022). An AI-enabled internet of things based autism care system for improving cognitive ability of children with autism spectrum disorders. *Computer Intelligence Neuroscience, 2022*, Article e2247675. https://doi.org/10.1155/2022/2247675
17. Gupta, S., Singh, A., & Ranjan, J. (2023). Multimodal, multiview and multitasking depression detection framework endorsed with auxiliary sentiment polarity and emotion detection. *International Journal of System Assurance Engineering Management, 14*, 337–352. https://doi.org/10.1007/s13198-023-01861-z
18. Gupta, S., Singh, A., & Kumar, V. (2023). Emoji, text, and sentiment polarity detection using natural language processing. *Information, 14*, 222. https://doi.org/10.3390/info14040222

Chapter 17
Indian Edtech Post COVID-19—Fertile Grounds for Investors

Mamta Chawla, Amit Verma, Supriya Lamba Sahdev, and Ali Hussein Zolait

Abstract COVID-19 resulted in acceptability and growth of e-learning worldwide. Due to pandemic-led restrictions, the scope of e-learning is widening worldwide. In India also there is an emergence of education technology "EdTech" sector, which is predicted to reach to US$30 billion industry by 2030. A number of players have reached the status of unicorn. Till June 2022, there are seven unicorns in the EdTech sector namely Byju's, Unacademy, Eruditus, Lead, Vedantu, Upgrad, and latest entry is PhysicsWallah. The New Education Policy, strengthen in 2023, offered structural reforms in the education sector by emphasizing the role of technology and flipped class. Thus, article provides an overview of EdTech sector in India—as a land of opportunities for the investors, details of major players along with the myriads of issues and challenges in the Indian education sector along with potential solutions that can be offered by different stakeholders in the Indian EdTech industry.

Keywords e-learning · EdTech industry · Growth · India · Unicorns

M. Chawla (✉)
AGBS, Noida, Amity University Uttar Pradesh, Noida, Uttar Pradesh, India
e-mail: mchawla@amity.edu

A. Verma
Kalicharan PG College, Lucknow, Uttar Pradesh, India
e-mail: amitvm2007@gmail.com

S. L. Sahdev
Office of International Affairs & Global, Partnerships and Associate Professor – Alliance School of Business, Alliance University, Anekal, India
e-mail: lamba.supriya9@gmail.com

A. H. Zolait
University of Bahrain, Zallaq, Kingdom of Bahrain

P. K. Kapur et al. (eds.), *Data-centric Approaches to Industrial Decisions*, Asset Analytics, https://doi.org/10.1007/978-981-96-7556-2_17

17.1 Introduction

The Indian EdTech industry emerged after COVID-19 strike India, which affected the whole country in the worst way, all the sectors were completely stuck and so was the case with the education industry [1, 2, 3]. This was the point when everyone started working and studying online; after a while all the students throughout the country started studying online. In this digital era knowledge imparting activity is not restricted to the traditional way of classrooms teaching. The encouragement by the governing agencies and the nations along with the evolving digital safety protocols have encouraged the institutions to utilize the new-age digital pedagogies for teaching learning. A consequence of these disruptions is welcomed by all the stakeholders involved in the knowledge imparting process; including—primary and secondary schools run by public, private, or jointly by both, the government, coaching institutions, teachers, learners, tutors, and academicians. Adoption of learning in the digital mode resulted in a boom in the form of technology-based education, and another segment emerged in the market as EdTech sector [4, 5].

In as early as 2021, Indian EdTech industry received funding of $16.1 billion by Venture Capitalists, which is 32 times surge from $ 500 million received in the year 2010 [6, 7, 41]. This surge is majorly by the segments called K-12, higher education, and upskilling categories. This is also combined with the intensifying popularity of courses offered in the name of distance education and courses launched by Massive Open Online Courses (MOOCs). In their education industry report, RBSA Advisors predicted that Indian EdTech industry will be of $30 billion by 2030. EdTech sector can cover the learning gap between technology's ability to eliminate geographical obstacles. Educators' adoption of digital mode is evident, even though they are also determined to retain a few incomparable paybacks of a traditional teaching modes such as one-on-one support from teachers, peer discussion, and generating opportunities to learn in collaboration e.g., group assignments and projects. According to RBSA Advisor report 2023 titled "Unleashing EdTech Potential in India," progress in the EdTech sector in India is supported by the rising requirements for education, skilling and coaching, which is adding up to the consolidation and growth of newer business models, encouraged by government to support quality education to the masses through digital mode. Latest provisions of digital education include immersive learning, on-demand learning, video-assisted remote learning, employing Artificial Intelligence and Virtual Reality, etc. falling under the broad term EdTech [8, 9].

However, accepting the latest trend means realization of skill gap by both the educators and the digital learners for the learning process to complete. Thus, making training to learn and practice different tools, techniques, and platforms for digital knowledge delivery. With the upsurge in the demand for digital education, further advancements on the technological front are predictive to be integrated with the teaching in classrooms; termed as Hybrid mode of education.

In comparison to classroom mode of education in school or college, digital education is economical in terms of time and money, making it comparatively affordable.

According to one of the msn.com (2023) industry reports education from first grade to highest twelfth grade enrollment has increased by almost 6.3 times in 2022 from the base year 2019. Learners from different economic status and different social classes have been able to access quality education via digital platforms due to better access, affordability, and being more flexible [10].

17.1.1 Indian Edtech Industry Outlook

Due to outbreak of COVID-19 gross domestic production (GDP) of India shrunk to 7.3 percentage during the year 2020–2021. The only exception as outlier was EdTech. The sector registered growth and increase of private equity, or technically termed as venture capital funding, from $ 325 million in the year 2019 to almost three times amounting to $1833 million in the year 2020. Following Fig. 17.1 clearly demonstrates an increase. According to a report on Indian EdTech sector by RBSA Advisors it is forecasted that it will be of $ 30 billion in one decade. BARC India in collaboration with Nielsen reported 30 percent hike in the EdTech related apps, in the amount of time spent during initial first three months of the lockdown in the country [11, 12].

On the top of all this background the supportive start-up ecosystem in India also contributed by way of generating remarkable employment opportunities during challenging times. Thus, the EdTech sector has developed in the form of a substantial contributors toward employment generation. According to 2022 survey organized by the Indian EdTech Consortium (IEC) which is founded under the aegis of IAMAI (Internet and Mobile Association of India) Indian EdTech sector has produced more than 75 thousand jobs during last five years [13, 14, 15].

Edtech start-ups are gaining popularity. Annually more than 15 million students appear in competitive exams in India which is driving a vast market for EdTech start-ups. Because of this boom training and coaching for the competitive examinations is getting affordable. Students now-a-days face growing competition to explore better openings in various arenas like medical and engineering (at K-12 stage), getting government- and public-sector employment e.g., Short-Service Commission (popular as SSC) and Bank PO (Probationary Officer), and clearing different Common Admission/Entrance Test (CAT) to get admission in different courses. Major start-ups funding destination is dominated by only a few (See Fig. 17.2).

EdTech start-ups run on lesser costs and even on low bandwidth, and these are further advances in technology adaption as compared to traditional offline coaching hubs. Due to these drivers, these start-ups are relatively popular in the Indian Tier II and III cities. But there is a lack of skilled trainers and teachers in these cities. Several brick-and-mortar training and coaching institutes are also offering mobile-apps along with their core services to increase their online footprint. The popularity of tech companies' apps such as WhatsApp, Telegram and Amazon India also provides opportunities in this segment which is more appealing to the EdTech start-ups. The major challenge is targeting audience and funding [15, 37].

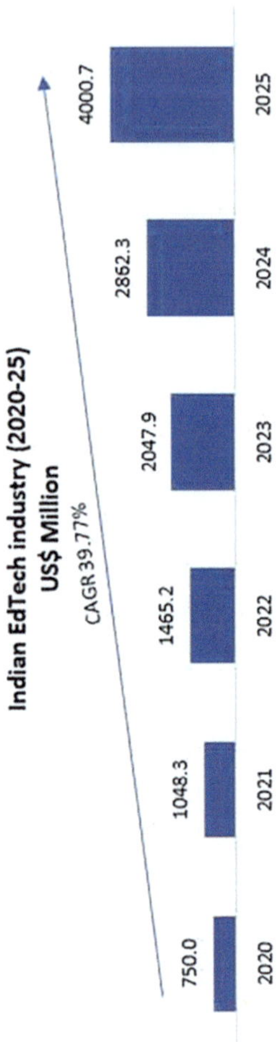

Fig. 17.1 Indian EdTech industry. (*Source* https://www.ibef.org/blogs/india-to-become-the-edtech-capital-of-the-world)

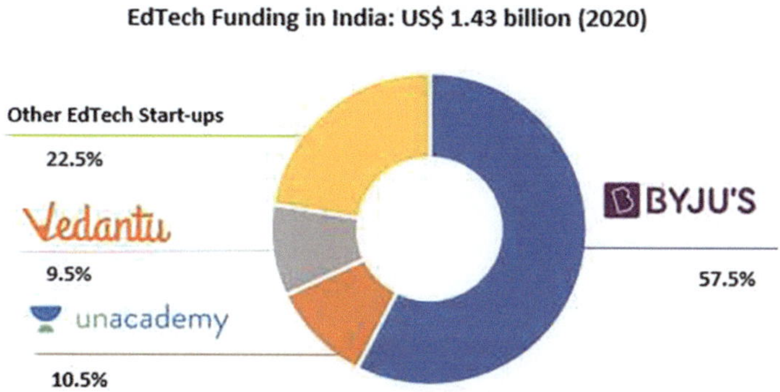

Fig. 17.2 Indian EdTech industry (Funding)

17.1.2 Major Players in the EdTech Sector

Fastest growing, as of 2022, Indian EdTech start-ups have been listed in the Fig. 17.3. These 40 players focused on their consumer and business side. Following statistics highlights the landscape of Indian EdTech market: EdTech offerings for classes first to twelfth are projected to increase 6.3 times by the end of 2023 and the post K-12 market is expected to grow 3.7 times to touch $1.8 billion.

17.2 Classification in EdTech

In this section, Indian EdTech sector categorization has been discussed. EdTech sector in India can be classified into: K-12 segment, Test-preparation segment, Skill development segment, Online certification segment, and Enterprise solution. Growth drivers comprise of huge growing target market, convenience of internet-based services, price advantage, along with pandemic-led disruptions and mandate, aided with drastically evolving information and communication technology capabilities along with improvements in the infrastructure [16, 17, 18].

17.2.1 K-12 Market

This is the biggest and most appealing segment among all. In 2022 total enrollment is more than 250 million in India. This segment of EdTech Sector can be further sub-segmented into smart class solutions, online preparation for exams, online tutoring, STEM learning incorporating simulation and virtual reality (VR), Artificial Intelligence and robotics, and online assessment. Major players in India are Byju's, Whitehat Jr, Cuemath, STEPapp, and CampK12.

1. PlayShifu
2. Flinto Learning
3. Skillmatics
4. Miko
5. Dost Education
6. Byju's
7. Vedantu
8. Practically
9. CountingWell
10. Doubtnut
11. Cuemath
12. ConveGenius
13. Edufiq
14. Testbook
15. Unacademy
16. Toppr
17. College Dunia
18. Leverage Edu
19. Blue Learn
20. Meet University
21. Kirger Campus
22. Lead School
23. Questt
24. Bit Class
25. No Paperforms
26. Teachmint
27. Classplus
28. Yellow Class
29. Front Row
30. Qshala
31. Hello English
32. Knudge me
33. Language curry
34. Upgrad
35. Edutius Executive Edu
36. Alma Better
37. My Captain
38. Pesto Tech
39. Avanse Financials
40. Eduvans

Fig. 17.3 Top Edtech start-ups in India. (*Source* Researchers compiled)

17.2.2 Entrance Exam Preparation Segment

The second most rising segment, in the past few years is for coaching or test preparation for competitive exams. A substantial number of users count on platforms like CAT, GRE, IAS, JEE, NEET, and GMAT. Main motivation for the growth of this cost effectiveness and convenience. Institutions, like Times group-T.I.M.E, Career Launcher, major business schools like IIM- Kozhikode and IIM- Ahmedabad to local famous business schools have also started hybrid courses which are a combination of online as well as offline classes for preparation. The major competitors in India include—Byju's, Embibe and Gradeup, Toppr, Unacademy, and Vedantu.

17.2.3 Online Certification Segment

The key growth driver for online professional education is its stress on improved employability and career growth. It includes various courses of mobile app development, usage of Big Data, Analytics, digital marketing, computing, and project management. These are mostly undertaken by the working professionals for upskilling and reskilling. The most active key players in India includes: Eruditus, Coursera, LinkedIn, Jigsaw Academy, upGrad, Simplilearn, etc [19, 20].

17.2.4 Skill Development Segment

Edtech start-ups in India are also enabling skill development effectively, upskilling, and even reskilling workers and entry-level employees having technical roles. Additionally, its target is to upskill and reskill science and engineers, software developers, to accelerate adaption of most recent technologies which are most used. Prominent participants are—Disprz, Edureka, InterviewBit, Crio, Quizizz, and Do.

17.2.5 Enterprise Solution Segment

This segment in the Indian EdTech sector has witnessed maximum growth due to the background of various companies' efforts to integrate with such solutions that they can stay sustainable and agile in the prevailing dynamic digital world. In this Indian EdTech market the competitors are Byndr, Classplus, Liqvid, and Foradian (Table 17.1).

Table 17.1 Indian EdTech market size (in US$ Billion)

Segment	2020	2025 (Projection)
K-12	US$1.16 billion	US$4.3 billion
Test preparation	US$0.8 billion	US$3.99 billion
Online certification	US$0.58 billion	US$1.38 billion
Skill development	US$0.26 billion	US$0.73 billion
Total	US$2.8 billion	US$10.4 billion

Source https://www.india-briefing.com/news/profiling-indian-edtech-industry-us-10-billion-dollar-opportunity-24013.html/

17.3 Growth Drivers, Challenges, and Future

The Indian Edtech industry has experienced tremendous growth in recent years, with a compound annual growth rate of around 44% since 2016. The industry has been fueled by a combination of factors, including a rapidly growing population, rising disposable incomes, and the government support to promote digital education. According to a report by RedSeer Consulting, the Indian Edtech market is expected to reach $10.4 billion by 2025, up from $2.8 billion in 2020.

17.3.1 EdTech Sector Growth Drivers

One of the main drivers of the industry growth has been the COVID-19 pandemic, which has led to a surge in demand for online education. The pandemic has disrupted traditional classroom-based learning and forced students and teachers to adapt to online learning. As a result, Edtech companies have seen a surge in demand for their products and services.

Another key factor driving the growth of the Edtech industry in India is the increasing availability of affordable smartphones and high-speed internet. The proliferation of mobile devices has made it easier for students to access online courses and study materials, while high-speed internet has made it possible to stream videos and attend live classes without any interruptions [21, 22, 23].

The Indian government has also played a significant role in promoting digital education in the country. In 2018, the government launched the Digital India campaign to promote the adoption of digital technologies across the country. As part of this initiative, the government has launched several programs to promote digital education and provide online learning resources to students.

The Edtech industry in India is dominated by a few major players, including Byju's, Unacademy, and Vedantu. Byju's is currently the most valuable Edtech company in the world, with a valuation of $16.5 billion. The company offers a wide range of online courses and study materials for students from kindergarten to 12th grade and has more than 100 million registered users.

Unacademy is another major player in the Indian Edtech industry, offering online courses for students preparing for competitive exams such as the Joint Entrance Exam (JEE) and the Union Public Service Commission (UPSC) exam. The company has raised over $1 billion in funding and has a valuation of around $4.5 billion. Vedantu is a live online tutoring platform that offers one-on-one and small group tutoring sessions for students. The company has raised over $200 million in funding and has more than one million registered students.

17.3.2 Challenges for EdTech Sector

Indian Edtech industry faces number of challenges-

17.3.2.1 Internet Connectivity Poses a Significant Challenge for Indian EdTech, Impacting the Efficacy of Digital Learning

In rural and remote areas in India, internet infrastructure is either non-existent or insufficient, resulting into no to limited access to online educational resources [4]. Even in urban areas, inconsistent internet speed, fluctuation and high costs obstruct uninterrupted learning experiences [2]. These connectivity concerns generate a digital divide, exacerbating educational inequities by handicapping learners without reliable internet access [25]. Students are unable to attend live classes, submit assignments on time, or access supplementary learning materials, leading to gaps in knowledge and skill acquisition [7]. Teachers also struggle to deliver interactive and engaging lessons, reducing the overall quality of education [13]. Consequently, the potential of EdTech to democratize education remains unrealized, perpetuating existing disparities in educational outcomes.

To address these issues, it is imperative for policymakers and stakeholders to invest in improving internet infrastructure, particularly in underserved regions, and make internet access more affordable and reliable [1, 24].

17.3.2.2 Devices, Infrastructure and Required Technology Are Critical yet Hindering Factors in the Growth of Indian EdTech

The lack of robust infrastructure, including adequate internet connectivity and sufficient technological resources, impedes the adoption and efficacy of digital education platforms [4]. Many schools, particularly in rural areas, lack basic facilities such as reliable electricity and internet access, making it difficult to implement and sustain EdTech solutions [11]. Additionally, the high cost of advanced technological tools and devices is prohibitive for a significant portion of the population, limiting access to digital learning. These infrastructural and technological barriers have significant impacts. Students in some areas and regions are completely deprived of the online quality education and resources and unable to take the benefits of interactive and engaging digital learning environments, exacerbating educational inequities [8]. On the other side due to inadequate training and resources educators also face difficulties in delivering effective online lessons, further compromising improvements in educational quality [13]. Consequently, the potential of EdTech to democratize education and attainment of learning outcomes remains largely untapped.

To overcome these challenges, it is essential for the government to invest in improving infrastructure, provide affordable technological tools, and ensure widespread internet access. Government and private sector collaboration can play a pivotal role in addressing these issues [1].

17.3.2.3 Internet Language Barriers Significantly Obstruct the Growth of Indian EdTech, as India is a Linguistically Diverse Country with Over 1600 Languages Spoken

Most EdTech platforms primarily use English, which limits accessibility for non-English speaking students and teachers. This language disparity means that a large portion of the population, especially in rural areas, cannot fully benefit from digital educational resources [24]. The impact of these language barriers is intense. Students who are not proficient in English face difficulties understanding the content, leading to disengagement and lower academic performance [27]. This issue also applies to the educators, who may tussle to deliver engage and deliver content effectively in a language that is not their own [13]. Consequently, the educational divide between urban and rural areas, as well as between different linguistic groups, widens.

To address these challenges, it is crucial for EdTech platforms to offer multilingual support and content tailored to regional languages. This inclusivity can enhance learning outcomes and make digital education more accessible to a broader audience [1].

17.3.2.4 Teaching Pedagogy and Learning Outcomes Development Present Another Substantial Challenge for Indian EdTech

Deeply rooted traditional teaching methodology in the Indian education system, often do not translate well into digital formats, leading to ineffective instructional delivery [18]. The lack of adapted pedagogical approaches as per the requirement of online learning means that many EdTech platforms fail to engage students effectively to meet the diverse learning requirements [13]. Furthermore, there is a lack of trained educators who can appropriately include digital tools to enhance learning experiences [11]. The impact of these challenges is significant. Students often experience a disjointed learning process, where the digital content does not align well with their cognitive and emotional needs, leading to lower engagement and poor retention of information [26]. This results in suboptimal learning outcomes, with students failing to achieve the desired educational objectives [16]. Moreover, the lack of effective assessment tools and personalized feedback mechanisms in many EdTech solutions further exacerbates this issue, as students do not receive the necessary support to address their learning gaps [1].

To address these challenges, EdTech providers must develop innovative pedagogical models tailored to digital platforms and invest in training educators to use these technologies effectively [18].

17.3.2.5 Affordability and Accessibility Are Significant Challenges for Indian EdTech

The high cost of digital devices, internet connectivity, and subscription fees for educational platforms limits access for a large segment of the population, especially in rural and economically disadvantaged areas. Many families cannot afford the necessary technology, resulting in a digital divide that prevents equal access to educational opportunities [11].

These barriers have substantial impacts. Students from low-income households are disproportionately affected, leading to unequal learning opportunities and widening the educational gap between different socio-economic groups [23]. This lack of access hampers their academic progress and future prospects, perpetuating the cycle of poverty. Additionally, teachers in underfunded schools face challenges in integrating EdTech into their curriculum, further diminishing the quality of education [13].

To address these challenges, it is crucial to implement policies that subsidize digital devices and internet services for low-income families. Public–private partnerships can also play a role in making EdTech more affordable and accessible [1]. Ensuring that EdTech solutions are inclusive and equitable is also essential for bridging the digital divide and promoting educational equity in India [26, 27, 28].

17.3.2.6 Teacher Training and Capacity Building Pose Significant Challenges for Indian EdTech Due to Several Factors

Firstly, the existing educational infrastructure in many regions is underdeveloped, limiting access to quality training programs. Secondly, there is a shortage of qualified trainers proficient in integrating technology with pedagogy, making it difficult to upskill teachers effectively. Moreover, the diverse linguistic and cultural landscape of India requires tailored training programs, which are often lacking.

These challenges impact the effectiveness of EdTech solutions. Teachers who are not adequately trained struggle to utilize digital tools to their full potential, leading to suboptimal student engagement and learning outcomes. Additionally, the digital divide exacerbates disparities, as students in rural or underprivileged areas may not benefit equally from EdTech advancements due to their insufficient trained educators. This undermines the goal of equitable education through technology.

Ultimately, addressing these issues requires concerted efforts to develop comprehensive and localized training programs, improve infrastructure, and ensure continuous professional development for teachers [29, 30].

17.3.2.7 Student Engagement and Retention Present Major Challenges for Indian EdTech Due to Several Factors

Firstly, the lack of personalized learning experiences can make it difficult to maintain student interest. Many EdTech platforms offer a one-size-fits-all approach, which fails to cater to the diverse learning needs and preferences of students. Secondly, inadequate access to reliable internet and digital devices, particularly in rural and economically disadvantaged areas, hampers continuous and effective engagement with EdTech resources. Additionally, cultural and language barriers can also play a role, as many platforms may not offer content in regional languages or in a culturally relevant manner. The impact of these challenges is significant. Low engagement levels can lead to poor academic performance and reduced motivation among students. Consequently, this can result in higher dropout rates, negating the potential benefits. Moreover, inconsistent engagement and retention rates can undermine the scalability and sustainability of EdTech initiatives, limiting their long-term effectiveness in improving educational outcomes across India [31, 32, 33].

17.3.2.8 Content Quality and Credibility Are Significant Concerns for Indian EdTech Due to Several Reasons

Firstly, the rapid proliferation of EdTech platforms has led to a wide variation in the quality of educational content. Many platforms may prioritize quantity over quality to attract more users, leading to content that is poorly researched, outdated, or not aligned with the curriculum. Secondly, the lack of stringent regulatory frameworks and quality control mechanisms allows the dissemination of substandard or incorrect information. This can confuse students and undermine their learning process.

The impact of these issues is profound. Low-quality content can hamper students' academic progress by providing them with inaccurate or incomplete knowledge. This is especially detrimental in subjects that require precise understanding, such as science and mathematics. Furthermore, the credibility of EdTech platforms suffers when students and parents encounter unreliable content, leading to a loss of trust and decreased adoption rates. In the long run, this can hinder the overall growth and effectiveness of the EdTech sector in India, which aims to democratize access to quality education.

To address these concerns, it is crucial for EdTech companies to invest in rigorous content creation and review processes, collaborate with educational experts, and establish transparent quality assurance standards [35].

17.3.2.9 Data Privacy and Security Are Major Concerns for Indian EdTech Due to the Sensitive Nature of the Data Collected

EdTech platforms often gather extensive personal information from students, including their academic records, personal details, and usage data. The lack of robust data protection laws and regulations in India exacerbates these concerns, making it easier for unauthorized access and misuse of data to occur. Additionally, many EdTech companies may not have the necessary infrastructure to secure this data adequately, increasing the risk of data breaches and cyberattacks.

The impact of these issues is significant. Compromised data privacy can lead to identity theft, financial fraud, and other forms of cybercrime, putting students and their families at risk. Moreover, the lack of trust in data security can deter users from adopting EdTech solutions, thereby limiting the potential benefits of digital education. Ensuring data privacy and security is crucial for building trust and fostering a safe digital learning environment.

To mitigate these risks, EdTech companies must implement stringent data protection measures, comply with international data security standards, and advocate for stronger regulatory frameworks.

17.3.2.10 Competition and Market Saturation Are Significant Concerns for Indian EdTech Due to the Rapid Influx of Numerous Platforms Vying for Market Share

This intense competition leads to a crowded marketplace, making it challenging for individual companies to distinguish. Many platforms resort to aggressive marketing and pricing strategies, which can result in unsustainable business models and a focus on short-term gains over long-term educational impact.

The impact of these challenges is multifaceted. For consumers, the overwhelming number of options can lead to confusion and difficulty in identifying high-quality, reliable EdTech solutions. This can result in inconsistent educational experiences for students. For EdTech companies, high competition can reduce profitability, limit growth, and stifle innovation as resources are diverted to competitive tactics rather than product improvement. Additionally, market saturation can lead to the collapse of weaker platforms, disrupting the learning process for their users.

To navigate these issues, EdTech companies must focus on delivering superior educational value, fostering innovation, and building strong, sustainable business models [36].

17.3.2.11 Government Regulations Pose Significant Concerns for Indian EdTech Due to the Evolving and Often Stringent Policies Governing the Sector

The Indian government has been increasingly active in regulating online education, aiming to ensure quality and protect student interests. However, the regulatory framework is still developing, leading to uncertainties and compliance challenges for EdTech companies.

Firstly, frequent changes in regulations can create an unstable business environment. EdTech companies may struggle to adapt to new rules quickly, leading to operational disruptions and increased compliance costs. Secondly, stringent regulations on data privacy, content quality, and teacher qualifications can be difficult for smaller EdTech firms to meet, potentially stifling innovation and growth. Additionally, regulatory requirements such as mandatory accreditation and adherence to specific curricula may limit the flexibility and diversity of educational offerings.

The impact of these regulatory challenges is significant. For EdTech companies, the cost of compliance can reduce profitability and divert resources from product development and innovation. Moreover, regulatory uncertainties can deter investment in the sector, slowing its overall growth. While regulations aim to protect students interests, stringent or poorly implemented policies can reduce access to diverse and innovative educational resources. To address these concerns, a balanced regulatory approach that ensures quality and safety without stifling innovation is essential [37, 38, 39, 40, 41].

17.4 Conclusion

In conclusion, the Indian Edtech industry has experienced tremendous growth in recent years, driven by a combination of factors such as the COVID-19 pandemic, increasing availability of affordable smartphones and high-speed internet, and the government push to promote digital education. Majorly growth has been further propelled by post COVID-19 disruptions and fast digitalization. As of June 2022, there are seven Edtech unicorns in the country and their future valuation is secured at US$30 billion. A hundred percent FDI in the education sector is permissible under the automatic route. An exponential upsurge in the number of paid Edtech users, is predicted to reach 37 million by 2025 [3]. Though the sector faces several challenges and there is surge in variation of products and services, as the competition is intensifying, pressurizing government to improve, standardize and safeguard interests of various stakeholders. With this background the sector is expected to grow at a rapid pace in the coming years and is likely to play a key role in shaping the future of education in India and can turn India into a knowledge point in the world map. Thus, Indian Edtech industry presents fertile ground for Indian as well as foreign investors.

Appendix 1

Ed tech challenges faced by the companies in India.

S.no	Challenge	Cause	Effect
1	Internet Connectivity	Limited infrastructure in rural areas	Hinders access to online learning platforms
2	Infrastructure and Technology	Lack of access to computers/smartphones	Impedes effective participation in digital education
3	Language Barriers	Limited content in regional languages	Reduces reach to a broader audience
4	Pedagogy and Learning Outcomes	Difficulty in designing effective pedagogy	Challenges in delivering measurable learning outcomes
5	Affordability and Accessibility	Financial constraints for digital devices/internet	Hinders access for certain socio-economic groups
6	Teacher Training and Capacity Building	Lack of adequately trained teachers	Limits effective integration of technology in education
7	Student Engagement and Retention	Struggles to maintain student interest and motivation	High dropout rates in online learning environments
8	Content Quality and Credibility	Difficulty in ensuring credible educational materials	Affects the overall learning experience
9	Data Privacy and Security	Handling sensitive student data and privacy	Breach of data privacy and potential legal issues
10	Competition and Market Saturation	Rapid growth leading to increased competition	Challenges in differentiating offerings and sustaining growth
11	Government Regulations	Compliance with education and data privacy regulations	Impacts business operations and expansion plans

Appendix 2

Edtech solutions to the above stated challenges with reference to the given industry conditions.

S.no	Challenge	EdTech Solutions
1	Internet Connectivity	1. Offline Access: Provide downloadable content that students can access offline, reducing dependency on constant internet connectivity

(continued)

(continued)

S.no	Challenge	EdTech Solutions
		2. Lite Versions: Develop lightweight versions of apps/platforms that consume less data and are accessible in areas with limited bandwidth
		3. Mobile Learning: Focus on mobile-friendly content delivery, as smartphones are more prevalent than computers in remote areas
2	Infrastructure and Technology	1. Collaborative Learning Centers: Establish learning centers equipped with computers and internet in underserved areas
		2. Device Support: Provide low-cost devices or technology bundles in partnership with hardware manufacturers to increase access to digital learning tools
3	Language Barriers	1. Multilingual Content: Develop educational content in various regional languages to cater to a diverse audience
		2. Translation Tools: Implement translation features to convert content from one language to another in real-time
		3. Localized Curriculum: Customize the curriculum based on regional needs and preferences
4	Pedagogy and Learning Outcomes	1. Interactive Content: Create engaging and interactive content that promotes active learning and better comprehension
		2. Adaptive Learning: Utilize AI algorithms to personalize learning paths for each student, catering to their individual strengths and weaknesses
		3. Regular Assessments: Incorporate regular quizzes and assessments to monitor progress and adjust teaching methods accordingly
5	Affordability and Accessibility	1. Freemium Models: Offer free basic access with premium features available at affordable prices, catering to students with different financial capabilities
		2. Scholarships and Discounts: Provide scholarships or discounts to underprivileged students or those with financial constraints
6	Teacher Training and Capacity Building	1. Professional Development: Offer online courses and training programs for teachers to enhance their digital teaching skills

(continued)

(continued)

S.no	Challenge	EdTech Solutions
		2. Virtual Workshops: Conduct virtual workshops and webinars on effective integration of technology in education
7	Student Engagement and Retention	1. Gamification: Integrate gamified elements into the learning process to increase student engagement and motivation
		2. Peer Interaction: Facilitate online discussion forums and group activities to foster collaboration and peer learning
8	Content Quality and Credibility	1. Curated Content: Partner with reputable educational content providers and subject matter experts to ensure the quality and credibility of the content
		2. User Reviews and Ratings: Implement user review systems to help learners assess the quality of the content
9	Data Privacy and Security	1. Robust Data Protection: Implement stringent data protection measures and adhere to privacy regulations to safeguard student data
		2. Transparent Policies: Clearly communicate data usage policies and seek explicit consent from users for data collection and processing
10	Competition and Market Saturation	1. Niche Offerings: Identify and cater to specific niches or underserved markets to differentiate from competitors
		2. Continuous Innovation: Regularly update and improve the platform with new features and functionalities to stay ahead in the market
11	Government Regulations	1. Compliance Team: Form a dedicated team to ensure adherence to education and data privacy regulations, keeping the company up-to-date with any changes in the regulatory landscape
		2. Engaging with Authorities: Establish open communication channels with relevant government authorities to address compliance-related queries and concerns proactively

Appendix 3

Ed tech challenges faced by the companies in India.

Player	Founders (origin)	Year of formation	Market valuation	No. of registered user	Annual paid subscribers (%)	Renewal rate (%)	Market catering
Byju's	Raveendran (Bangalore, India)	2011	$5.4 billion	100 million	6.5%	86%	Coaching for competitive entrance exams like IIT-JEE, CAT, UPSC, GMAT, GRE, Engineering & Medical, and courses of grades 6th to 12th
Unacademy	Gaurav Munjal, Dr. Roman Saini, and Hemesh Singh (Bengaluru India)	2015	$ 2000 million	40 million	2%	–	Coaching for compettetice exams like UPSC, SSC, IIT-JEE, CAT, And cources of grade 6 to 12
coursera	Andrew Ng Daphne Koller	2012	$ 4 Billion	92 Million	3.47%		Coaching and proper courses in collaboration with some of the best universities in the world Coaching is available for all grades of student
Vedantu	Vamsi Krishna, Saurabh Saxena, Pulkit Jain, Anand Prakash (Bengaluru India)	2014	$ 600 Million	35 million	250,000	–	K 1–K 12

(continued)

(continued)

Player	Founders (origin)	Year of formation	Market valuation	No. of registered user	Annual paid subscribers (%)	Renewal rate (%)	Market catering
Coding Labs	Yong Ning Foo (Singapore)	$ 29,738 m illion	–	–	–	–	Software testing courses. Learn Quality Assurance online. ... Learn coding from scratch. Online IT courses. ... Software testing courses. ... Learn coding from scratch
Toppr	Zeeshan hayath, Hemanth Goteti, (Maharashtra)	84.3 Crore +	30 million	49.5%	100% growth		Entrance exams in engineering, medicine, law and chartered accountancy

References

Journals and conferences

1. Agarwal, S., & Aggarwal, R. (2021). Bridging the digital divide: Effective pedagogies for online learning. *Journal of Education and Technology, 15*(3), 45–56.
2. Arumugam, U., & Manida, M. (2020). Covid-19: Transformation of Edtech in India from real to virtual: An analysis of issues and challenges by University Students in Tamilnadu. *ComFin Research, 10*(1), 80–91.
3. Augustine, R., & Reji, M. S. K. (2021). *Online teaching and learning during COVID-19: A study among selected university students* (pp. 22–24). Culture and Society.
4. Bhatia, R. (2021). Internet accessibility in India's Education Sector. *International Journal of Educational Research, 8*(2), 102–115.
5. Choudhary, R. (2020). Challenges in implementing educational technology in India. *Journal of Educational Technology & Society, 23*(4), 111–121.
6. Ghosh, A., & Arora, N. (2021). The challenges of data privacy and security in Indian EdTech. *Journal of Cyber Security and Mobility, 10*(2), 213–232.
7. Jha, M. (2021). Challenges of online learning in rural India. *Asian Journal of Distance Education, 19*(1), 77–89.

8. Joshi, A. (2021). Technological challenges in Indian EdTech. *Journal of Digital Learning, 10*(1), 65–78.
9. Khamis, T., Naseem, A., Khamis, A., & Petrucka, P. (2021). The COVID-19 pandemic: A catalyst for creativity and collaboration for online learning and work-based higher education systems and processes. *Journal of Work-applied Management, 13*(2), 184–196. https://doi.org/10.1108/jwam-01-2021-0010
10. Kim, J. H., Hong, J. A., & Song, D. H. (2019). The roles of academic engagement and digital readiness in students' achievements in university E-learning environment. *International Journal of Educational Technology in Higher Education.* https://doi.org/10.1186/s41239-019-0152-3
11. Kumar, V., & Sharma, D. (2022). The competitive landscape of Indian EdTech: Challenges and strategies. *International Journal of Educational Technology in Higher Education, 19*(1), 45–59.
12. Kurt, S. D. (2021). From virtual to participatory learning with technology during Covid-19. *E-Learning and Digital Media: Sage Journal.* https://doi.org/10.1177/20427530211022926
13. Mishra, L., Gupta, T., & Shree, A. (2020). Online teaching-learning in higher education during lockdown period of COVID-19 pandemic. *International Journal of Educational Research.* https://doi.org/10.1016/j.ijedro.2020.100012
14. Ngoasong, M. Z. (2021). Curriculum adaptation for blended learning in resource-scarce contexts. *Journal of Management Education, 46*(4), 622–655. https://doi.org/10.1177/10525629211047168
15. Palvia, S., Aeron, P., Gupta, P., Mahapatra, D., Parida, R., Rosner, R., & Sindhi, S. (2018). Online education: Worldwide status, challenges, trends and implications. *Journal of Global Information Technology Management, 21*, 233–241. https://doi.org/10.1080/1097198X.2018.1542262
16. Patel, R. (2020). Assessing learning outcomes in digital education. *Journal of Educational Assessment, 22*(3), 112–123.
17. Pusca, D., & Northwood, D. O. (2021). Teaching and learning engineering design: Creative methods for remote education. *World Transactions on Engineering and Technology Education, 19*(3), 306–312.
18. Rao, K., & Srinivasan, V. (2020). Adapting traditional pedagogy for digital platforms. *International Journal of Educational Technology, 8*(4), 77–89.
19. Rizvi, Y. S., & Nabi, A. (2021). Transformation of learning from real to virtual: an exploratory-descriptive analysis of issues and challenges. *Journal of Research in Innovative Teaching & Learning, 14*(1), 5–17. https://doi.org/10.1108/jrit-10-2020-0052
20. Sahu, A., & Samantaray, S. (2022). Digitalization of education: Rural India's potential to adapt to the digital transformation as new normality. In *Biologically Inspired Techniques in Many Criteria Decision Making: Proceedings of BITMDM 2021* (pp. 377–388). Singapore: Springer Nature Singapore.
21. Sathiyamoorthy, R., Ramesh, V. L., Chettri, M., Ziephru, S., & Kamath, R. (2023). Enablers and Barriers to online education in India: A systematic review. *International Journal of Educational Reform.* https://doi.org/10.1177/10567879231202485
22. Sharma, A. (2021). Student engagement and retention in Indian EdTech: Challenges and opportunities. *International Journal of Educational Technology in Higher Education, 18*(1), 98–112.
23. Sharma, S., & Gupta, R. (2021). The impact of economic disparities on digital learning. *Educational Technology & Society, 25*(1), 88–99.
24. Sharma, S., & Gupta, R. (2022). The impact of language on digital learning in Rural India. *Educational Technology & Society, 25*(1), 88–99.
25. Sharma, S., & Kumar, A. (2022). Digital divide in Indian education: Causes and solutions. *Educational Technology & Society, 25*(1), 112–125.
26. Singh, R., & Kaur, N. (2020). Student engagement in online learning environments. *Journal of Digital Education, 13*(2), 203–219.

27. Singh, R., & Kumar, A. (2020). The role of language in educational accessibility. *Journal of Education Policy, 22*(3), 104–119.
28. Singh, R., & Srivastava, S. (2020). The impact of internet connectivity on urban education in India. *Urban Education Review, 13*(2), 203–219.
29. Zhao, Y. (2019). The Future of Education: 2030. In S. Yu, H. Niemi, J. Mason (Eds.), *Shaping future schools with digital technology: An international handbook*, (pp. 1–24). Springer. https://doi.org/10.1007/978-981-13-9439-3

News article or online articles

30. Education in India: Higher Education Trends & Index | IBEF. (n.d.). India Brand Equity Foundation. Available at: https://www.ibef.org/industry/education-sector-india#:~:text=The%20Indian%20edtech%20market%20size,E%2Dlearning%20after%20the%20US.
31. Inc42 Media. (2020). The Future of India's $2 bn Edtech Opportunity Report 2020. https://aspirecircle.org/wp-content/uploads/2021/04/Inc42_Edtech_Report_updated.pdf
32. India Today. (2022). EdTech: The new growth catalyst of Indian education industry. Available at: https://www.indiatoday.in/education-today/featurephilia/story/edtech-the-new-growth-catalyst-of-indian-education-industry-1912116-2022-02-12
33. Naina Bhardwaj. (2022). Profiling India's EdTech Industry: A US$10 Billion Opportunity. India-Briefing. Available at: https://www.india-briefing.com/news/profiling-indian-edtech-industry-us-10-billion-dollar-opportunity-24013.html/
34. Online, F. (2022). India's EdTech Sector generates 75,000 jobs in the last 5 years. Financial Express. https://www.financialexpress.com/jobs-career/education-indias-edtech-sector-generates-75000-jobs-in-the-last-5-years-2428543/
35. Ramaswamy, N. (2022). *The rise of digital in Indian education sector - KPMG India.* Available at: https://home.kpmg/in/en/blogs/home/posts/2022/01/education-skill-nep-digital-mantra.html
36. Rest of World. (2022). *India's online education startups are struggling to survive in a post-pandemic world.* Available at: https://restofworld.org/2022/indias-online-education-startups-struggling-post-pandemic/
37. Sanghvi, S., & Westhoff, M. (2022). *Five trends to watch in the edtech industry.* McKinsey & Company. https://www.mckinsey.com/industries/education/our-insights/five-trends-to-watch-in-the-edtech-industry
38. Sarkar Gargi. (2022). *How Indian Edtech Startups Can Bounce Back After Slowdown Bump.* Available at: https://inc42.com/features/how-indian-edtech-startups-can-bounce-back-after-slowdown-bump/
39. The Economics Times. (2022). *Higher education, lifelong learning to drive edtech market growth.* Available at: https://economictimes.indiatimes.com/tech/technology/higher-education-lifelong-learning-to-drive-edtech-market-growth-says-redseer-report/articleshow/90475148.cms?utm_source=contentofinterest&utm_medium=text&utm_campaign=cppst
40. thestartuplab.in. (2021). *Future of Edtech India: Factors driving the growth of Edtech Industry in India.* Available at: https://thestartuplab.in/future-of-edtech-india-factors-driving-the-growth-of-edtech-industry-in-india/
41. Yelenevych, A. (2022). *The future of EdTech. Forbes.* Available at: https://www.forbes.com/sites/forbesbusinesscouncil/2022/12/26/the-future-of-edtech/

Chapter 18
Case Study of Brand Resilience with Special Reference to S7 Crisis of Samsung Electronics

Shweta Manchanda and Mamta Mohan

Abstract This paper discusses resilience strategies required to be demonstrated by the brands after a product harm crisis, taking into consideration Samsung, one of the leading brands, as a case study. Product harm crisis often bring an organization into a conflicting position. Not only does the crisis affect the reputation of the brand negatively, but also the media focus on the same allows the consumers to judge the brand in a manner which is not the usual course. Samsung had to face a similar situation when S7 launched by the company turned out to be failure due to multiple instances of exploding of the handset. At such a point of time, the brand not only requires to manage its reputation which is at stake at the time of product failure but is also required to demonstrate resilience and act agile. Consumer beliefs toward a brand also get distorted in the case of product failure that affects their future intention to purchase from the brand along with the WOM (Word of Mouth) which act as referrals. Therefore, it impacts the way how the future consumers would like to interact with the company. Limited research has been done on the factors that affect the brand resilience and the recovery steps taken by the brand to compensate for the loss of image caused due to product harm crisis. This research paper therefore tries to examine the strategies adopted by Samsung as a brand after the crises to regain consumer trust and act as a resilient brand.

Keywords Brand resilience · Product harm crises · Consumer future engagements · Samsung

S. Manchanda (✉) · M. Mohan
Amity International Business School, Noida, India
e-mail: shweta.manchanda53@gmail.com

M. Mohan
e-mail: mmohan@amity.edu

P. K. Kapur et al. (eds.), *Data-centric Approaches to Industrial Decisions*, Asset Analytics, https://doi.org/10.1007/978-981-96-7556-2_18

18.1 Introduction

In the highly volatile environment of today, a brand should be capable to manage and effectively face any kind of uncertainties, crisis and change [8]. The uncertainties arise from the environmental turbulence or may be a part of the internal risks faced by the brand. Any kind of turbulence can have positive or negative outcomes depending upon how the brand responds to the situation [6]. During the last decade, the focus of the firms has widely shifted from "profits" to "resilience" owing to increasing uncertainties in the environment [4]. For instance, when Toyota had to recall 8 million vehicles in 2009–2010 due to mechanical defects which led to unintended acceleration and was the cause of several accidents and deaths, the brand was able to bounce back within a period of two years. The brand was able to make a quick recovery because it had invested many years in delivering its brand promise and effectively communicating with its consumers. It was the slow built resilience over a number of years that led to the fast rebound of the brand [19].

18.1.1 Resilience-Meaning

Resilience refers to the brand's capacity to absorb, adapt, and bounce back from the internal and external threats by anticipating proactively, avoiding them and engaging in constructive actions to transform any kind of challenges as opportunities for growth [4]. This requires the brand to *reactively and proactively, offensively and defensively manage the internal as well as external turbulence over a period of time by being flexible and adaptable to any kind of risks* [4]. The need of the hour calls for a brand to be resilient to manage the different kinds of risks a brand may be exposed to [10].

18.1.2 Attributes of a Resilient Brand

A *brand may require certain attributes such as knowledge, ability to change, openness, availability of adequate resources, flexibility and a wide network of relationships to be resilient* [8]. Resilient brands are easily adaptable to changing business environments, are flexible enough to change their directions during adversities and make a comeback much stronger than before [9]. According to prior research, resilience may be conceptualized on the basis of time i.e., attitude of the firm before the occurrence of the event (proactiveness of the firm), during the event (how the firm responds) and after the event (bouncing back of the firm). The different attributes of resilience associated before the occurrence of the event include "alertness," "readiness" and "preparedness." During the occurrence of the event a firm have either an absorptive or adaptive resilience attitude. Absorptive resilience refers to the organization's capacity to "sustain" and "absorb" during the occurrence of the event. It may include an organization's capacity to "bounce back" during the crisis without bringing any major changes. Whereas, in adaptive resilience, the organization focusses on modifying its functions incrementally in the face of change [8].

18.1.3 Literature Survey

Prior researchers have focused their studies on brand crisis that have an adverse effect on their reputation and the responses of the brands in relation to those crises. The term "resilience" has gained a lot of attention and various definitions and views on "resilience" with respect to organizations, apart from the individuals have come into picture [4]. Recent researchers have expressed their opinion on the resilience of the firms during environmental turbulence and concepts related to recovery and reinvention [6]. Academia however has limited literature available on the interrelation between brands and resilience. To bridge this gap, the current study aims to determine by means of example of Samsung, a diversified electronics brand, who proved itself to be a resilient brand after undergoing the unexpected S7 crisis. The study aims to determine the factors of a resilient brand, taking the brand resilience framework by Edelman into consideration and how Samsung was able to withstand the blow by learning from the crisis and changing its future strategies.

18.2 Launch of the Flagship Phone—S7

18.2.1 Flagship Product of Samsung

In the year 2016, Samsung was planning to launch their much-awaited android phone Samsung's Galaxy Note 7. The Galaxy Note 7 was expected to provide many improvements over the previous series of Galaxy phones to provide a better service experience to its customers. The consumers expected a lot from this phone especially after the launch of S5 which acted as a best performer in this category. S5 had occupied the market as a leader immediately after it was introduced as a big phone and had all the technical features as compared to the other phones in the same category [11, 12]. S7 was expected to add on to the same thus making Samsung the leader in the category. "*This is because S7 being the flagship product of Samsung aimed at retaining the differentiation conscious consumer base.*" [1]. Secondly, phone was highly awaited due to the fact that Samsung aimed at launching the same with the best of features at a reasonable price. The term reasonable price means a price for the similar products as compared to the competitors in the industry [2].

18.2.2 Features of S7

S7 was a 4G, LTE Phone, comprising of a better stylus with iris scanning capabilities and a waterproof body to compete with Apple. It was equipped with a 3500 mAh battery with quick charging and wireless charging capabilities. The handset also had a "Secure folder" capability with an extra layer of privacy [3]. The features of S7 have been highlighted in Fig. 18.1.

18.2.3 Final Launch of S7

The long-awaited phone of Samsung was finally launched on August 19, 2016 in 10 countries including America, Australia, Korea, etc. It was considered to be a unique phone in its own category [1]. Due to its power packed features, as soon as the phone was launched, it immediately became a hit in its category. On the very first day more than 35,000 units had been sold which was really exciting for the company [3]. Samsung found it difficult to manage the high demand for this phone in the US and Asian Market as the company was experiencing a high growth [3].

18.3 Problems

18.3.1 Bursting into Flames

As per the reports of Wall Street Journal, a week after the mobile phone was launched on August 24, 2016, there was first complaint of the phone being burst into flames with more following [12, 15]. Even before the company could understand the reason for the same, complaints started multiplying. The United States Consumer Products Safety Commission advised the company to go for a recall of the phones. According to the CPSC, Samsung had received 92 reports of the batteries overheating in the United States, including 26 reports of burns and 55 reports of property damage, including fires in cars and a garage [12, 15, 16]. The detailed event timeline is featured in Fig. 18.2. The thing that made the situation all the worse was that such complaints were being shared on social media, harming the brand name. Also, since this was having a potential to cause human harm, the same was banned on the flights, eroding the image of Samsung badly [11].

Versions: N930 (USA); N930F (Europe); N930G (India)
Also known as Samsung Galaxy Note7 Duos with dual-SIM card slots

NETWORK	Technology	GSM / HSPA / LTE EXPAND ▾
LAUNCH	Announced	2016, August. Released 2016, September
	Status	Discontinued
BODY	Dimensions	153.5 x 73.9 x 7.9 mm (6.04 x 2.91 x 0.31 in)
	Weight	169 g (5.96 oz)
	Build	Corning Gorilla Glass 5 back panel
	SIM	Single SIM (Nano-SIM) or Dual SIM (Nano-SIM, dual stand-by)
		- IP68 certified - dust proof and water resistant over 1.5 meter and 30 minutes - Stylus - Samsung Pay (Visa, MasterCard certified)
DISPLAY	Type	Super AMOLED capacitive touchscreen, 16M colors
	Size	5.7 inches (~78.0% screen-to-body ratio)
	Resolution	1440 x 2560 pixels (~518 ppi pixel density)
	Multitouch	Yes
	Protection	Corning Gorilla Glass 5
		- Always-on display - TouchWiz UI
PLATFORM	OS	Android OS, v6.0.1 (Marshmallow)
	Chipset	Exynos 8890 Octa
	CPU	Octa-core (4x2.3 GHz Mongoose & 4x1.6 GHz Cortex-A53)
	GPU	Mali-T880 MP12
MEMORY	Card slot	microSD, up to 256 GB (dedicated slot) - single-SIM model microSD, up to 256 GB (uses SIM 2 slot) - dual-SIM model
	Internal	64 GB, 4 GB RAM
CAMERA	Primary	12 MP, f/1.7, 26mm, phase detection autofocus, OIS, LED flash, check quality
	Features	1/2.5" sensor size, 1.4 µm pixel size, geo-tagging, simultaneous 4K video and 9MP image recording, touch focus, face/smile detection, Auto HDR, panorama
	Video	2160p@30fps, 1080p@60fps, 720p@240fps, HDR, dual-video rec., check quality
	Secondary	5 MP, f/1.7, 22mm, dual video call, Auto HDR
SOUND	Alert types	Vibration; MP3, WAV ringtones
	Loudspeaker	Yes
	3.5mm jack	Yes
		- 24-bit/192kHz audio - Active noise cancellation with dedicated mic
COMMS	WLAN	Wi-Fi 802.11 a/b/g/n/ac, dual-band, Wi-Fi Direct, hotspot
	Bluetooth	v4.2, A2DP, EDR, LE
	GPS	Yes, with A-GPS, GLONASS, BDS, GALILEO
	NFC	Yes
	Radio	No
	USB	v3.1, Type-C 1.0 reversible connector
FEATURES	Sensors	Iris scanner, fingerprint (front-mounted), accelerometer, gyro, proximity, compass, barometer, heart rate, SpO2
	Messaging	SMS(threaded view), MMS, Email, Push Mail, IM
	Browser	HTML5
	Java	No
		- Fast battery charging - Qi wireless charging (market dependent) - ANT+ support - S-Voice natural language commands and dictation - MP4/DivX/XviD/WMV/H.265 player - MP3/WAV/WMA/eAAC+/FLAC player - Photo/video editor - Document editor
BATTERY		Non-removable Li-Ion 3500 mAh battery
MISC	Colors	Blue Coral, Gold Platinum, Silver Titanium, Black Onyx
	SAR EU	0.25 W/kg (head) 1.28 W/kg (body)
	Price group	9/10
TESTS	Performance	Basemark OS II 2.0: 2676 / Basemark X: 32648
	Display	Contrast ratio: Infinite (nominal), 4.247 (sunlight)
	Camera	Photo / Video
	Loudspeaker	Voice 69dB / Noise 69dB / Ring 72dB
	Audio quality	Noise -92.8dB / Crosstalk -92.9dB
	Battery life	Endurance rating 90h

Disclaimer. We can not guarantee that the information on this page is 100% correct. Read more

Fig. 18.1 (Samsung Galaxy Note7 - Full phone specifications). (*Source*: GSM Arena, 2017)

Aug-19	Note 7 launches in 10 countries including South Korea, U.S. and China
Aug-24	First case of Note 7 fire reported in Korea, with more following
Sep-02	Samsung confirms battery defect, initiates recall
Sep-15	Samsung, US CPSC announces official recall in US
Sep-19	Replaced Note 7 equipped with new batteries released in Korea
Sep-21	Replaced Note 7 equipped with new batteries released in US
Sep-26	Case of replaced Note 7 fire reported in China
Oct-01	Case of replaced Note 7 fire reported in Korea
Oct 2-5	Samsung, KTL say new cases of Note 7 fire in Korea were caused by "External Pressure" after testing
Oct-06	Replaced Note 7 catches fire aboard Southwest airlines flight in Louisville, Kentucky, US
Oct 7-9	Additional cases of replaced Note 7 fires reported in Taiwan and other US cities
Oct-09	Major US Carriers voluntarily halt Note 7 sales
Oct-10	Samsung halts production of Note 7 devices
Oct-11	Samsung suspends Note 7 sales globally

Fig. 18.2 Timeline of Samsung Galaxy Note 7 debacle. (*Source* https://www.patentlyapple.com/patently-apple/2016/10/samsung-caves-production-of-the-note-7-has-been-permanently-discontinued.html)

18.3.2 Samsung Strikes Action

Finally, the company had to make its first recall on September 2. The recall program by Samsung required the customers to refund their note 7 by offering a refund or replacement phone. The replacement phones were to be made available by September 21. It was expected that the new phones would be safe. Contrary to the expectations, the replacement phone also started catching fire during charging. In one of the instances the phone caught fire while in car charging. This was highly unexpected as car chargers are comparatively slower than the electric chargers [1].

Although the company tried to identify the issue but due to increasing complaints and not being able to fix the issue, in the best interest of its consumers, it went for a complete recall of approximately 2.5 million phones. Taking their customer's safety as their ultimate priority, Samsung decided to halt sales and production of Galaxy Note 7 on 11th October, the same year [11–16]

Samsung, although one of the biggest brands in Electronics and Mobile phones was faced with the following challenges:

18.3.2.1 Financial Crisis

As per the reports of Wall Street Journal, Samsung declared a fall in the net profit by 16.8% to 4.54 trillion in the third quarter due to the disastrous event of recall of Samsung Note 7. The operating profit of the company fell by 30%. This was denoted to be the least quarterly profit since the launch of mobile phone Galaxy series for more than 6 years ago. The recall of the mobile phones negatively impacted the market value of Samsung. Samsung shares fell by 8% which was considered to be the biggest one-day decline in eight years [15]. Samsung's discontinuation of the phone from the market along with the global recall, cost the company $ 17 billion in sales [20].

18.3.2.2 Reputational Impact of Crises

This unforeseen disastrous event also led to the downfall in the brand image of Samsung. The brand image refers to the consumers' perception of the company and its products and is measured by its intangible assets such as management expertise, innovation and intellectual property. Samsung carried a greater brand credibility and was considered to be the biggest competitor of Apple. The simple reason for this was that Samsung as a brand was excelling in delivering customer services along with their expectations and was considered as a socially responsible brand. As per Samsung's newsroom of 2016, Samsung had been ranked 7th by Interbrand "Best Global Brands" [1]. The customers, therefore, expected an immensely high-quality phone from Samsung. The last two iPhones that were launched during that phase were not having any major technological changes. In order to compete with the major player in the market, Apple, Samsung thus wanted to launch a phone that was slim and packed with all the latest features to meet customer's expectations. Batteries from two different manufacturers had flaws. Moreover, the phone was banned on the flights and in public places thereby leading to a downfall in the name of Samsung [11]. It takes years for a brand to build trust in the minds of the consumers and being named and shamed in such a way was causing Samsung to lose not only its image but also the trust that it had created over the years in the minds of the consumers.

The main questions that arise here is:

a) What kind of resilient strategies were followed by the brand in order to save its brand image?
b) To identify factors affecting brand resilience
c) Do the resilient strategies help to regain consumer trust and help in positive future engagements of consumers with the brand.

18.4 Corrective Resilient Actions

4.1. Although it was the first kind of incident in the smart phone category, and was highly unexpected, Samsung took an initiative of quick voluntary recall of phones. Without any delay, Samsung detected a battery defect and started its first recall on September 2nd followed by a complete recall of 2.5 million phones later. Although difficult for Samsung to understand the root cause initially, Samsung took the responsibility and, in a press conference acknowledged that the phones were exploding [1, 3]

4.2. For the concern of people's safety Samsung immediately sent alert emails and messages to its customers for the recall of phones. They completely stopped the manufacturing process and sales of Galaxy Note 7. They tied up with the telecom companies for software updates to disable the charging function thereby turning the phone useless [1, 3, 7, 12–16].

4.3. The company started with the replacement offer whereby the consumers could opt for taking S7 edge along with cash refund for the balance amount or a complete refund if they wanted to. Samsung also decided to give the Gear VR Headset and an Oculus App coupon worth 50 $ free to customers who opt for company phones [23]. In order to compensate its customers, Samsung offered up to 100$ in bill credit to those consumers who were willing to exchange their Note 7 for any Samsung Smart phone in US. For customers who were willing to buy any other brand's smart phone or required a full refund, Samsung offered to pay 25$ [21]. The firm also decided to compensate its suppliers who incurred losses due to manufacturing various parts of S7, their losses estimated up to $ 1.7 billion [22]. The decision of voluntary recall was highly appreciated by the consumers reflecting the commitment of the brand toward the consumers safety and also limited their damage to the reputation [1, 3, 7, 12–18].

4.4. For a thorough check, Samsung built a lab where 200,000 phones were tested by 700 researchers in order to determine the cause of the battery exploding. The phones were tested under extreme conditions where by the ultimate cause of this disaster was found. The problem was with the batteries. The principal root cause of the first manufacturer's battery problem (Battery A) was negative electrode deflections. The second manufacturer's product (Battery B), suffered from abnormal ultrasonic welding burrs. In fact, as per the reports it was the slim design of the phone that had given rise to this unfortunate event. There was no space for batteries to contract or expand as a result of which it led to overheating and then exploding. As per the report of 3 external investigators appointed by Samsung the fusing was not done in an effective manner. Hence the batteries malfunctioned and ultimately exploded [1, 3, 7].

4.5. Although Samsung shifted the responsibility to its own subsidiary Samsung SDI for the production of default batteries, yet it could not prevent its image from being eroded as the quality should have been thoroughly checked by Samsung before the launch of the phone. The batteries were produced by its own Subsidiary SDI and Samsung being over ambitious to launch this phone

10 days earlier than its usual date failed in its quality check procedure, thus giving rise to this contingency [7].

18.5 Factors Affecting Resilience

As per Edelman, any resilient brand should have the following set of attributes which are presented in Fig. 18.3:

Although the set of attributes was given by Edelman in the context of COVID-19, but the questions that seemed to be following it are very much relevant to the case of Samsung S7 Crisis.

- The thought process that a brand needs to enter while its marketing activities are on hold.
- How to regain the trust of the consumers?
- Steps required to be followed by the brand to modify current marketing and communications?
- Focusing on the near-term needs while planning for an unknown longer-term reality.

18.5.1 Resilient Approach Followed by Samsung

Any organization's ability to face such an unforeseen event or crises depends upon its robustness and also the organization's ability to respond i.e., Agility. Resilience refers

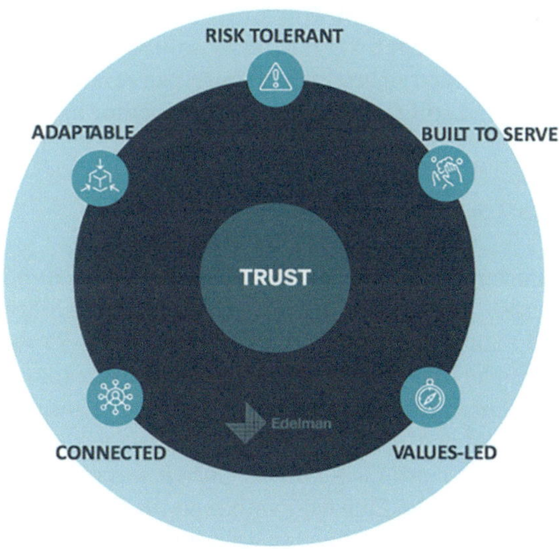

Fig. 18.3 Edelman's framework of Resilience. (*Source* https://www.edelman.com/insights/brand-resilience-building-future-ready-brands-covid19-context)

to the brands' capacity to bounce back, specifically on anticipating, preparing, and effectively avoiding threats, and possibly converting turbulence-induced challenges as opportunities for growth [4]. Samsung's approach toward being a resilient brand can be related to the Edelman's attributes of a resilient brand.

18.5.1.1 Adaptability

The organization's ability to respond to any kind of internal or external turbulence will not only depend on its strength but also on its ability to be agile. Agility depends on three factors i.e., corporate mindset, organizational flexibility and resource fluidity [5]. Samsung proved itself to be adaptable to the crises. It immediately went for a voluntary recall as soon as complaints started logging in regarding the exploding of batteries. Keeping consumer safety as their priority area, after getting multiple complaints even with the replacement sets the company didn't take much time to stop the production process and remove the handsets from market completely. Being agile, the company decided to focus on the other products in its portfolio and achieved relatively stable sales of semi-conductors and display panels. They also planned to come up with their new flagship phone Galaxy S8 which would have an artificial intelligence system just like SIRI offered by Apple [13].

18.5.1.2 Risk Tolerant

In the above-mentioned crisis, Samsung is considered to be highly risk tolerant brand. Samsung had to face a huge financial crisis and its profits dipped badly with the crisis. As per the reports of Wall Street Journal, overall operating profit for the three months ended Sept. 30 fell 30% to 5.2 trillion won from nearly 7.4 trillion won a year earlier, while revenue fell 7.5% to 47.82 trillion won. But it was still considered to be a risk tolerant brand since although impacted severely, Samsung was able to rise itself once again and still standing tall as a competitor to Apple [15].

18.5.1.3 Built to Serve

Samsung as a brand has been able to establish its position as a dominant player in the market. The solo reason for this is Samsung's ability to keep its customers as a priority and their willingness to serve them always. S7 was launched with the intention of giving an exceptional quality phone to its customers. Barely did Samsung know that the phone would be creating such a major disaster. However, once the phone was launched and started exploding, Samsung was proactive to recall the phones even before the reason was known. Once it was realized that the replacement phones were also exploding, Samsung within no time, halted the sales and production of the S7. This was because Samsung was committed to serving its customers and did not want them to face any unpleasant consequences. "Taking our customer's safety as our highest priority, we have decided to halt sales and production of the Galaxy Note 7," the company said [12, 13].

18.5.1.4 Values Led

Immediately after the crises was reported, Samsung built a lab where 200,000 phones were tested by 700 researchers. Samsung opted for a global recall along with offering its customers a complete refund or replacement with a similar high tier S7 + voucher for the balance amount. Samsung published a full page apology in the major newspapers like The Wall Street journal, The New York Times and The Washington Post [13].

18.5.1.5 Connected

Once the phones started exploding and the complaints were registered, Samsung did its best efforts to stay connected to its consumers and retailers. They issued e-mails and messages for recall of phones. Thereafter once the plan had been formulated regarding total halt of production process Samsung assured the customers of complete refund or replacement. They were highly apologetic of this and stayed connected with the consumers throughout this process. When the phones were banned on the flights by Department of Telecom (DOT), Samsung also sent their team members on the airport to help consumers and collect phones [13].

18.5.1.6 Trust

Trust is something which the brand establishes in the form of Brand Equity over the years. The company went through a product harm crisis which caused a reputational damage to the firm. In order to gain the consumer's trust, Samsung focused on its future course of action and took some additional time to launch its next phone S8 after conducting all the necessary quality checks. They were sure that they would be able to regain the consumer trust.DJ Koh, President Mobile Communications Business for Samsung Electronics, said in an interview with industry analysts, "Technology innovation is important to Samsung, but our customer's safety is more important. We want to reinstate trust in the brand." To regain trust, Samsung had to identify the root cause of the issues with the Galaxy Note 7 and apply these learning to future product designs [12–18].

18.6 Steps for Revival and Regaining Customer Trust

Samsung is considered to be one of the major leaders in the electronic industry. It had been able to revive itself after this fiasco and regain the trust of the consumers. Samsung thoroughly focused on its weaknesses to eradicate them and capitalized on its strengths to become a leading brand once again.

Samsung took the crisis constructively i.e., it focused on its faults and improved the quality processes and came up with an 8-point battery safety system which comprised of Durability test, Visual Inspection, X Ray Test, Disassembling Test, TVOC test, ΔOCV Test, Charge and discharge test and accelerated usage test. Such a multi-layer safety implied that the customers would be getting better phones in the future [13].

Secondly, Samsung had a great brand image even before it launched S7 Note. It had a large number of loyal customers. As per the reports of Harvard Business Review, Samsung sold over 78 million smartphones in the second quarter of 2016. In the initial days of recall, when the customers were given a choice for refund or another Note 7, the choice was latter. Consumers had a strong faith in Samsung products. This is termed as "Brand Insulation effect." A large and loyal customer base insulated Samsung from the huge fiasco loss [14].

Thirdly, Samsung is considered as a national pride for the South Korean Customers. As per the reports, if the brand belongs to a particular heritage, then the consumers are likely to forgive and trust the brand easily (like in case of Volkswagen emission case). Therefore, the brand was forgiven and given a second choice, which spilled over to the other customers [14].

Fourthly, Samsung did not escape from the scene after the crisis. Instead, they held a press conference where the mobile president came up with apology for the exploding batteries and also claimed that Samsung would recall 2.5 million Note 7 devices that it had sold. Also, since the recall of such a huge number would take a lot of time, they designed an app update which would limit the maximum battery charge to 60% to avoid any further incidents. Thereafter the company decided to bring the production process to stop completely [12, 14, 16].

All this was considered to be sincere efforts of recall which made the consumers trust Samsung once again which led to positive future engagements of the consumers with the brand.

18.7 Conclusion

Samsung had to undergo a major crisis when it had to go with a major recall of its most awaited phone which was totally unexpected. However, Samsung, as a brand is agile and acted as a resilient organization and turned its negative points into positive ones thereby regaining consumer trust once again. In order to achieve this, Samsung came up with new quality assurance processes which were to be followed by Samsung as well as its component manufacturers. It initiated a multi-layer safety protocol at the level of product planning and also an 8-point battery safety system which included various tests at all stages to avoid any kind of failure in the future. Samsung as a brand has emerged to be stronger than before.

References

1. Shamsi, A. F., Ali, M. H., & Kazmi, S. S. (2017). *Samsung Note 7—an unprecedented recall that created history: exploding phones recovered–exploded trust?*
2. Falayi, O. R., & Adedokun, A. J. (2014). The demand for smartphones among students in university of Ibadan. *Economics, 1*(1), 30–34.
3. Zhang, W. (2020). Crisis PR study on Samsung Galaxy Note 7 explosion incident. *International Journal of Business and Management, 15*(7), 68–76.
4. Conz, E., & Magnani, G. (2020). A dynamic perspective on the resilience of firms: A systematic literature review and a framework for future research. *European Management Journal, 38*(3), 400–412.
5. Doz, Y. L., & Kosonen, M. (2010). Embedding strategic agility: A leadership agenda for accelerating business model renewal. *Long range planning, 43*(2–3), 370–382.
6. Rego, L., Brady, M., Leone, R., Roberts, J., Srivastava, C., & Srivastava, R. (2022). Brand response to environmental turbulence: A framework and propositions for resistance, recovery and reinvention. *International Journal of Research in Marketing, 39*(2), 583–602.
7. Lim, J., & Kim, J. H. (2018). Examining risk communication of Samsung by analysing customer reaction on social media: from Galaxy Note 7 explosion to Galaxy S8 Release. In *CS & IT Conference Proceedings* (Vol. 8, No. 3). CS & IT Conference Proceedings.
8. Miceli, A., Hagen, B., Riccardi, M. P., Sotti, F., & Settembre-Blundo, D. (2021). Thriving, not just surviving in changing times: How sustainability, agility and digitalization intertwine with organizational resilience. *Sustainability, 13*(4), 2052.
9. Lewis, J. D. (2017). *Brand vs. wild: Building resilient brands for harsh business environments.* Taylor & Francis.
10. Copulsky, J. R. (2011). *Brand resilience: Managing risk and recovery in a high-speed world.* Martin's Press.
11. https://continuitycentral.com/index.php/news/business-continuity-news/2684-crisis-recovery-case-study-samsung-18-months-on-from-the-explosive-s7-note
12. https://www.wsj.com/articles/samsung-to-permanently-discontinue-galaxy-note-7-smartphone-1476177331
13. https://www.forbes.com/sites/maribellopez/2017/01/22/samsung-reveals-cause-of-note-7-issue-turns-crisis-into-opportunity/?sh=76b7a27424f1
14. https://hbr.org/2016/10/why-samsungs-note-7-crisis-wont-hurt-its-brand-long-term
15. https://www.wsj.com/articles/samsung-profit-falls-on-galaxy-note-7-recall-1477526701
16. https://www.wsj.com/articles/samsungs-management-of-recall-wounds-companys-image-1473928872
17. https://www.wsj.com/articles/u-s-consumer-safety-agency-plans-recall-of-samsung-phone-1473968157?tesla=y
18. https://www.theverge.com/circuitbreaker/2016/9/9/12865966/samsung-note-7-recall-battery-mark-blue-s
19. https://lisamerriam.com/the-toyota-brand-bounces-back/
20. https://www.androidauthority.com/samsung-galaxy-note-7-permanently-discontinued-721283/
21. https://www.theguardian.com/technology/2016/oct/13/samsung-to-pay-galaxy-note-7-owners-to-buy-another-brands-phone

22. https://www.ndtv.com/world-news/samsung-to-compensate-suppliers-hit-by-note-7-crisis-1475686
23. https://indianexpress.com/article/technology/tech-news-technology/samsung-to-compensate-galaxy-note-7-parts-suppliers/

Chapter 19
Understanding the Paradox of Employer–Employee Relationship Using Text Analytics

Poorvi Prakash, Rashmi Srinivasan, Tripti Mahara, and V. L. Helen Josephine

Abstract Indian companies have recently paid exorbitant salaries to acquire and retain the country's top digital professionals. However, when Russia invaded Ukraine in February 2022, the scenario changed. The global economy's impact of geopolitical uncertainty, as well as poor markets, high inflation, and the fear of a long-term global recession, has tempered the fresh zeal in startup land in India and elsewhere. The paradigm has evolved from "growth at any cost" to "profit at any cost" for both large and small firms, as investors' fear of missing out (FOMO) has given way to traditional belt-tightening. The year of "loud" layoffs has arrived, and the devastation is likely to last until 2023. This leads to the employer–employee relationship paradox: despite massive layoffs, hiring is still a concern. An unsupervised statistical method, Topic Modelling, is used to find "themes" that exist inside a collection of documents in order to understand and evaluate hidden trends utilizing annotated data and specified topics. As a result, four topics were generated that give the overall summary of the articles, i.e., "Corporate Reorganization", "Impact of Macroeconomic Condition on Hiring", "Unethical Retrenchments", and "Government Intervention on Industrial Revolution ". There is a path to recovery, but no dates have been specified. However, thousands of employees may lose their jobs in upcoming years, and there might be no quick recovery.

Keywords Text analytics · Topic modelling · Employer · Employee · Paradox

---

P. Prakash · R. Srinivasan · V. L. H. Josephine
Christ (Deemed to Be University), Bangalore, Karnataka, India
e-mail: poorviprakash1998@gmail.com

R. Srinivasan
e-mail: 2703rashmi@gmail.com

V. L. H. Josephine
e-mail: helen.josephine@christuniversity.in

T. Mahara (✉)
Prin. L. N. Welingkar Institute of Management Development and Research, Bengaluru, Karnataka, India
e-mail: tripti.mahara@welingkar.org

P. K. Kapur et al. (eds.), *Data-centric Approaches to Industrial Decisions*, Asset Analytics, https://doi.org/10.1007/978-981-96-7556-2_19

19.1 Introduction

The employee–employer paradox refers to the tension that exists between the interests of employees and employers. Employers seek to maximize profits and maintain stability in their operations, often at the expense of employee well-being. On the other hand, employees seek meaningful work, fair compensation, and job security, often at the expense of the employer's bottom line. This paradoxical relationship is particularly evident during times of economic upheaval, such as the Great Resignation and the mass layoffs.

The enormous trend of workers quitting their employment in historically high numbers in pursuit of better professional possibilities during the COVID-19 pandemic is referred to as "The Great Resignation", as it was first proposed by Professor Anthony Klotz. Numerous factors have contributed to this phenomena, such as the trend towards remote work, the desire for a better work-life balance, and a more thorough evaluation of one's own and one's career's top objectives.

Employers now face significant hurdles as a result of the Great Resignation, including coming up with plans to hold onto their best workers in a very competitive labour market. Many employees are quitting their current jobs in search of ones that provide greater pay, perks, and more room for advancement. Employers have been compelled by this circumstance to review their benefits packages and chances for career progression in order to stay competitive. Employers face the difficult task of balancing the need to retain top talent with the imperative to maintain financial stability. As a result, companies have had to make tough decisions regarding compensation, benefits, and job security. This balancing act has led to a range of strategies aimed at retaining employees, such as increasing salaries, offering enhanced benefits packages, and providing more opportunities for professional development and career advancement.

The impact of the Great Resignation has been felt across various industries, prompting a surge in hiring as businesses scramble to fill the gaps left by departing employees. The need to bridge the skills gap has led to an intensified focus on recruitment and retention efforts. Companies are not only competing to attract new talent but also striving to create a work environment that encourages existing employees to stay. The Great Resignation has significantly reshaped the job market, compelling employers to adapt their strategies to attract and retain employees in a time of heightened mobility and shifting priorities. This period of increased job turnover has highlighted the importance of providing competitive compensation, fostering a positive work culture, and offering clear paths for career progression.

However, the picture changed in February 2022, when Russia invaded Ukraine. The impact of geopolitical instability on the global economy, along with weak markets, excessive inflation, and the dread of a long-term global recession, has dampened the fresh zeal amongst businesses across globe and in India also. For both large and small organizations, the paradigm has shifted from "expansion at any cost" to "profit at any cost", as investors' fear of missing out (FOMO) has given place to traditional belt-tightening. The year of mass layoffs announced vociferously through all communication channels and social media has arrived and this carnage is expected to continue into 2023. Statistics reveal, that as of December 8, 2022, many Indian firms, including a slew of unicorns like BYJU'S, MPL, ChargeBee, Meesho, Cars24, LEAD, Ola, Innovaccer, Udaan, Unacademy, Vedantu, Meesho, etc. have gone for a loud layoff of more than 20,000 employees on the whole [1]. According to data compiled by Moneycontrol, startups have laid-off about 2100 employees in the initial three weeks of 2023. The primary causes of layoffs are the oncoming recession and the geopolitical scenario, which has resulted in inflation, affecting numerous businesses. The other reasons include funding crunch, organization restructuring, employee performance, cost cutting etc. [2]. The message of mass layoffs is out and clear there which further highlights the employee–employer paradox, as businesses struggle to balance the need for profitability with their responsibility to their employees. While layoffs may be necessary to maintain a business's financial stability, they can also result in significant hardship for affected employees and their families.

This wave of mass layoffs underscores the ongoing employee–employer paradox. Companies are faced with the challenge of balancing the need to remain profitable while also upholding their responsibilities to their workforce. On one hand, layoffs might be deemed necessary to ensure a company's financial health and long-term sustainability. They can be a strategic move to reduce costs during economic downturns or when the business environment is particularly volatile. On the other hand, these layoffs can cause significant distress for the affected employees and their families, leading to financial insecurity and emotional turmoil.

The dilemma arises from the competing demands of preserving corporate operations and guaranteeing worker well-being. Although layoffs might help a business weather hard circumstance, they can also undermine the confidence of the remaining staff members and ruin the company's reputation. This delicate balance emphasizes how companies must come up with more long-term solutions to safeguard their bottom line without negatively affecting their workers. Instead of laying off employees, businesses should look into other options including voluntary leave packages, temporary salary reductions, or shortened workweeks. The tendency of widespread layoffs eventually acts as a sobering reminder of the difficulties of running a company during difficult times. It necessitates a more thorough investigation of how companies might more effectively manage financial difficulties while helping their employees.

Traditionally, Indian firms have recently given absurd wages to attract and keep the most remarkable tech talent in the nation. However, a wave of mass layoffs has caused professionals to doubt their ability to advance and the viability of employers. This leads to the paradox of employer–employee hiring that despite massive layoffs, hiring is still a problem [3]. Thus, in this research, an attempt is made to understand and analyse this problem using analytics techniques to identify the hidden trends for this relationship using annotated data and predefined topics.

19.1.1 Background Study

Notable research was done by Yu et al. [4] to comprehend and validate the employee–employer conundrum. In a similar vein, surveys have been used in additional research by Nelissen et al. [5], Devos et al. [6] and Acikgoz et al. [7] to investigate this phenomenon and uncover several elements impacting the paradox. To obtain information, this research, however, mostly rely on survey data. To completely understand the root causes of this contradiction, none has made full use of the enormous amounts of unstructured secondary data included in reports and newspaper stories. Making use of such unstructured data could yield more insightful results because it includes a range of viewpoints from different stakeholders. This method would enable a more thorough examination of the elements causing the paradox, guaranteeing that the conclusions are thorough and pertinent to the given situation. Future studies could provide a more sophisticated view of the employee–employer relationship by examining this underutilized data and capturing the complexity and dynamics that conventional survey methods could overlook.

Thus, this research focuses on analysing the secondary data to understand the paradox. Text analytics is a useful tool for understanding unstructured data and is used in research. Studies have been found where Great Resignation and Mass Layoffs have been studied using text analytics. For example, a study by Wang et al. [8] used text analytics to analyse employee reviews on Glassdoor and found that factors such as job security, work-life balance, and employee benefits were important drivers of the Great Resignation. Similarly, a study by Yang and Kim [9] used text analytics to analyse news articles and found that the COVID-19 pandemic and the resulting economic downturn were the primary drivers of Mass Layoffs. The majority of available research examines the Great Resignation and widespread layoffs in isolation rather than the influence on the employer–employee relationship. Furthermore, there is a scarcity of research that employs text analytics to uncover the fundamental causes of the conundrum that lead to the foundation of this research.

19.2 Methodology

19.2.1 Data Collection

The aim is to explore the fundamental reasons and realities behind the complex relationship between employees and employers. To accomplish this, a broad collection of articles was gathered from various sources, such as Money Control, Forbes, Fox Business, Live Mint, Business Insider, NY Times, India Today, and Times of India. A total of 150 articles were systematically gathered using Python scripts. This process involved not only scraping the articles but also collecting their URLs, ensuring a thorough and organized compilation. The data was then compiled into an Excel file for detailed analysis. This extensive dataset is intended to offer a deeper insight into the intricate dynamics between employees and employers, highlighting the core reasons and realities that shape this paradoxical relationship.

19.2.2 Data Cleaning

Each row in the Excel file contains both the URL and the full article content. These articles often include tables, figures, and numerical data that are not relevant to our analysis. To refine the dataset, we meticulously cleaned the articles by removing this extraneous information. This cleaning process ensured that only the pertinent text remained. Each cleaned article was then saved back into the file, making the dataset more focused and valuable for analysis. This step was crucial to ensure that the information used in the study was both relevant and concise, facilitating a deeper understanding of the underlying reasons and realities of the employee–employer relationship.

19.2.3 Text Preprocessing

The text preprocessing steps undertaken include several key transformations to ensure the data is clean and suitable for analysis. Initially, the text is converted to lower case to maintain uniformity. Punctuation marks, which do not contribute meaningful information, are then removed [10]. Following this, the text undergoes tokenization, a process that breaks down the text into individual words or tokens.

Next, stop words, which are frequently occurring words that offer little informational value, are removed. These words, such as "and", "the", and "is", are filtered out to focus on the more significant terms within the text. After stop words removal, lemmatization is performed. Lemmatization involves transforming words into their base or root form, known as the lemma [11]. This process is essential as it reduces different inflected forms of a word to a common base form, enhancing the consistency and accuracy of the analysis. For instance, words like "sustainability" and "sustainable" are both converted to their root form "sustainable".

In addition to these steps, bigram and trigram models are constructed. Bigrams and trigrams are sequences of two or three words, respectively, that frequently appear together in the text. These models help in understanding the context and identifying common phrases within the dataset. Likewise, a dictionary and corpus are created. The dictionary consists of all the unique words in the dataset, while the corpus is a collection of all the processed documents. These components are crucial for various text analysis tasks, including topic modelling and sentiment analysis. Overall, these comprehensive preprocessing steps are designed to enhance the quality and utility of the text data, ensuring a robust foundation for subsequent analysis.

19.2.4 Model Building

To find the topics and subjects in big text data sets, topic modelling is a crucial text analytics technique. When examining the complex employer–employee relations during times of economic instability and transition, this method is especially helpful. Topic modelling can identify the underlying themes and issues causing employee discontent and turnover by examining large volumes of text data. Using this method gives businesses a better grasp of the important concerns influencing their employees. For example, it can pinpoint important issues that are frequently on employees' concerns, like work-life balance, job security, and workplace culture. Businesses can address particular issues and put plans in place to increase staff retention and satisfaction by emphasizing these themes.

The capacity of topic modelling to handle and analyse enormous amounts of text data is one of its main benefits. With so much input and communication in a business setting, this technology can cut through the clutter and identify the most important problems. This simplifies the data and makes it easier to obtain and assess, enabling decision-makers to make better-informed choices. Additionally, subject modelling can offer useful information that is essential for strategic planning. For instance, if management notices that a recurring issue is a lack of satisfaction with work-life balance, they might take action to introduce more flexible work schedules or remote work opportunities. In a similar vein, employers may prioritize open communication about employment stability and future potential if job security becomes a major worry. To sum up, topic modelling is an effective technique for comprehending and managing the intricate relationships that exist between companies and employees. Businesses can utilize this information to make data-driven decisions that improve workplace satisfaction and productivity by identifying the underlying reasons of employee dissatisfaction. Businesses may better handle economic downturns by utilizing this technology, which also ensures a more resilient and satisfied workforce.

Latent Dirichlet Allocation (LDA) is a widely used unsupervised machine learning algorithm for topic modelling. This generative probabilistic model identifies relevant topics within a collection of documents. In LDA, each document is viewed as a combination of several latent topics, and each word within the document is associated with one of these topics. A topic, in this context, is essentially a collection of words that frequently appear together.

The process begins by the researcher specifying the number of topics to be identified. LDA then works to distribute the words in each document across these topics based on their co-occurrence patterns. This allocation is done in such a way that documents sharing similar topics will have a similar distribution of words. In essence, LDA seeks to discover the underlying structure of topics by analysing how words group together across the provided documents.

First, every word in the papers is randomly assigned to one of the predetermined subjects using Latent Dirichlet Allocation (LDA). The cyclical procedure that follows has its beginning point in this first random allocation. Next, taking into account the word's context within the document, LDA reassigns each word in each document to a subject depending on the likelihood that the word corresponds to each topic. To fine-tune the topic allocations, this reassignment phase is repeated numerous times. The word distributions within subjects and the themes inside documents are changed in real time to reflect the most recent assignments as these iterations continue. The repeated procedure persists until the assignments attain stability and cease to undergo major fluctuations, signifying that the topic distributions have attained an ideal condition. Ultimately, LDA reveals topics by identifying groups of words that frequently appear together across the documents. This method allows researchers to understand the thematic structure within large text datasets, providing insights into the common topics discussed in the documents [12].

Algorithm of LDA:

Step 1: Set the number of topics, k.

Step 2: For each document, compute the proportion of words assigned to each topic, P(topic t | document d).

Step 3: For each word in the corpus, compute the proportion of assignments to each topic, P(word w | topic t).

Step 4: For each word in each document, update the probability of the word belonging to each topic using the following formula in step 5.

Step 5: P(word w with topic t) = P(topic t | document d) * P(word w | topic t).

Step 6: Repeat step 4 for all words in the corpus.

Step 7: After all words have been updated, repeat steps 2–5 until convergence.

Step 8: Assign the topic with the highest probability to each word in the document.

Step 1: Set the number of topics, k. The first step in LDA is to define the number of topics (k) that you want to uncover in your document set. This is a predefined parameter that guides the entire algorithm.

Step 2: Compute the proportion of words assigned to each topic for each document. For each document, calculate the probability distribution of topics, denoted as P(topic t|document d)P(\text{topic} t \mid \text{document} d)P(topic t|document d). This represents the proportion of words in the document that are assigned to each topic.

Step 3: Compute the proportion of assignments to each topic for each word in the corpus. Next, determine the probability distribution of words within each topic, P(word w|topic t)P(\text{word} w \mid \text{topic} t)P(word w|topic t). This step involves calculating how frequently each word is assigned to each topic across the entire corpus.

Step 4: Update the probability of each word belonging to each topic. For each word in each document, update its topic assignment probabilities using the formula provided in Step 5.

Step 5: Calculate the probability of the word with the topic. The probability of word www being associated with topic 't' in document 'd' is computed as.

P(word 'w' with topic 't') = P(topic 't' | document 'd') * P(word 'w' | topic 't').

This product combines the likelihood of the topic given the document and the likelihood of the word given the topic.

Step 6: Repeat the update for all words in the corpus. This step involves iteratively reassigning each word in every document to a topic based on the updated probabilities.

Step 7: Iterate until convergence. After updating all the words, repeat Steps 2 to 5 iteratively until the assignments stabilize, meaning the topic distributions no longer change significantly.

Step 8: Assign the highest probability topic to each word. Once the algorithm converges, assign each word in the document to the topic with the highest probability. This final step results in a clear topic assignment for each word, thus uncovering the latent topics within the documents.

These procedures help LDA find patterns and word clusters that appear together frequently, which helps uncover the document set's hidden thematic structure. This facilitates comprehension of the recurring themes and how they are distributed throughout the texts.

Python is used to implement the LDA algorithm in the Gensim package. The algorithm assigns each word in the document a unique id. It is a mapping between word id and word frequency used as input to the LDA model. Alpha (auto) and Beta (β) are hyperparameters representing document-topic and topic-word densities, respectively. The chunk size, set at 50 in this case, represents the quantity of documents utilized in each training batch. The model is updated after every frequency and is executed for maximum 20 iterations.

19.2.5 *Model Evaluation*

Perplexity score and Coherence score are used to evaluate the model (Xu et al. 2019; Tresnasari et al. 2020) [7]. The perplexity score is a statistical metric that indicates how effectively a probability model predicts a sample. This is achieved by determining the number of optimal topics and is computed using log-likelihood of documents. It is used to compare different models' and the model with maximum log- likelihood score is a better fit. The log-likelihood is evaluated as follows:

$$L(w) = (w|\Phi, \alpha) = \sum_{d} (w_d|\Phi, \alpha)$$

where:

Φ = topic matrix.

Wd = unseen document.

$\boldsymbol{\alpha}$ = alpha parameter (topic distribution of documents).

That is, log-likelihood of a set of unseen documents wd given the topics Φ.

In contrast, Coherence Score is used to evaluate the semantic similarity between the words on a particular topic. Thus, it determines the quality of the learned topics. The score determines the number of topics that must be included in the model. Coherence score is defined as follows:

$$\boldsymbol{c(t, W_t) = \sum_{w1, w2 \in W_t} loglog \frac{d(w1, w2) + \in}{d(w1)}}$$

where:

WT denotes a topic t defined by a set of top words.

d(w1) = the number of documents that contain word **w1.**

d(w1,w2) = the number of documents that contain words w1 AND w2 (co- occur together).

ε = smoothing count set to 1 or 0.01.

19.2.6 Post Processing

Once the keywords corresponding to the topics are extracted, this step involves assigning the label to the topics and interpreting the topics.

19.3 Results and Analysis

The Table 19.1 depicts the perplexity and the coherence score computed for different number of topics. Here, lower the perplexity better the model, and the higher the topic coherence, higher is the interpretability of the topic. Therefore, model with 4 topics that has the least Perplexity score and High Coherence Score with no overlapping of the topics is considered the best model amongst all.

For $n = 4$, the relevant clusters corresponding to 4 topics are depicted in the Fig. 19.1 as represented in Gensim. All the four clusters are entirely separated and there is no overlapping. The size of each cluster corresponds to the number of the documents incorporated in it. The larger the cluster in size, a greater number of documents it contains.

Table 19.1 Model comparison based on perplexity and coherence score

No. of Topics	Perplexity score	Coherence score
2	−7.390	0.309
3	−7.390	0.306
4	−7.400	0.309
5	−7.2	0.302
6	−7.23	0.305

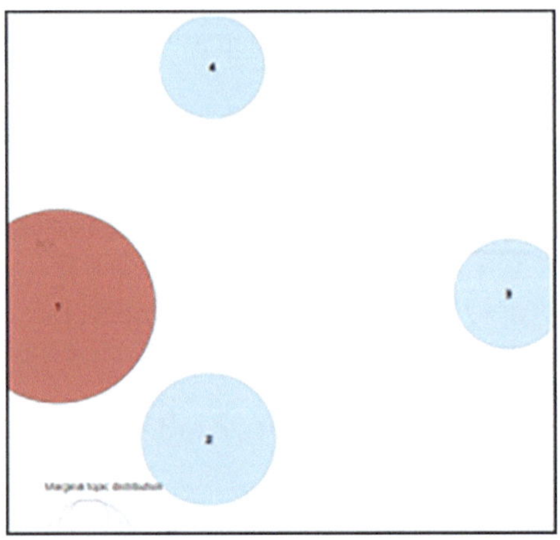

Fig. 19.1 Cluster for n = 4 topics

Figure 19.2 depicts the top 10 keywords for each topic and Table 19.2 defines the topic by labelling it along with the percentage of the documents that has mention of these topics.

The first topic has keywords like "Company", "tech", "employee", "cut", "meta", "revenue", "lay", "global", "pandemic", "inflation", "job_cut", "slow", and is labelled as "Corporate Reorganization". In a free-market setting, the reorganization of corporate policies and culture is crucial, as they profoundly influence the relationship between human resource management and other departments of the organization. The changes that employees in the company experience during the reorganization process include increasing macroeconomic uncertainty, job loss fears, global mass layoffs, and financial insecurity resulting in a significant increase in stress and anxiety-related disorders in the last 6–8 months. Unemployment and recession go hand in hand and reinforce one another. In a downturn, unemployment rises swiftly but falls slowly, and

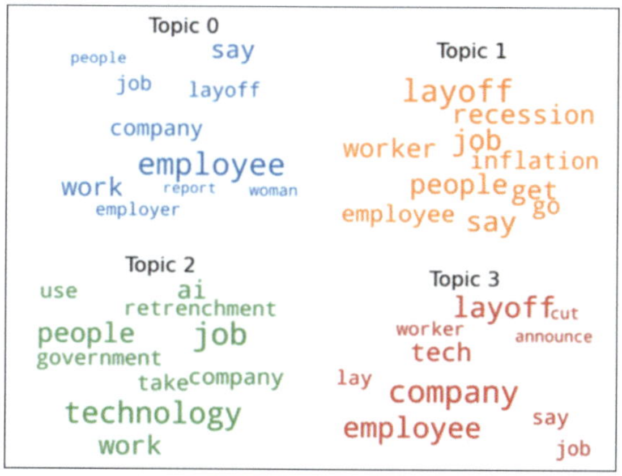

Fig. 19.2 Visualization of top 10 keywords in each topic

Table 19.2 Analysing the topics based on keywords and tokens

Topic name	Keywords
Topic 0: Corporate Reorganization	"Company", "tech", "employee", "cut", "meta", "revenue", "lay", "global", "pandemic", "inflation", "job_cut", "slow"
Topic 1: Unethical Retrenchments	"woman", "employer", "workplace", "uber", "unethical", "visa", "Wipro", "moonlighting", "recruitment", "elimination", "resign"
Topic 2: Government Intervention on Industrial Revolution	"retrenchments", "replace", "chatgpt", "id_act", "ai", "workman", "government", "technology", "lay off", "illegal", "revolution", "automation", "industrial establishment"
Topic 3: Impact of Macro-economic Condition on hiring	"student", "yohanne", "inflation", "linkdin", "index", "rent", "regional", "economy", "diamond", "feeling"

the long-term consequences are costly. There was an unusual discrepancy between economic growth (quantified by GDP) and unemployment during the latest recession of 2022.

The second topic is labelled as "Unethical Retrenchments" as the keywords focus on unethical issues. The problem of moonlighting gripped India's much-vaunted IT sector, dividing it due to differing views of several key leaders on the subject. Moonlighting is the practice of a company's full-time employee taking on an extra or a side job, usually without the employer's knowledge, and earning money from both sources. Some organizations including TCS, Wipro, and Infosys, have debated side gigs for several weeks. For various reasons, many businesses have claimed that several staff have started moonlighting after the Covid-19 pandemic. The Minister of State for Electronics and Information Technology endorsed moonlighting, but advised employees to refrain from breaching their contractual obligations. According to him, this is the age of employee entrepreneurs and corporations. Businesses must recognize that there has been a significant shift in the views and attitudes of young Indian IT workers. When a company goes through layoffs, women and minority employees are not always treated equally.

The third topic is labelled as "Government Intervention on Industrial Revolution " as the keywords are focusing on various government interventions. Establishments with 100 or more employees are obliged under the ID Act to obtain prior approval from the appropriate authorities before closing, retrenchment, or layoff. The ID Act also allows for the right of laid-off and retrenched workers to be compensated and retrenched workers re-employment. According to the Act's provisions, the central and state governments try to address the difficulties of workers and defend their interests based on their respective jurisdictions as delineated in the ID Act. In his response, the minister stated that any retrenchment or layoff not carried out following the provisions of the ID Act is illegal. Economists have come to believe that, as artificial intelligence and robotics improve manual labour and other blue-collar occupations will be largely eliminated from the economy. Unlike the Industrial Revolution (IR) technologies that increase the overall number of jobs through deskilling, the most recent upheaval, the introduction of information and communications technologies, is defined by skill bias. As a result, in recent decades in the United States, more occupations requiring a high level of knowledge, such as programmers and data scientists, have been created, whereas unskilled work has been eliminated.

The fourth topic focusses on hiring and has been labelled as "Impact of Macroeconomic Condition on hiring". The recurrent crises intensify the deterioration of the status of the labour market's youngest members. Failure to integrate the young generation means a loss of production, productivity, and most likely, the innovative potential at the macroeconomic level. The condition of these people is heavily influenced by globalization processes, employment availability, credentials, and economic prosperity. The continual transformation processes and socioeconomic changes necessitate continuous monitoring of this topic and updating relevant knowledge.

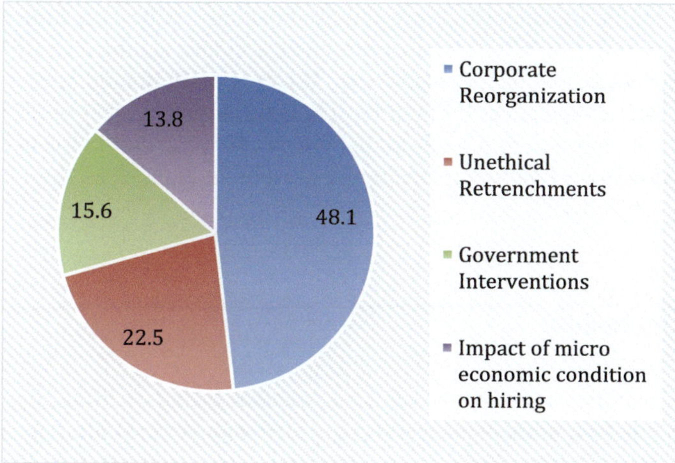

Fig. 19.3 Occurrences of Topics in the articles

The Fig. 19.3 shows the percentage of articles having a topic. The most occurring topic is Corporate Reorganization and the least discussed topic is Impact of micro economic conditions.

19.4 Implications

Although there are no fixed timelines, a path to recuperation exists. The perception of employees within the organization has undergone a noteworthy transformation. They are increasingly regarded as capital investments rather than merely a cog in the company's operations. At the same time, external circumstances such as mass layoffs, economic volatility, and general unrest have made it challenging to keep this valued talent.

HR Managers of companies must be taught to handle adverse employee reactions and retaliation and should have sound legal advice and contingency plans in place for all worst-case scenarios. Managers and the people function should also be trained to handle people's emotions at work, as mental health issues can reflect decreased productivity and destructive behaviour when under excessive stress. Balancing the interests of two key stakeholders—investors and employees—is crucial. Investors must be convinced of long-term gains because financial losses are only transitory. Instead of layoffs, employees should be encouraged to advance their careers and work extra hours to assist the company in getting through tough times. When faced with a Moonlighting dilemma in the firm, HR managers may seek to adopt regulations to minimize the adverse effect of moonlighting while reaping the few benefits.

Creating a conflict-of-interest policy and establishing a policy for further employment consent will aid. Consider incorporating a clause stating that an employee must obtain permission to work elsewhere. The employees should be provided with reasonable salaries and perks for their level of contributions so they are less prone to contemplate moonlighting. A well-crafted policy can assist the company in creating an open and constructive workplace while preventing employees from concealing their simple objectives.

19.5 Conclusion

Apart from avoiding pre-emptive choices, whether hiring or firing, organizations must have annual recruitment plans and engage in employee upskilling and reskilling. Having clarity from the start can be beneficial. The company must prioritize its culture, vision, and strategy to recruit the top individuals. When aiming to impress a potential new hire during the courting stage, companies frequently roll out the red carpet and portray themselves in the best possible light. These are challenging times. Following an encouraging first half of 2023, there are hints of stagnation and even a downturn due to various macroeconomic issues. Going back to the basics of hiring and engaging staff is the only way to manage the hiring-firing quandary successfully. IT is particularly hit as automation eliminates low-skilled jobs. According to recruitment specialists and engineering institutes, while top campuses such as Indian Institutes of Technology (IITs) may not be affected, lower-tier campuses may observe a reduction in recruitment. It could be a chance for developing organizations, startups, and unicorns to strengthen and expand their talent pool, source laid-off people, and invest in talent pipelines.

References

1. Kashyap, H. (2023). *Indian Startup Layoff Tracker: 37,260+ Employees Laid Off By 130+ Startups Since 2022.* https://inc42.com/features/indian-startup-layoffs-tracker/
2. Mujtaba, B. G., & Senathip, T. (2020). Layoffs and downsizing implications for the leadership role of human resources. *Journal of Service Science and Management, 13*, 209–228.
3. Bhatnagar, V. (2022). *The employee-employer paradox: Layoffs everywhere, but hiring still a problem.* The Ken. Retrieved February 5, 2023, from https://the-ken.com/the-employee-employer-paradox-layoffs-everywhere-but-hiring-still-a-problem/
4. Yu, H., Yan, C., Zhang, L., Dong, Z., Cheng, L., Zheng, X., & Zhao, Z. (2021). Does the employability paradox exist or not? An inverted U-shaped model. *Frontiers of Psychology.* https://doi.org/10.3389/fpsyg.2021.588793
5. Nelissen, J., Forrier, A., & Verbruggen, M. (2017). Employee development and voluntary turnover: Testing the employability paradox. *Human Resource Management., 27*, 152–168.
6. DeVos, A., Forrier, A., Der Heijden, B. V., & De Cuyper, N. (2017). Keep the expert! Occupational expertise, perceived employability and job search: A study across age groups. *Career Development International, 22*, 318–332.

7. Acikgoz, Y., Sumer, H., & Sumer, N. (2016). Do employees leave just because they can? examining the perceived employability–turnover intentions relationship. *Journal of Psychology, 150*, 666–683.
8. Wang, H., Xie, Y., & Xie, Y. (2021). Analyzing the great resignation through employee reviews: A text mining study. *Journal of Business Research, 136*, 697–704.
9. Yang, S. A., & Kim, J. H. (2021). News analysis of corporate layoffs during COVID-19 pandemic: A machine learning approach. *Sustainability, 13*(6), 3466.
10. Josephine, V. H., & Duraisamy, S. (2018). Novel pre-processing framework to improve classification accuracy in opinion mining. *International Journal of Computing, 17*, 234–242.
11. Mahara, T., Josephine, V. L. H., Srinivasan, R., Prakash, P., Algarni, A. D., & Verma, O. P. (2023). Deep vs. shallow: A comparative study of machine learning and deep learning approaches for fake health news detection. *IEEE Access, 11*, 79330–79340. https://doi.org/10.1109/ACCESS.2023.3298441
12. Kulshrestha, R. (2019). *Latent Dirichlet allocation (LDA)*. Medium. Retrieved December 21, 2022, from https://towardsdatascience.com/latent-dirichletallocation-lda-9d1cd064ffa2

Chapter 20
Modelling Technology Diffusion Process Under Dynamic Market Scenario

Ompal Singh, Hitesh Kumar, Adarsh Anand, and P. K. Kapur

Abstract Adoption process plays an important role in defining the success of any product. But usually, it is preceded by many other processes, like awareness and motivation. Therefore, in order to talk about adoption, one should inculcate the role of awareness and motivation as separate processes. Because awareness and motivation attributes are continuously changing, one has to think about the impact they have in creating a dynamic market size. In this work, authors have studied the role of awareness and motivation in the adoption of a product simultaneously. The numerical method technique has been used meticulously, in the absence of a closed-form solution. Model validation done on different data sets shows the acceptability of the models in terms of accuracy obtained.

Keywords Consumer behaviour · Diffusion of innovation · Dynamic market · Market expansion · Numerical methods

20.1 Introduction

The performance of the product in the market mainly depends on two factors which include the promotional efforts made by the company for the product and the feedback given by consumers for the product [1]. These two factors play a significant role in promoting the product to potential customers. In this way, the process of reaching the potential adopters of the newly introduced innovation in the market can be understood under the diffusion of innovation. "Diffusion of innovation" is a theory under which the method of adopting innovation by its potential adopters is studied.

O. Singh · H. Kumar (✉) · A. Anand
Department of Operational Research, University of Delhi, Delhi 110007, India
e-mail: hiteshmeena00155@gmail.com

H. Kumar
Department of Mathematics, Zakir Hussain Delhi College, University of Delhi, Delhi, India

P. K. Kapur
Amity Centre of Interdisciplinary Research, Amity University, Noida, UP, India

P. K. Kapur et al. (eds.), *Data-centric Approaches to Industrial Decisions*, Asset Analytics, https://doi.org/10.1007/978-981-96-7556-2_20

Innovation diffusion theory has been widely accepted and utilized to study the diffusion phenomenon of the new product [2–4]. This is because it can be used for better understanding of customer behaviour and the acceptance of product, especially in the consumer goods industry.

In the technological advance market, the information about the product spreads quickly among the potential market. After getting information about the existence of the product the customer wants to get fully aware about the efficiency, accuracy, the social impact of the product, usability of the product, etc. [5]. This awareness of the product comes into picture prior the eventual adoption of the product by the customer. In real market scenario, it is not the case that the customer straightforwardly purchases the product. For a clear understanding of the adoption process, it can be studied in different stages. These stages might include customer awareness, generating interest, conducting an evaluation and trial, and implementing the product [6]. Each stage is designed to assist the adopter in gaining a deeper understanding of the product and its potential benefits. The awareness campaigns assist the information to potential adopter about the existence of the product. After getting information, the customers try and evaluate the product which motivates the potential customer for innovation adoption and eventually adoption of the product takes place, which indicates that the adoption process can be studied in three stages. Some researchers have also examined the product adoption process in five stages, namely knowledge, persuasion, decision, implementation, and confirmation [7]. Due to the impact of technological development and the increasing influence of promotional strategy, the above stages can also be studied in three stages: awareness, motivation, and eventual adoption. The awareness stage comprises knowledge and persuasion; secondly, the decision and implementation can be studied under the motivation stage and lastly, the adoption stage incorporates the confirmation of the purchasing decision [8]. In short, the awareness stage is where potential customers become aware of the existence of innovation in the market and start gathering information about the innovation. In the next stage, known as the motivational stage, potential customers, after getting information, create some perception about the product. In this stage, potential customers decide to adopt or not adopt the innovation. In the final stage, potential customer adopts the innovation according to their demands and needs.

In order to increase the adoption of a product, it is very crucial for a firm to understand the behaviour of its potential adopters. Based on the potential adopter's behaviour the firms can optimize the expenditures incurred on promotion, advertisement and can help them to alter their messages more efficiently and effectively [9, 10]. Moreover, companies can adjust their promotional strategies accordingly which will result in an increased profit margin. For instance, a firm may use customer segmentation to target their advertisements to specific customer segments [11]. This will increase the likelihood that potential customers will convert from one stage to the next stage.

Further, an approximate estimation of the number of potentially aware and motivated customers at each level of the product adoption not only helps company management in creating a proper marketing strategy but also reduces costs associated with it [12]. For instance, consider a case of a firm that produces larger quantities irrespective of having less number of motivated customers and on the contrary if the motivated customers are in large numbers but the produced quality is less as compared to the demand, in both the cases the firm can experience monetary loss as well as can result in loss of goodwill of the company as there might be a possibility that due to the unavailability of the product the customers might leave the system, respectively [13, 14]. Which indicated the importance of estimating the count of potentially aware and motivated customers.

Plenty of research work has been conducted in the field of diffusion of innovation [15–17]. However, with the technological advancement and dynamic market the studies conducted prior seems to be outdated with time which makes the availability of different attributes to research on such as the effect of price, advertisement, budgetary constraints, promotional efforts, etc. over the diffusion process. In 1969, researcher Frank M. Bass presented the most well-known mathematical framework for understanding the sales of products [18]. But the framework gained recognition in 1987 when it was utilized to understand product sales across multi-generations [19]. Frank M. Bass (1969) divided the adopter categories into two categories: innovators and imitators on the basis of getting information through different media exposure. Innovators are the adopters who adopt the product by the effect of companies' promotional and advertising strategies. Adopters who adopt the product by the influence of word-of-mouth from previous adopter are termed as imitators. Due to the complexity of differentiating between innovators and imitators in 2004, an alternative form of the Bass model was proposed [20]. Most of the researchers assumed the constant potential market size [21, 22] which is not true in real life. The government policies, environmental factors, wants and needs of potential customers are some of the factors that keep on changing which can lead to expansion or reduction of potential market size [23]. Only a few researchers have presented their work considering the dynamic nature of market size. In 2012, the market expansion was studied considering the repeat purchase under stochastic phenomena [24]. Some of the work has been done considering different dynamic behaviour of potential market [6, 22]. The diffusion of innovation has also been studied using the multi-stage adoption process considering the different stages with constant market size [8, 25]. The novelty of the proposed article is the relaxation of the assumption in multi-stage adoption process, i.e., constant market size, which is not a real world marketplace scenario. The market expansion can incur due to first-time purchasers or it can be due to the repeat purchasing behaviour of consumers. The authors have considered the adoption process as three-stage process (awareness stage, motivation stage, and eventual adoption) incorporating the market expansion concept.

Section 20.1 includes the introduction and the review of the existing literature for the proposed modelling framework. In this section, the authors talk about the outline of the chapter and the significance of the study, incorporating the research gap and novelty of the proposed framework in the field of academics and marketing. Section 20.2 includes the notations and assumptions that follow the modelling framework. These basic assumptions are crucial in understanding the modelling framework. Precise definitions of different variables and parameters have been provided, ensuring readers have a comprehensive grasp of the model's components.

In Sect. 20.3, the authors explain the key terminologies necessary to understand the article's building. This section provides a basic definition of market expansion and its reason. It also offers different types of market expansion. Following this, a mathematical modelling framework with different market expansion cases has been introduced. This modelling framework is not limited to predicting sales, although using this framework, the behaviour of potential customers in different stages. Section 20.4 has solution methodology, data analysis, and model validation. The authors employed the numerical methods in the absence of an analytical solution. Also, different numerical methods should be compared to find the most appropriate method. After that, they provide a detailed analysis of data, demonstrating how the proposed modelling framework is useful in real-life scenarios. Model validation indicates the accuracy of the proposed modelling framework through various tests on actual sales. In Sect. 20.6, managerial implications help decision-making from an organizational point of view. Section 20.7 includes the future scope and applications of models in different fields. Section 20.8 has the conclusion of the chapters, which indicates the outcome of this discussed chapter. The final sections of the article include acknowledgements and references. The acknowledgements section pays tribute to individuals and institutions that supported the research. At the same time, the references provide a comprehensive list of all sources cited throughout the paper, ensuring the academic field and allowing readers to trace the study's intellectual lineage.

20.2 Notations and Assumptions

This section comprises notations and the assumptions that have been used for modelling the proposed modelling framework.

20.2.1 Notations

The notations used in the proposed model are as follow:

$m(t)$ = Dynamic market size of potential adopters.

$N_{AW}(t)$ = Count of aware potential adopters till time t.

$N_{MO}(t)$ = Count of motivated potential adopters till time t.

$N_{AD}(t)$ = Number of adopters till time t.

b_1 = Rate by which awareness is spread.

b_2 = Rate of conversion in motivated potential adopters from aware potential adopters.

b_3 = Rate of conversion in adopters from aware motivated potential adopters.

β_1 = learning parameter for spreading awareness.

β_2 = learning parameters for motivation among aware potential customers,

β_2 = learning parameters for motivation adoption.

$N_{AW'}(t) =$ Number of aware and non-aware potential customers at time t but not motivated.

$N_{MO'}(t) =$ Number of motivated potential customers but not adopters.

$\alpha =$ Rate by increment in first-time adopters.

$\lambda =$ proportion of repeat purchasers.

20.2.2 Assumption

The fundamental assumptions considered for modelling the proposed model are given below.

1. Before final adoption, each potential customer passes through all stages of product adoption, i.e., Potential customers first become aware, then they become motivated, and then they become adopters.
2. By influence of first-time adopters, the market can expand either exponentially or linearly.
3. Rate of the spread of awareness, rate of conversion from awareness to motivation and adoption rate from motivators have been assumed to be logistic in behaviour.

20.3 Mathematical Modelling

Before modelling the framework, one must understand some important activities followed by the market for better understanding of the model. The relevant literature related to the proposed study has been provided in further subsection.

20.3.1 Market Expansion

Market expansion can be defined as an increase in the potential market size for a product [23]. According to the saying, "One should either create the product based on the needs of the market, or create the market demand for the product first, then provide the solution based on the need". Both ways are intended to increase the demand for the product, and to do so, the company tries to increase the number of potential customers. Therefore, a situation where the number of potential customers of the product increase in comparison to the initial potential customers can be defined as a market expansion [25]. Market expansion is primarily driven by two types of customers. Firstly, the potential customers who are buying the product for the first time, and secondly, the potential customers who are adopting the product again. Despite the fact that both types of potential customers behave differently, most researchers have studied these market expansions in a similar way. Firms can implement different types of market expansion strategies to cater to these distinct market expansions. Moreover, the implementation of market expansion scenarios has been discussed in detail in further subsections.

20.3.1.1 Market Expansion by First-Time Purchasers

In this type of expansion, the potential product customer purchases the product for the first time. For example, drones have been used in army operations; however, these days, their use and application have been extended beyond military operations. Drones are now widely used in various fields, such as delivery partners, agriculture for crop monitoring, real estate for aerial photography, etc. So, the drone has a meagre count of potential customers in the initial time phase. Still, as it finds different applications in different fields, the count of potential customers increases. Smartphones are another example. They were initially innovated for businesses and professionals who needed emails and access to the Internet; however, their uses and applications have extended. Today, smartphones have become an integral part of human life, and they are used for social media, educational purposes, gaming purposes, health consultations, etc. For these types of customers, firms initially creates awareness about the product, and after that companies try to motivate these potentially aware customers to make the final purchase. In literature, market expansion considering the first-time purchaser is generally studied in two scenarios [26]:

i. Exponential market expansion
ii. Linear market expansion

Exponential Market Expansion

In exponential market expansion scenario market expansion increased exponentially. If m be the initial market size and α be the rate by which potential market size increases then potential market size at time t can be given as $me^{\alpha t}$.

Linear Market Expansion

Market expansion has been considered linearly in this section. If m be the initial market size and α be the rate by which potential market size increases then potential market size at time t can be given as $m(1 + \alpha t)$.

Model 1 Adoption Process Inculcating Market Expansion by First-Time Purchasers.

Stage 1 To get the attention of potential adopters during the launch of a new product, the company uses different marketing strategies to spread awareness of the innovations. The awareness rate of the product's existence in the market is proportional to the number of potential adopters who remain aware of the product's existence. The differential equation for the given scenario can be represented as Eq. (20.1)

$$\frac{dN_{AW}(t)}{dt} = \frac{b_1}{1 + \beta_1 e^{-b_1 t}}(m(t) - N_{AW}(t)); \quad N_{AW}(0) = 0 \tag{20.1}$$

where $m(t)$ denotes the market expansion; it may follow any scenario of market expansion. In this article, the authors have considered the most commonly used forms of market expansion, i.e., the exponential market expansion (Model 1(a)) and the linear market expansion (Model 1(b)).

Stage 2 In this stage, the authors have considered the situation when the aware potential customers want to know about the product which motivates them to gather information about product capacity, usability and other features and services offered by the firms. Based on these factors, potential customers become motivated towards the product, and based on their knowledge and needs, they form a positive perception towards the product. Moreover, the rate of motivation depends on the remaining number of aware potential customers which can be seen from Eq. (20.2).

$$\frac{dN_{MO}(t)}{dt} = \frac{b_2}{1 + \beta_2 e^{-b_2 t}}(N_{AW}(t) - N_{MO}(t)); \quad N_{MO}(0) = 0 \tag{20.2}$$

It is a crucial aim for Marketers to set marketing strategies to influence and attract potential adopters towards product adoption. Understanding the count of aware and non-aware potential customers is essential for this process. By using Eqs. (20.1) and (20.2), it is not an easy task to estimate the count of aware and unaware potential customers. Therefore, the number of aware and unaware potential customers can be given by the difference between motivated potential customers and the potential non-aware potential customers which can be represented as Eq. (20.3).

$$N_{AW'}(t) = m(t) - N_{MO}(t) \tag{20.3}$$

With the help of the Eq. (20.3), managers can easily estimate the number of known and unknown potential customers. On the basis of these numbers, company management implements necessary and require marketing strategies.

Stage 3 This stage incorporates the marketing scenario where the motivated potential customers will purchase the product on the basis of their needs. This also implies that motivation does not imply adoption; both can be differentiated as individual stages. Further, the rate of adoption depends on the remaining number of motivated potential customers. The number of adopters can be easily estimated using Eq. (20.4).

$$\frac{dN_{AD}(t)}{dt} = \frac{b_3}{1 + \beta_3 e^{-b_3 t}}(N_{MO}(t) - N_{AD}(t)); \quad N_{AD}(0) = 0 \tag{20.4}$$

Undoubtedly, the sales of the product can be predicted with the help of the above equation. Moreover, it is essential for the firms to estimate the number of potential customers who are motivated to make the purchase of the product based on product's performance in the market but due to some reasons couldn't buy it. By knowing the

Table 20.1 Multi-stage adoption process in different market scenarios without repeat purchasing

Type of market expansions	Multi-stage adoption process
a. Exponential Market Expansion	$\frac{dN_{AW}(t)}{dt} = \frac{b_1}{1+\beta_1 e^{-b_1 t}}\left(me^{\alpha t} - N_{AW}(t)\right); N_{AW}(0) = 0$
	$\frac{dN_{MO}(t)}{dt} = \frac{b_2}{1+\beta_2 e^{-b_2 t}}(N_{AW}(t) - N_{MO}(t)); N_{MO}(0) = 0$
	$\frac{dN_{AD}(t)}{dt} = \frac{b_3}{1+\beta_3 e^{-b_3 t}}(N_{MO}(t) - N_{AD}(t)); N_{AD}(0) = 0$
b. Linear Market Expansion	$\frac{dN_{AW}(t)}{dt} = \frac{b_1}{1+\beta_1 e^{-b_1 t}}(m(1+\alpha t) - N_{AW}(t)); N_{AW}(0) = 0$
	$\frac{dN_{MO}(t)}{dt} = \frac{b_2}{1+\beta_2 e^{-b_2 t}}(N_{AW}(t) - N_{MO}(t)); N_{MO}(0) = 0$
	$\frac{dN_{AD}(t)}{dt} = \frac{b_3}{1+\beta_3 e^{-b_3 t}}(N_{MO}(t) - N_{AD}(t)); N_{AD}(0) = 0$

estimated count of such potential customers, the company can increase its product sales to a significant extent by converting such potential customers into adopters of the product. The differential Eq. (20.5) can provide the estimated potential customers number who were motivated to make the final purchase but could not adopt it.

$$N_{MO'}(t) = N_{MO} - N_{AD}(t) \tag{20.5}$$

With the help of the Eqs. (20.3), (20.4), and (20.5), the sales process of any product can be easily understood, which is very helpful for management in understanding the sales of newly introduced products. Considering different market scenarios multi-stage adoption process can be given as follows (Table 20.1).

20.3.1.2 Market Expansion by Repeat Purchasing

This section discusses consumer repeat purchasing behaviour. Customers who have already purchased a product are generally more familiar with its use, capabilities, capacity, and the services the firm offers. This familiarity makes it easier for firms to motivate repeat customers to buy products than first-time buyers.

For example, consider the smartphone brand iPhone. Most iPhone users re-adopt the product because of its brand value and social influence. A similar pattern can be observed in the retail industry, where customers who regularly purchase clothing from the same brand and company are more likely to continue doing so.

Every company aims to improve its repeat sales, employing various strategies to achieve this goal. The importance of repeat purchasing lies in the fact that companies have a clear understanding of their potential customers. By analysing feedback from previous adopters, companies gain insights into the needs of potential customers and adjust their marketing strategies accordingly. Companies often send emails and messages to encourage customers to re-adopt their products, offering significant discounts on repeat purchases.

One effective marketing strategy to boost innovation sales is the use of membership cards and online reward points. These incentives attract previous adopters to make repeat purchases. For instance, H&M, a popular clothing brand, employs this strategy by providing membership cards. The company collects customer contact information through these cards and sends regular emails with discounts. Additionally, they offer reward points on every purchase, encouraging repeat sales.

Model 2 The Adoption Process Inculcates Market Expansion by First-Time Purchasers as well as Repeat Purchasing

Stage 1 On the similar note of Model 1, this stage models the potential non-aware adopters who become aware of the product's existence in the market by internal or external factors, for which the rate of spreading awareness can be understood by Eq. (20.6).

$$\frac{dN_{AW}(t)}{dt} = \frac{b_1}{1+\beta_1 e^{-b_1 t}}(m(t) - N_{AW}(t)); \quad N_{AW}(0) = 0 \tag{20.6}$$

Here, $m(t)$ denotes the market expansion as discussed in Model 1. For Model 2, the authors have considered exponential market expansion (Model 2(a)) and linear market expansion (Model 2(b)) forms of market expansion.

Stage 2 As discussed in the model (1), in this stage aware potential customers gather information about the product and convert it into motivation. But an important question arises here, "who can be motivated towards the product?" the Answer to this question is very easy; aware potential customers as well as some proportion of adopters can be motivated towards product adoption. Thus, the potentially motivated customers can be defined as $N_1(t) + \lambda N_3(t)$ and rate of motivation can be represented as Eq. (20.7).

$$\frac{dN_{MO}(t)}{dt} = \frac{b_2}{1+\beta_2 e^{-b_2 t}}(N_{AW}(t) + \lambda N_{AD}(t) - N_{MO}(t)); \quad N_{MO}(0) = 0 \tag{20.7}$$

Since the company's objective is to constantly increase the sales of the marketed product, the company wants to know the approximate number of potentially motivated adopters who can adopt the product. In a fundamental way, non-aware potential customers, aware potential customers, and a portion of previous adopters may be motivated to re-adopt the product. Aware potential adopters may be motivated to purchase the product because of the brand recognition or the quality of the product. Previous adopters may be motivated to re-adopt the product due to loyalty and a positive experience with the product. Indirectly, repeat potential adopters might be considered aware potential adopters. Thus, the number of non-aware and aware adopters for the product at any fixed time can be given as Eq. (20.9).

$$N_{AW'}(t) = m(t) + \lambda N_{AD}(t) - N_{MO}(t) \tag{20.8}$$

Table 20.2 Multi-stage adoption process in different market scenarios with repeat purchasing

Type of market expansions	Multi-stage adoption Process
a. Exponential Market Expansion	$\frac{dN_{AW}(t)}{dt} = \frac{b_1}{1+\beta_1 e^{-b_1 t}}(me^{\alpha t} - N_{AW}(t)); N_{AW}(0) = 0$ $\frac{dN_{MO}(t)}{dt} = \frac{b_2}{1+\beta_2 e^{-b_2 t}}(N_1(t) + \lambda N_3(t) - N_{MO}(t)); N_{MO}(0) = 0$ $\frac{dN_{AD}(t)}{dt} = \frac{b_3}{1+\beta_3 e^{-b_3 t}}(N_{MO}(t) - N_{AD}(t)); N_{AD}(0) = 0$
b. Linear Market Expansion	$\frac{dN_{AW}(t)}{dt} = \frac{b_1}{1+\beta_1 e^{-b_1 t}}(m(1 + \alpha t) - N_{AW}(t)); N_{AW}(0) = 0$ $\frac{dN_{MO}(t)}{dt} = \frac{b_2}{1+\beta_2 e^{-b_2 t}}(N_1(t) + \lambda N_3(t) - N_{MO}(t)); N_{MO}(0) = 0$ $\frac{dN_{AD}(t)}{dt} = \frac{b_3}{1+\beta_3 e^{-b_3 t}}(N_{MO}(t) - N_{AD}(t)); N_{AD}(0) = 0$

Stage 3 In this stage, motivated potential customers purchase the product whenever they need it. The final adoptions of the product can be understood using Eq. (20.9).

$$\frac{dN_{AD}(t)}{dt} = \frac{b_3}{1 + \beta_3 e^{-b_3 t}}(N_{MO}(t) - N_{AD}(t)); \quad N_{AD}(0) = 0 \tag{20.9}$$

Companies also want to know the approximate number of motivated potential customers so that they can implement their different marketing strategies to sell the product to motivated potential customers. That's why marketers need to know the approximate number of motivated potential customers that can be given by Eq. (20.10).

$$N_{MO'}(t) = N_{MO} - N_{AD}(t) \tag{20.10}$$

At last, the Eq. (20.8) can be utilized to obtain the approximated number of aware and non-aware potential customers and similarly Eq. (20.10) can be used to give approximated motivated potential customers and Eq. (20.9) can be utilized to predict the sales of the product. Considering different market scenarios multi-stage adoption process considering repeat purchasing can be given as follows (Table 20.2).

20.4 Solution Methodology

As outlined in the proposed mathematical modelling framework, the authors focus on the framework as simultaneous ordinary differential equations (ODEs). However, an explicit solution for these simultaneous ODEs must still be included. In other words, an explicit or analytical solution of these simultaneous ODEs is not obtainable using traditional methods. Numerical methods can be applied to obtain a numerical solution to overcome this issue. Numerical methods are techniques used to find numerical solutions to problems without actually solving them analytically.

The literature provides various numerical methods for different types of problems [27]. These methods are used to find the root of any equation, perform integration, calculate derivatives, determine areas and volumes, and perform interpolation and extrapolation. Some of the most widely used numerical methods to deal with ordinary and simultaneous differential equations include the Picard method, Taylor series method, Euler's methods, improved Euler's method, modified Euler's method, Runge–Kutta method, Milne-Multistep method, Adams–Bashforth methods, and predictor–corrector methods [28–30] (Table 20.3).

In the proposed mathematical modelling framework, the author has chosen the Runge–Kutta method of order four for numerical illustrations. This decision has been made based on the balance of high accuracy and reasonable computational complexity. The Runge–Kutta method is particularly advantageous because it minimizes truncation errors while providing a robust solution without needing more than one boundary condition. By employing this method, an effectively approximate numerical solution of the proposed modelling framework can be obtained, ensuring the reliability and precision of their numerical illustration [31].

20.5 Data Analysis and Model Validation

To show the usefulness and relativity of the proposed modelling framework in real life, the authors used two real life and technological products data sets. The first data set (DS1) is of apple iPhone sales data[1] from Q3 of 2007 to Q2 of 2021, and the second data set is of ACER personal computer sales data[2] from Q3 of 2009 to Q4 of 2022. As mentioned earlier, the proposed frameworks need to yield analytical solutions. Consequently, their parameters cannot be estimated using traditional methods. Instead, numerical methods are employed to predict sales and estimate parameters effectively. To calculate the values of different parameters, initial guesses can be considered by observing the datasets. Initial values from the dataset help find the adoption rate and learning parameters.

The peak value and nature of the modelling framework assist in estimating the initial guess of the market size. If the modelling framework assumes a constant market size, then the initial values of the market size must be considered more significant than the last value in the dataset. Conversely, if the authors assume market expansion, the initial market size can be less than the final sales value. Another approach to finding the initial guess of the market size is to consider that if the market size is constant and product sales are not increasing at some final points, then the initial market size can be regarded as equal to the final value of the dataset. Similarly, if the final values of

---

[1] https://dazeinfo.com/2018/08/24/apple-iphone-sales-worldwide-by-quarter-graphfarm/.

[2] https://www.statista.com/statistics/298947/pc-shipments-worldwide-acer-group/#:~:text=In%20the%20fourth%20quarter%20of,the%20company%20sixth%20among%20vendors.

Table 20.3 Comparison between different Numerical methods to deal ODEs

Method	Comparison	Advantage	Disadvantage	Applicability
Taylor's Series method	Expand the function into Taylor series around a point	High Accuracy for small intervals	Complexity increases with higher-order terms	Suitable for small intervals where high precision is required
Picard's method	Iterative approach that successively approximate the solution	Useful for proving existence and uniqueness	Slow convergence rate, impractical for complex problem	Theoretical interest, limited practical use due to slow convergence
Euler's method	Steps forward using the derivative at the current point	Easy to implement	Significant truncation errors, less accurate	Simple problem or educational purpose, not ideal for high accuracy
Improve Euler's method	Enhance Euler's method by averaging slopes at beginning and end of interval	Reduces truncations errors compared to Euler's method	More complex than Euler's method	Provides a good balance of simplicity and improved accuracy
Modified Euler's method	Further refines Improve Euler by iterating on predicted values and recalculating slopes	Better accuracy by iterative corrections	Increased computational efforts	Suitable when higher accuracy is needed without full complexity of Runge–Kutta of order 4
Heun's method	Similar to Improved Euler's method, it adjusts initial step using average slopes	Significantly reduce truncation error	Slightly more complex than basic Euler's method	Useful for problems needing moderate accuracy improvements
Runge–Kutta (2nd Order)	Evaluate the derivative at multiple points within each step	Balances accuracy and computational cost	Less accurate than higher-order methods	Suitable for moderate accuracy needs
Runge–Kutta (3rd Order)	Similar to second order but with high accuracy	Provide better accuracy than 2nd order method	More Computationally intensive than 2nd order	Appropriate when higher accuracy is required

(continued)

Table 20.3 (continued)

Method	Comparison	Advantage	Disadvantage	Applicability
Runge–Kutta (4th Order)	Evaluate the derivative at four points within each step to provide high accuracy	High accuracy and efficiency	More complex than lower-order methods	Wieldy used due to its accuracy and reasonable computational efforts
Predictor–corrector methods	Use an initial prediction step followed by a correction step to refine the solution	High accurate	Required both initial and boundary conditions	Best for problems with well-defined initial and boundary conditions
Adams–Bashforth methods	Explicit multistep methods using past values to predict values	Efficient for known previous points	Needs multiple initial conditions to start	Suitable for large dataset with known past values and needing efficiency

the dataset show increasing patterns, then the initial values of the market size should be sufficiently large.

After guessing the initial values, the objective is to maximize the value of R-square and minimize the values of SSE, RMSE Bias, Mean Absolute Error, Variance, and so on. To optimize the values of different statistical measurements, authors need to adjust the values of different parameters (Tables 20.4, 20.5).

Table 20.4 The approximated parameter values for DS1

Parameters/ Model	m	α	b_1	β_1	b_2	β_2	λ	b_3	β_2
Model 1(a)	1095	0.0165	0.42	5	0.32	4.2	–	0.102	23
Model 1(b)	1248.36	0.0172	0.756	4.32	0.482	3.6	–	0.097	28.25
Model 2(a)	1055.5	0.017	0.658	15	0.372	20.35	0.0806	0.084	11.36
Model 2(b)	1240.25	0.01505	0.536	8.32	0.356	7.6	0.0628	0.103	29.12

Table 20.5 The approximated parameter values for DS2

Parameters/ Model	m	α	b_1	β_1	b_2	β_2	λ	b_3	β_2
Model 1(a)	175	0.016	0.882	0.2	0.582	0.35	–	0.155	0.365
Model 1(b)	188.02	0.0179	0.818	0.355	0.452	0.256	–	0.126	0.028
Model 2(a)	160.36	0.0170	0.885	0.202	0.556	0.38	0.0825	0.149	0.089
Model 2(b)	165.02	0.0235	0.892	0.332	0.539	0.0125	0.0282	0.139	0.0256

In the above tables of parameters, the impact of repeat purchasing can be directly seen in potential market size. The potential market size of "Model 1(a)" is greater than the potential market size of "Model 2(a)". Similarly, the potential market size of "Model 1(b)" is greater than the potential market size of "Model 2(b)".

To show goodness of fit of proposed modelling framework, authors use different statistical measurements, in which authors include R^2, MSE, MAE, variance, and RMPSE. High value of R^2 but less value of MSE, MAE variance, and RMPSE indicate good fitting of modelling framework [32] (Tables 20.6, 20.7).

By observing Tables 20.3 and 20.4, values of R^2, MSE, MAE, Variance, and RMPSE indicate that proposed modelling framework shows very good and acceptable results. By observing Tables 20.3 and 20.4, one can see that error in predictions of sales provided by "Model 2(a)" is less as compared to error in predictions of sales provided by "Model 1(a)". Similar results can be seen for "Model 2(b)" and "Model 1(b)". These results show the importance the study of repeat purchasing behaviour during study of market expansions.

Furthermore, for more clarity of proposed modelling framework goodness of fit curves can be used, in which actual sales and sales predicted by proposed model can be plotted in a graph (Figs. 20.1, 20.2, 20.3, 20.4).

Above figures indicate the goodness of fit of proposed modelling framework. If we see, there are a little gap between predicted and actual sales for Model 1(a) and Model 1(b), but this difference is less as compare to Model 2(a) and Model 2(b). Now author provides the graphs for predicted sales and actual sales for DS2 for all models, which are given below (Figs. 20.5, 20.6, 20.7, 20.8):

Above figure also indicate that Model 2(a) and Model 2(b) have predicted sales as good as compared to Model 1(a) and Model 1(b), respectively.

The presented modelling framework has supplied acceptable results for estimating the sales of product, but the authors intend to understand the product's future status

Table 20.6 Values of different statistical errors of DS1

	R^2	MSE	MAE	Variance	RMPSE
Model 1(a)	0.998	459.140	14.736	20.626	21.4275
Model 1(b)	0.998	703.393	19.982	21.999	26.521
Model 2(a)	0.999	336.146	14.980	18.234	18.334
Model 2(b)	0.999	282.744	12.148	14.299	16.815

Table 20.7 Values of different statistical errors of DS2

	R^2	MSE	MAE	Variance	RMPSE
Model 1(a)	0.996	26.82458	3.934834	5.086761	5.179245
Model 1(b)	0.993	58.34067	5.854488	7.151366	7.638107
Model 2(a)	0.999	14.05556	2.311667	3.420682	3.749075
Model 2(b)	0.999	24.72077	3.88955	9.767408	9.841409

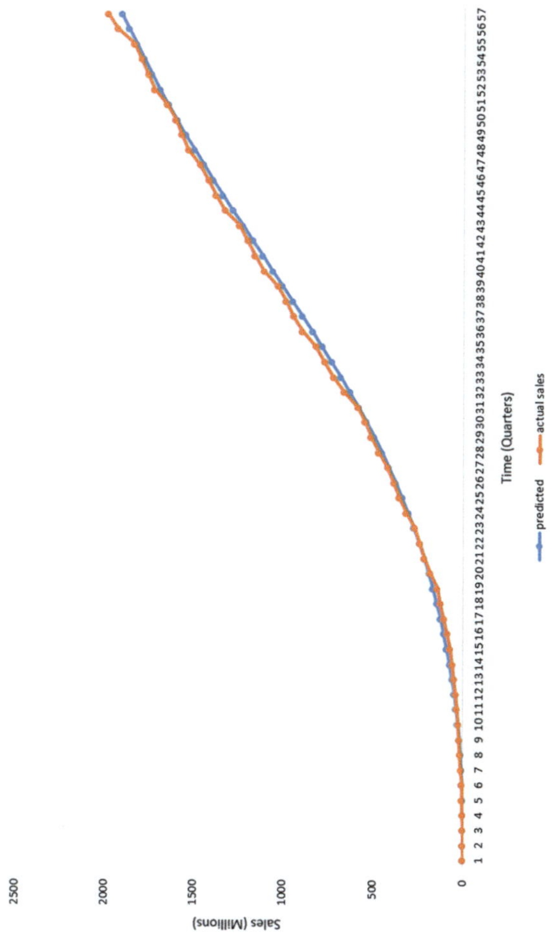

Fig. 20.1 Goodness of fit for Model 1(a) for DS1

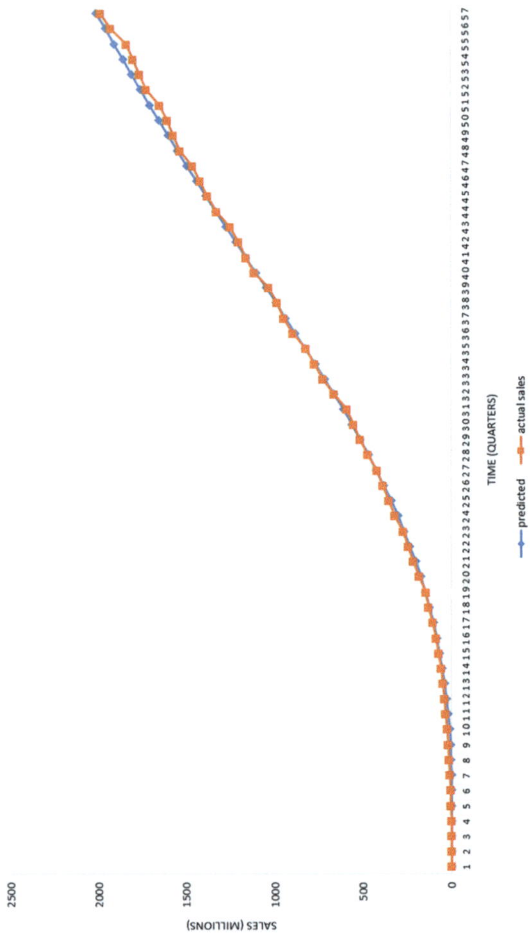

Fig. 20.2 Goodness of fit for Model 1(b) for DS1

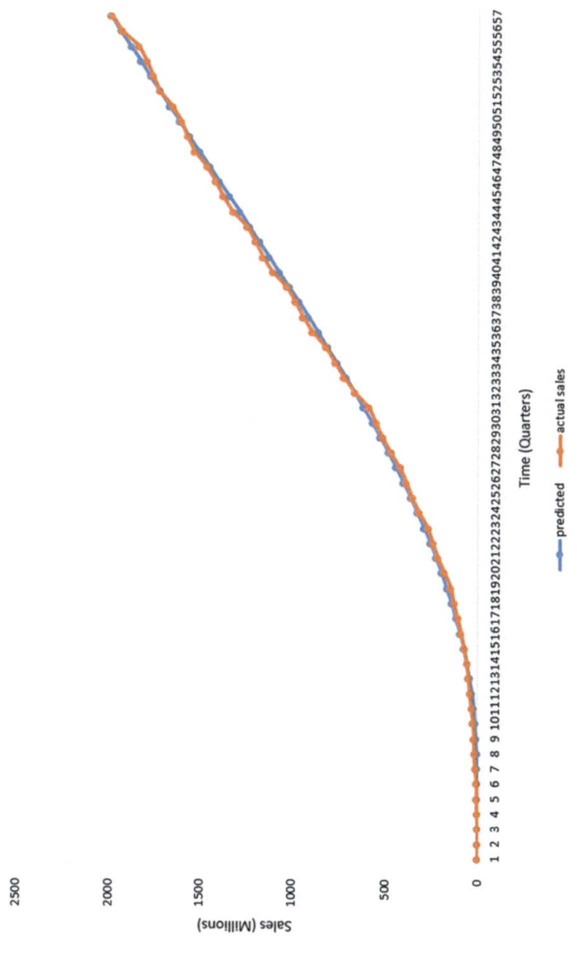

Fig. 20.3 Goodness of fit for Model 2(a) for DS1

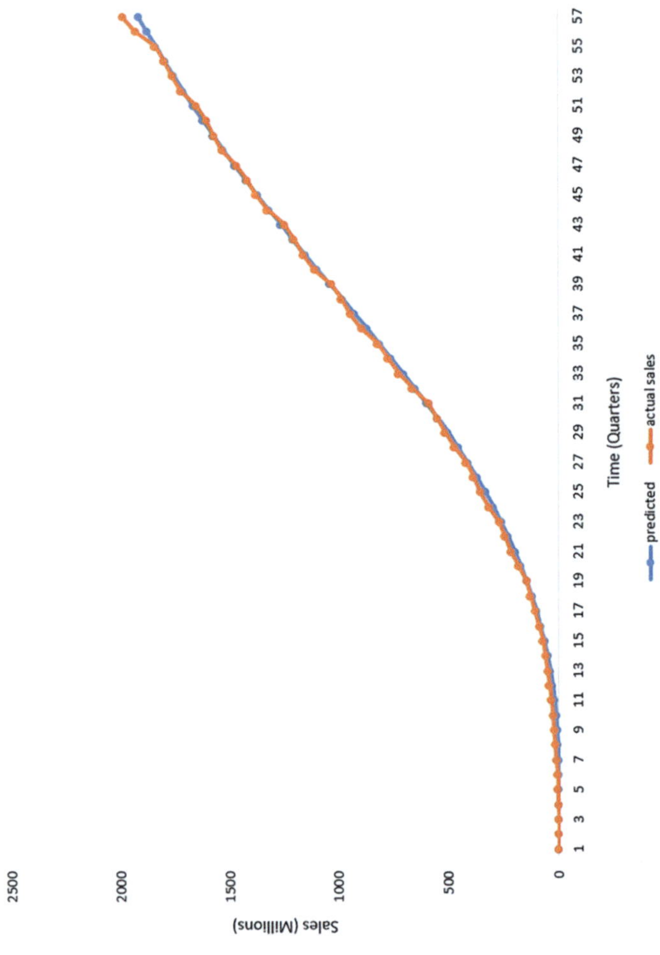

Fig. 20.4 Goodness of fit for Model 2(b) for DS1

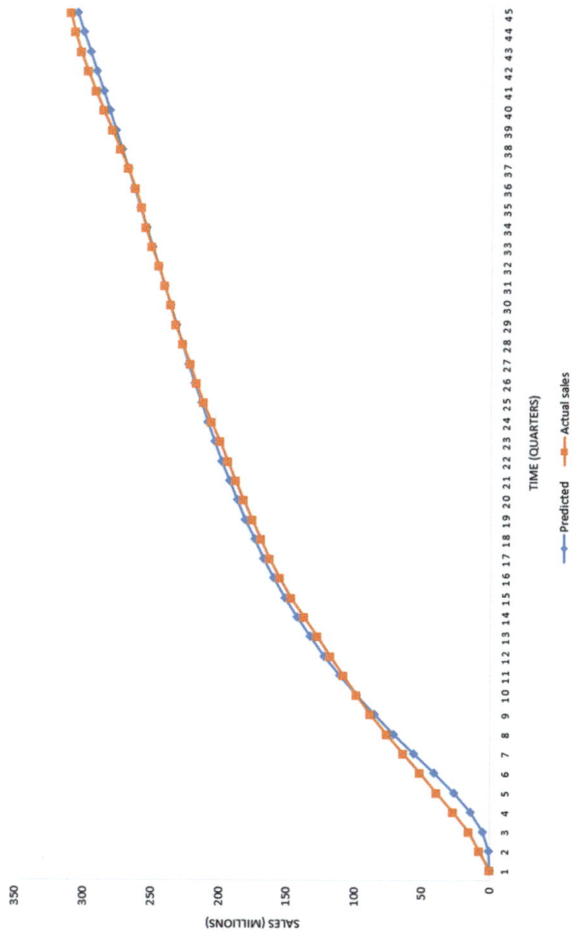

Fig. 20.5 Goodness of fit for Model 1(a) for DS2

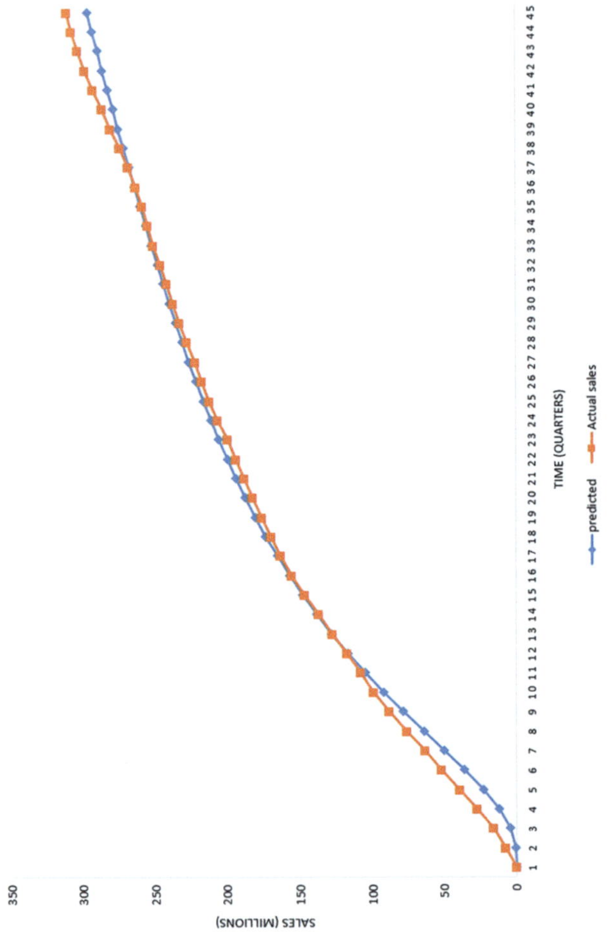

Fig. 20.6 Goodness of fit for Model 1(b) for DS2

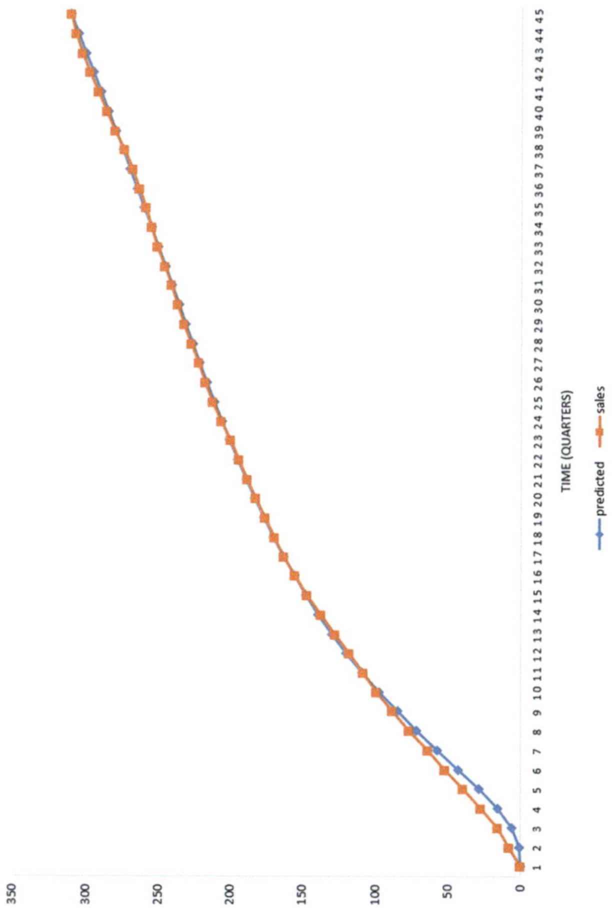

Fig. 20.7 Goodness of fit for Model 2(a) for DS2

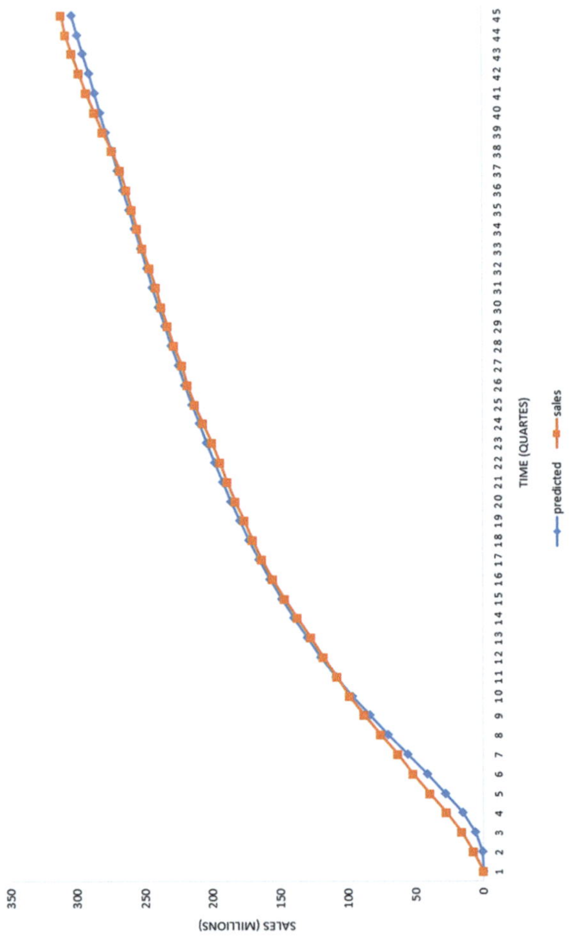

Fig. 20.8 Goodness of fit for Model 2(b) for DS2

along with its sales through this article. For this reason, it is essential to know the status of potential customers at every stage, i.e., first of all, the company needs to determine the count of aware and non-aware potential customers. After that, the company wants to determine the count of such attractive potential customers. This is because they like the performance of the product in the market but have not adopted it due to some factors. In other words, companies want to be aware of the number of motivated but not adopters potential adopters.

For Model 1, Eq. (20.3) helps to determine number of aware and non-aware potential adopters, Eq. (20.5) helps to determine motivated potential adopters and Eq. (20.4) helps to predicted sales of the product. Graph given below (Fig. 20.9), displayed numbers of aware and non-aware, motivated and predicted sales (Figs. 20.10, 20.11, 20.12).

With the help of above models, one can know count of potential customers at different stages. For example, suppose, if manager want to know status of market population in different stages in case of exponential market expansion at 25th time point for DS1, from Fig. 20.5 of Model 1(a), it can be seen at 25th time point, there are 147.864 million have aware or non-aware by the existence of product in market, 1138.31 million are motivated towards to product but not adopt the product yet, and 340.85 million has been estimated sales. Similarly, suppose the manager wants to know the number of potentially aware and non-aware potential adopters for the DS2 at time point 10. In that case, the above graph shows about 20 million potentially aware and non-aware adopters and 105 million potentially motivated adopters who did not adopt the product until this point. Predicted adopters at this time have been about 97 million.

For Model 2, aware and non-aware potential customers can be determined by Eq. (20.8) and motivated potential customers can be determined by Eq. (20.10) and prediction of sales can be done by Eq. (20.9). With the help of these equation, role of customers can be given as given in graph below (Figs. 20.13, 20.14, 20.15, 20.16):

By above graphs, one can determine the aware non-aware, motivated, and sales of the product at any given time, for example, suppose a manager of a company want to determine the status of population in different stages at time $t = 30$ in case of exponential market expansion for DS2, in this case, 24.43 million are aware or non-aware potential customers, 32.48 million are motivated potential customers, and 236.256 are predicted sales and $236.256*0.0282 = 6.6624$ million are satisfied by the use of product and they can adopt product in future. Similarly, at other time points, it is possible to accurately and effectively predict the categories of individuals who are aware but not yet motivated, those who are unaware, those who are motivated, and the cumulative number of adopters.

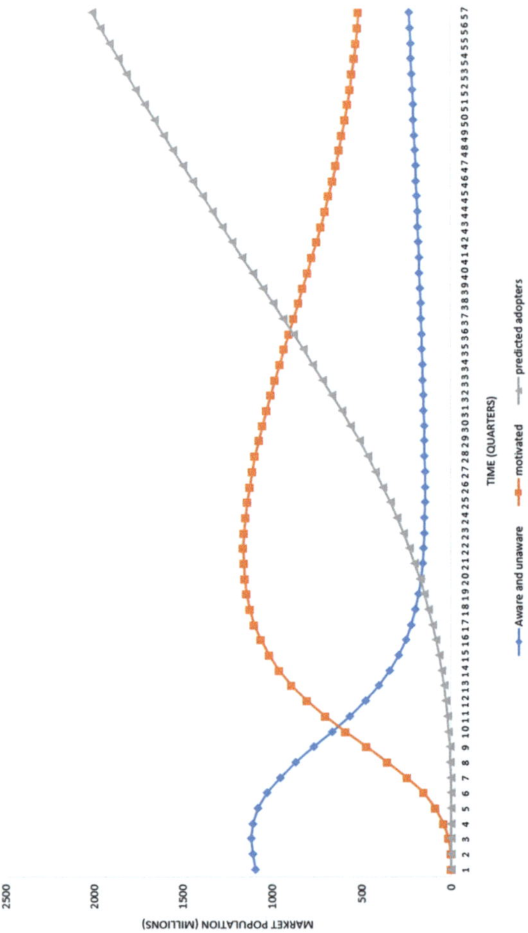

Fig. 20.9 Potential adopters in different role for Model 1(a) for DS1

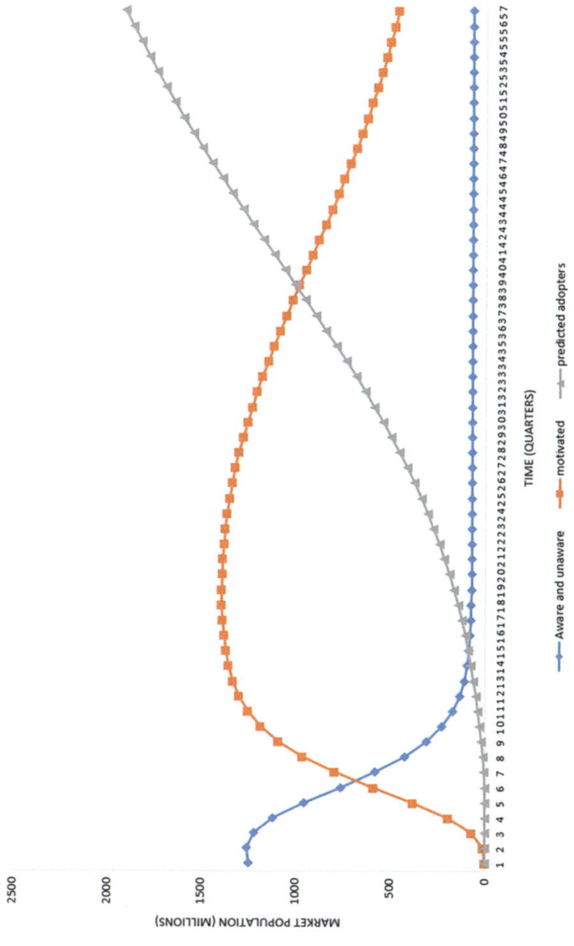

Fig. 20.10 Potential adopters in different role for Model 1(b) for DS1

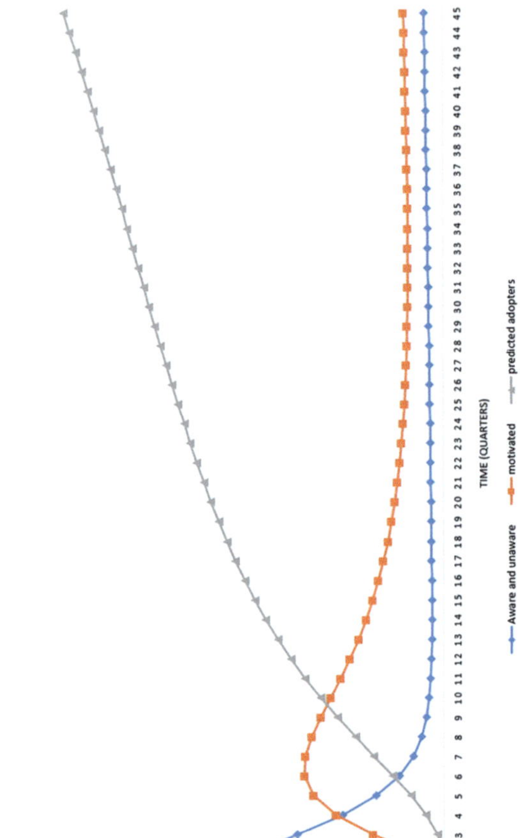

Fig. 20.11 Potential adopters in different role for Model 1(a) for DS2

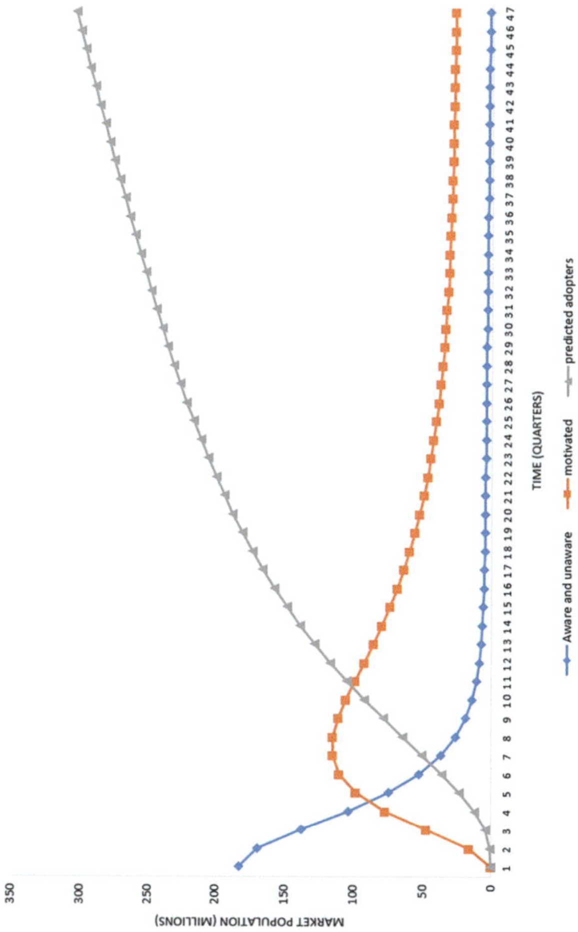

Fig. 20.12 Potential adopters in different role for Model 1(b) for DS2

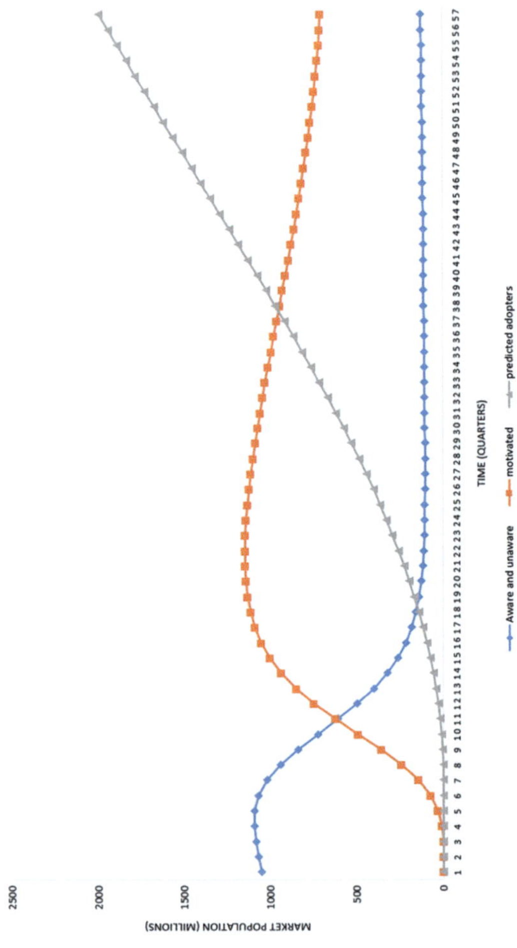

Fig. 20.13 Customers in different role for Model 2(a) for DS1

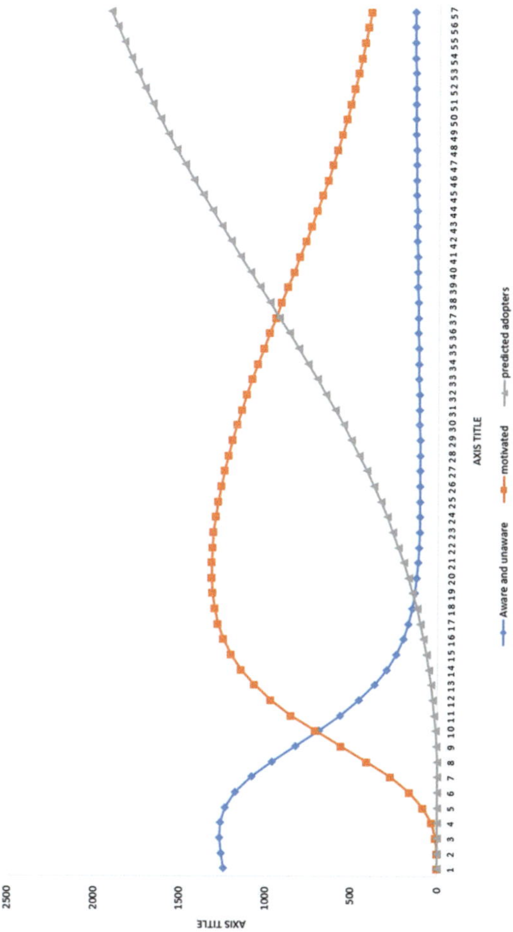

Fig. 20.14 Customers in different role for Model 2(b) for DS1

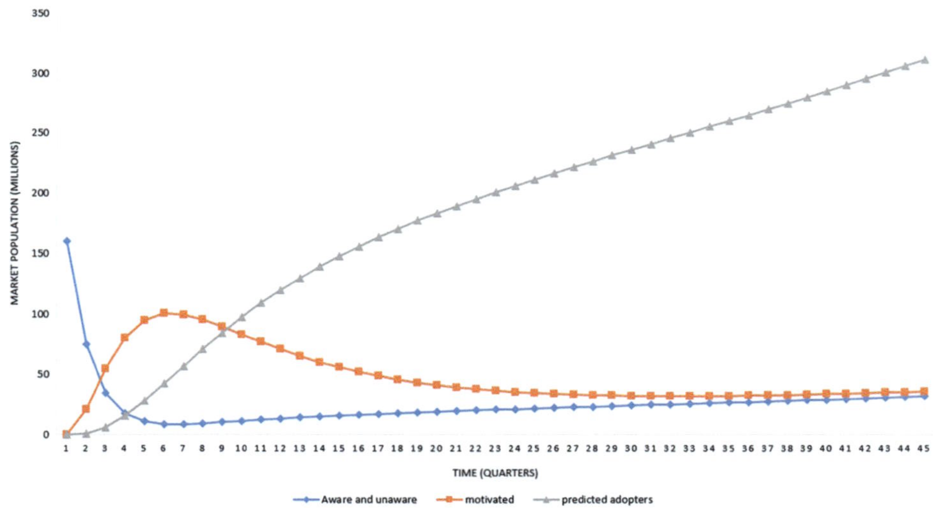

Fig. 20.15 Customers in different role for Model 2(a) for DS2

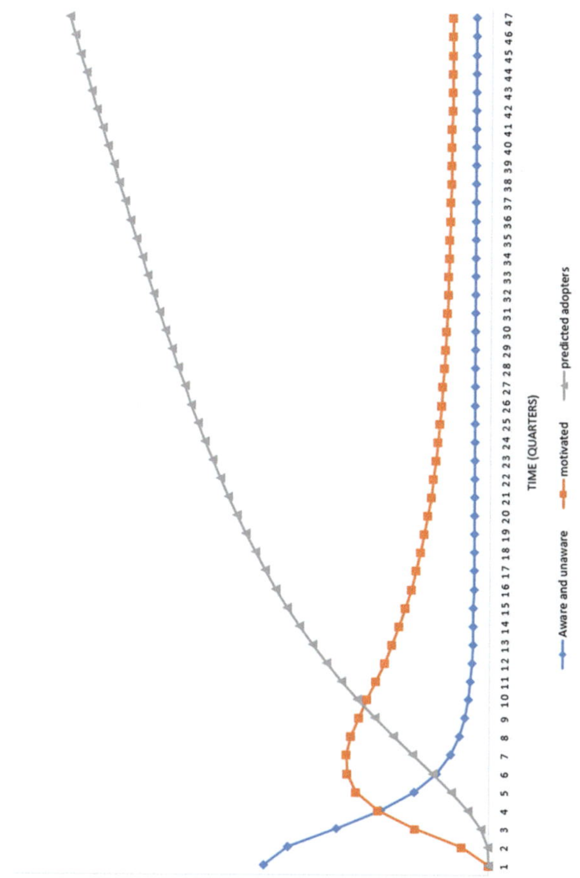

Fig. 20.16 Customers in different role for Model 2(b) for DS2

20.6 Managerial Implications

The concept discussed in this article helps to provide full-flagged information about the market population at every stage. The modelling framework discussed in this article helps the company manager implement different marketing strategies after determining the number of potential product customers at each stage. Knowing the number of potential customers at each stage will give the manager an idea of how much of the product needs to be sold and the quantity the company needs to produce in the market. With the help of this article, management can take the sales of the product to the desired level without incurring much cost.

Furthermore, detailed information about the dynamic market population allows targeted marketing strategies. Manager can derive their marketing strategies and efforts to different segments of the market, ensuring that promotional activities affect the right person at the right time. With this, marketing strategies can be set effectively with a high conversion rate. In summary, this chapter's modelling framework helps understand the market population at each stage, optimize production and inventory management, identify new growth opportunities, and tailor promotional efforts to specific customer segments. Ultimately, the methodology used in this chapter is helpful for management to achieve the desired sales level efficiently and cost-effectively, driving overall business success.

20.7 Future Scope

In this article, the authors assume that every aware potential customer must convert into a motivated one, and every motivated potential customer must adopt the product. However, these assumptions might not hold in real life. For example, for any product, some potential customers became aware of the product, but after gathering information, they may be motivated towards the product adoption. Similarly, some potential customers became motivated towards product adoption. Still, due to some constraints like financial issues and better alternatives, they do not need to proceed to accept the product. These problems can be considered in the future, and a more realistic modelling framework can be developed for these situations.

20.8 Conclusion

In this study, the authors suggest an effective and important framework to understand the adoption behaviour of potential adopters in the multi-stage adoption process in considering the dynamic market scenarios. The solution methodology used in this article can be established in case of a lack of closed-form solutions of ODE-based modelling frameworks. Furthermore, the proposed modelling framework evaluated on real-life data sets. Results of the proposed models indicate that market expansion due to first increment in the first-time potential adopters and repeat purchasers should be studied together.

Additionally, the proposed model framework provides valuable aspects related to real-life scenarios. The approach used in this chapter is not only valid and empirical in marketing but also crucial in different fields and scenarios. The finding of actual sales and the count of potential customers in various stages highlight the complexities and nuances of market behaviour. Results obtained from different data sets indicate that the proposed modelling framework is theoretically valid and provides actionable insights for marketers to optimize their marketing strategies and improve goodwill and terminology in the market.

References

1. Ref, O., & Shapira, Z. U. R. (2017). Entering new markets: The effect of performance feedback near aspiration and well below and above it. *Strategic Management Journal, 38*(7), 1416–1434.
2. Aggarwal, R., Singh, O., Anand, A., & Kapur, P. K. (2019). Modeling innovation adoption incorporating time lag between awareness and adoption process. *International Journal of System Assurance Engineering and Management, 10*, 83–90.
3. Anand, A., Agarwal, M., Bansal, G., & Garmabaki, A. H. S. (2016). Studying product diffusion based on market coverage. *Journal of Marketing Analytics, 4*, 135–146.
4. Agarwal, M., Anand, A., & Kumar, H. (2023). Determination of launch time for a multi-generational product: A fuzzy perspective. In *Advances in Soft Computing Applications* (pp. 105–119). River Publishers.
5. De Luca, L. M., & Atuahene-Gima, K. (2007). Market knowledge dimensions and cross-functional collaboration: Examining the different routes to product innovation performance. *Journal of marketing, 71*(1), 95–112.
6. Anand, A., Singhal, S., & Singh, O. (2016). Multi-stage diffusion dynamics based on optimal control theory. In *2016 5th International conference on reliability, infocom technologies and optimization (trends and future directions) (ICRITO)* (pp. 100–106). IEEE.
7. Anand, A., Agarwal, M., Aggrawal, D., & Singh, O. (2016). Unified approach for modeling innovation adoption and optimal model selection for the diffusion process. *Journal of Advances in Management Research, 13*, 154.
8. Anand, A., Singh, O., Agarwal, M., & Aggarwal, R. (2014). Modeling adoption process based on awareness and motivation of consumers. In *Proceedings of 3rd international conference on reliability, infocom technologies and optimization* (pp. 1–6). IEEE.
9. Hossain, M. S., & Rahman, M. F. (2022). Detection of potential customers' empathy behavior towards customers' reviews. *Journal of Retailing and Consumer Services, 65*, Article 102881.
10. Chang, H. J., Hung, L. P., & Ho, C. L. (2007). An anticipation model of potential customers' purchasing behavior based on clustering analysis and association rules analysis. *Expert systems with applications, 32*(3), 753–764.
11. Dibb, S. (1998). Market segmentation: Strategies for success. *Marketing Intelligence & Planning, 16*(7), 394–406.
12. Salim, S.A., Sedera, D., Sawang, S., & Alarifi, A. (2014). Technology adoption as a multi-stage process: Differing levels of significant. *ACIS*.
13. De Bruyn, A., & Lilien, G. L. (2008). A multi-stage model of word-of-mouth influence through viral marketing. *International journal of research in marketing, 25*(3), 151–163.
14. Bator, F. (1995). *The anatomy of market failure* (pp. 129–158). Macmillan Education UK.

15. Singhal, S., Anand, A., & Singh, O. (2019). SDE based generalized innovation diffusion modeling. *International Journal of Mathematical, Engineering and Management Sciences, 4*(3), 697.
16. Mehta, S., Chaudhary, K., & Kumar, V. (2020). Optimal promotional effort policy in innovation diffusion model incorporating dynamic market size in segment specific market. *International Journal of Mathematical, Engineering and Management Sciences, 5*(4), 682.
17. Meade, N., & Islam, T. (2006). Modelling and forecasting the diffusion of innovation—A 25-year review. *International Journal of Forecasting, 22*(3), 519–545.
18. Bass, F. M. (1969). A new product growth for model consumer durables. *Management Science, 15*(5), 215–227.
19. Norton, J. A., & Bass, F. M. (1987). A diffusion theory model of adoption and substitution for successive generations of high-technology products. *Management Science, 33*(9), 1069–1086.
20. Jha, P. C., Gupta, A., & Kapur, P. K. (2008). Bass model revisited. *Journal of Statistics and Management Systems, 11*(3), 413–437.
21. Singh, O., Singhal, S., & Anand, A. (2021). Multigenerational modelling incorporating time lag in innovation adoption. *International Journal of Services Operations and Informatics, 11*(2–3), 332–351.
22. Anand, A., Agarwal, M., Aggrawal, D., & Singh, O. (2018). Innovation diffusion modeling considering the time lag between awareness and eventual adoption. *Journal of Advances in Management Research, 15*(1), 4–16.
23. Singh, J., Kumar, H., Singh, O., Anand, A., & Bisht, M. (2022). Innovation adoption modeling incorporating market expansion and change point attribute. *Mathematics in Engineering, Science & Aerospace (MESA), 13*(2).
24. Singh, O., Anand, A., Kapur, P. K., & Aggrawal, D. (2012). Consumer behaviour-based innovation diffusion modelling using stochastic differential equation incorporating change in adoption rate. *International Journal of Technology Marketing, 7*(4), 346–360.
25. Singhal, S., Anand, A., & Singh, O. (2020). Studying dynamic market size-based adoption modeling & product diffusion under stochastic environment. *Technological Forecasting and Social Change, 161*, Article 120285.
26. Anand, A., Aggrawal, D., & Agarwal, M. (2019). *Market assessment with OR applications.* CRC Press.
27. Atkinson, K., Han, W., & Stewart, D. E. (2011). *Numerical solution of ordinary differential equations.* John Wiley & Sons.
28. Gerald, C.F. (2004). Applied numerical analysis. Pearson Education India.
29. Stroud, A.H. (2012). *Numerical quadrature and solution of ordinary differential equations: A textbook for a beginning course in numerical analysis* (Vol. 10). Springer Science & Business Media.
30. Kumar, H., Singh, O., Anand, A., & Irshad, M. S. (2023). Studying multi-stage diffusion dynamics using epidemic modeling framework. *International Journal of Mathematical, Engineering and Management Sciences, 8*(1), 105–119. https://doi.org/10.33889/IJMEMS.2023.8.1.007
31. Aggrawal, D., Anand, A., Kumar, H., Singh, O., & Singh, I. (2024). Multi-stage multi-generation based product adoption: An analysis using numerical methods. *Nonlinear Studies, 31*(2), 357–373.
32. Anand, A., Agarwal, M., & Aggrawal, D. (2022). *Multiple criteria decision-making methods: applications for managerial discretion* (Vol. 14). Walter de Gruyter GmbH & Co KG.

Chapter 21
Fuzzy Logic-Based Quantification of Usability Expectation for Novice User in Mobile Games

Manish Mishra and Reena Dadhich

Abstract Usability is one of the important characteristics of any mobile game application. This characteristic is very important because users directly interact with a mobile application, so the opinion of any end user is important for the success of a mobile game application. Hence, it is a big challenge to understand the end user's perspective on a mobile game, especially who interact with a mobile game first time, i.e., novice users. Novice users should look for a game that is easy to learn, fun to play and has tutorials or instructions to help them get started. Players are looking for a game that is entertaining, engaging, and provide a unique and enjoyable experience. Mobile games should also be easy to use, have intuitive controls, and be visually appealing. Additionally, mobile games should be optimized for mobile devices to ensure that they run smoothly and quickly. Challenge and engagement are fundamental and core parts of the enjoyment of mobile games and are both dependent on usability. Thus, it is required to explore sub-characteristics of usability, design its matrices, and evaluate it numerically. Unlike most other mobile applications, the focus of this article proposes a fuzzy logic-based usability evaluation model for novice users. This paper explores sub-characteristics of usability along with the baseline model ISO 9126, designs the matrices, and evaluates based on novice users' responses. These evaluations are expressed as imprecise, vague linguistic expressions using a five-point Likert scale, very high to very low. The theory of fuzzy logic has been chosen as the most appropriate tool to deal with such kind of uncertainty.

Keywords Usability · ISO 9126 · Five point likert scale · Fuzzy logic · Triangular fuzzy set · etc.

M. Mishra (✉)
University of Kota, Kota, Rajasthan, India
e-mail: manish.rpapers@gmail.com

R. Dadhich
Department of Computer Science and Informatics, UOK, Kota, Rajasthan, India
e-mail: profrdadhich@uok.ac.in

P. K. Kapur et al. (eds.), *Data-centric Approaches to Industrial Decisions*, Asset Analytics, https://doi.org/10.1007/978-981-96-7556-2_21

21.1 Introduction

Over the last several years, there has been an increase in the use of mobile apps as a means of connectivity. Users are seeking for methods to meet their requirements in connection to the application itself, such as mobile gaming, mobile commerce, mobile-based education, and other forms of mobile apps. These applications serve a range of tasks, and customers are looking for ways to satisfy their needs on their mobile devices. This is done in order to increase the likelihood that consumers will return and to ensure that they continue to be satisfied. Nevertheless, after the application has been published, it becomes a difficult task to assess the expectations of end-users. In the mobile application development sector, developers, testers, β-users, and decision-makers are continuously engaged in the process of exploring applications whose quality is to be evaluated and compared. Stakeholders, investors, and quality managers are all examples of people who may be considered decision-makers. In order to make the execution of this attempt more manageable, it will be beneficial to assess qualitative characteristics that are reliant on software quality metrics. It is for this reason that selecting the suitable method of evaluation is of the highest significance.

The degree to which a software program satisfies both the expectations of its users and the needs of its developers is referred to as the software's quality. In other words, it is a measurement of how effectively a program accomplishes the tasks for which it was designed and how simple it is to use. The quality of anything is essential since it may have an impact on how easily it can be used and how satisfied the end user is. When it comes to the cost of generating and maintaining the software, it might also have an impact. When it comes to quality, it is a subjective measurement that might change based on the requirements and anticipations of the consumer.

The quantification of the usability factor is not only highly significant for the end user but it also plays a vital part in becoming an asset for the popularity of mobile gaming applications. As described in Boehm's Model [1], Dromey's Model [2], and ISO/IEC 9126 Model [3], many methodologies have emerged that define software quality from a variety of viewpoints. These methodologies have been developed throughout time. There are two parts that make up quality: functional aspects, which refer to what a system does, and non-functional aspects, which refer to how the system is constructed and how it functions. However, these models are recommended for desktop applications. When compared to desktop computers, mobile devices have their own set of limitations, including the following:

1. Limited Storage Space: The majority of smartphones come with a limited amount of storage space, which makes it difficult to store huge files and programs.
2. The battery life of smartphones is less than that of conventional mobile phones, which means that they need charging more often than other types of mobile phones.
3. The Tiny Screen Size: The Tiny Screen Size of Smartphones Makes It Difficult to Examine Material in an Accuracy that is Satisfactory.

4. Insufficient Signal Strength Smartphones are dependent on a strong signal in order to connect to the internet; however, the signal strength is often inadequate in regions that are very distant or that have a great deal of interference.
5. Concerns Regarding Security Smartphones Are Susceptible to Viruses and Malware, Both of Which Have the Potential to Alter Data and Pose Significant Levels of Security Threats.

As a result, there is a need for the development of an appropriate model that incorporates a number of well-known standards in addition to features that are supporting in order to overcome the limitations of mobile.

From the perspective of the end user, usability is one of the most significant characteristics one may possess. The ease with which a user can learn and utilize a product, as well as the level of contentment they feel with their experience, are both factors that are taken into consideration.

The current study creates measurements, assesses on the basis of metrics with the assistance of beginner users, formulates it in terms of fuzzy calculation, and ultimately quantifies the quality attributes of usability. It does all of this by encapsulating sub-characteristics of ISO 9126 along with certain necessary sub-characteristics.

21.2 Background Work

When it comes to any sort of goods, quality is one of the most significant terms that is linked with it. As a result of the fact that software is a kind of product, the issue arises as to how one can explain the quality of a software product. Boehm's model [1] and Dromey's model [2] are two examples of the fundamental models that have been presented by the majority of scholars that study in this field and strive to justify quality. There are several different types of quality models that are established by the International Organization for Standardization (ISO). One of these models is ISO 9126 [3]; however, all of these models are general in nature. The following question is: how can the quantity qualities be quantified? There are several researchers that are now moving in this area. There is a discussion in Srivastava et al. (2010) regarding the measurement of software quality by the use of fuzzy multi-criteria method [4]. Using a multi-criteria decision approach, Sanjay Dubey and Disha Sharma (2015) suggest a method to quantify software quality [5]. On the other hand, Reena Dadhich and Bhavesh Mathur (2012) propose a method to assess the dependability of aspect-oriented software with the assistance of fuzzy logic [6]. A methodology was developed by Maryoly et al. (2003) to assess software products and generate a quality model [7]. Slaughter et al. (1998) evaluated the cost of quality software in order to produce and deliver high-quality products on time and within budget [8]. Despite the fact that Srivastava and Kumar (2009) presented project management, developer, and tester perspectives and extended the study to the next level [9]. Through the use of a survey, Nitze and Schmietendorf examine the perceptions and expectations of mobile users on the quality of software applications [10]. In their 2013 study,

Ali et al. extended the ISO 9126 model to include mobile environments [11]. Some models of software quality criteria and the link between them for mobile applications are discussed in Hotra and Roodolfo's (2012) review [12]. In paper [13], an extension of ISO/IEC 9126 is presented in order to develop a conceptual framework for mobile learning. Additionally, several matrices are proposed that will be of assistance in determining the quality of a mobile learning application. As a result of its relationship with the expectations of the end user, usability is considered to be one of the most essential quality attributes. The quantitative assessment of usability was the primary emphasis of Ghazizadeh and Vafadar's (2017) research [14], and this work likewise designs an empirical investigation. A paradigm was suggested by Barnett et al. (2018) for the purpose to access usability expectation of first-time user and proposed a model named as FTUE for mobile games [15]. Mobile gaming is one of the more complex sectors, and a significant amount of work has been put in to ensure that mobile game applications are user-friendly. According to Cui et al. (2017), a methodology is proposed to determine the ideal user interface among several options by using user interviews and questionnaires, as well as by satisfying user expectations [16]. Darin et al. (2019) provides a model PLUMB for a multi-model video game that is meant for blind persons [17]. This model is developed with the assistance of an expert opinion poll.

21.3 ISO/IEC 9126 Model

ISO 9126 is a quality model that is used for the purpose of supplying software with its features. Because its fundamental components continue to be applicable to the evaluation of software quality, it is being used today. The fact that it has been available since the early 1990s indicates that its usefulness has been well established. This model is considered to be quite mature. Because it offers a comprehensive and extensive set of criteria that may be used to assess software, it is a model that is widely used for evaluating the quality of software. The model has been revised over the course of several years to guarantee that it continues to be applicable and efficient. This makes it a tool that is easily accessible for evaluating the quality of software since it is not only simple to use but also simple to comprehend.

Providing a comprehensive description of software quality in terms of six primary criteria for software evaluation. The ISO standard also known as International Organization for Standardization produced a standard that is now known as the ISO/IEC 9126 Model [3]. Functionality, efficiency, maintainability, portability, reliability, and usability are the properties that are associated with this factor. This model encompasses almost all of the characteristics that have been addressed in models that have been offered in the past, such as Boehm's model [1] and Dromey's model [2], among others. From the point of view of the end user, usability is one of them; yet, these models are unable to explain how to measure its qualities and sub-characteristics. A discussion of the sub-characteristics of usability may be found in Table 21.1.

Table 21.1 Sub-characteristics of usability as per ISO 9126

Characteristic	Definition	Sub-characteristic
Usability	Usability is the ease of use and learn ability of a product, service, or system. It is a measure of how well users can interact with a product, service, or system to achieve their goals and complete their tasks	Understandability is the degree to which a product, service, experience, or communication can be understood by a customer or user
		Learnability is the quality of being easy to understand and use. It is a measure of the ease with which a user can learn to use a product
		Operability is a measure of the quality of a product that defines how easy it is to use and operate. It is an important factor in ensuring customer satisfaction
		Attractiveness is a quality that makes someone or something appealing. It can refer to designing of front-end
		Usability Compliance determines whether the software adheres to the compliance standards of usability or not

21.4 Multi-Criteria Fuzzy Approach

The multi-criteria fuzzy approach is a decision-making strategy that integrates the use of fuzzy logic and various criteria in order to analyze and prioritize choices. Taking this method is beneficial in circumstances in which there are several criteria, each of which has a varying degree of significance and ambiguity, and the conclusion that is wanted is not entirely specific. Uncertain or imprecise data may be input into fuzzy logic, which then offers a quantitative assessment of the choice. Fuzzy logic allows for the input of such data. The fuzzy multi-criteria method helps to quantify the decision-making process, which makes it simpler to compare and assess the many possibilities that are available.

A method known as the triangle fuzzy function has been used in this particular piece of writing. There is a sort of function known as a triangular fuzzy function that is used in fuzzy logic systems for the purpose of mapping an input value to a membership value. The input value is mapped to a value between 0 and 1, where 0 represents the lowest possible value and 1 represents the highest possible value. This graph is in the form of a triangle. The following parameters, in the form of fuzzy weight and fuzzy rating, are what this study uses as its parameters. Table 21.2 proposed two fuzzy inputs FW and FR, which calculates overall rating of six sub characteristics.

Table 21.2 Triangular fuzzy values for fuzzy weight and fuzzy ratings as per criteria

Linguistic variable	F-W	F-R
L1	(0.0, 0.0, 0.25)	(0.0, 0.1, 0.3)
L 2	(0.0, 0.25, 0.50)	(0.1, 0.3, 0.5)
M	(0.25, 0.50, 0.75)	(0.3, 0.5, 0.7)
H1	(0.50, 0.75, 1.0)	(0.5, 0.7, 0.9)
H2	(0.75, 1.0, 1.0)	(0.7, 0.9, 1.0)

21.5 Fuzzification and Defuzzification

The term "fuzzification" refers to the process of converting a numerical input into a fuzzy set. This is accomplished by mapping a precise input value to a fuzzy set and granting the input the capacity to assume any value between the two. The purpose of fuzzification is to enhance the precision of data representation by allowing for a greater degree of flexibility in the processing of inputs.

The process of fuzzy value conversion involves the conversion of matrices of usability sub-characteristics into fuzzy values through the process of fuzzification. For instance, the criterion may be defined by examining the error and recovery matrices, which are as follows (Table 21.3):

During the process of defuzzification, a set of fuzzy values is converted into a singular value that is distinct and unambiguous. The process of defuzzification is the inverse of the process of fuzzification, wherein a crisp value is transformed into a fuzzy set. This process is the antithesis of the process of fuzzification. The triangular fuzzy set is defuzzified using the centroid approach, which is used in this article.

$$\text{Centroid formula } z* = \frac{\int \mu(z) \cdot zdz}{\int \mu(z)dz} \tag{21.1}$$

In this context, z* represents the crisp value after defuzzification.

Table 21.3 Process of Fuzzification as per criteria

Matric	Description	Probability	criteria	Fuzzy rating
Probability (P) to evaluate recoverable errors	$P = \frac{\text{recoverable errors}}{\text{total number of errors}}$	0 < P < 0.2	VL	(0.0, 0.1, 0.3)
		0.2 < P < 0.4	L	(0.1, 0.3, 0.5)
		0.4 < P < 0.6	M	(0.3, 0.5, 0.7)
		0.6 < P < 0.8	H	(0.5, 0.7, 0.9)
		0.8 < P < 1	VH	(0.7, 0.9, 1.0)

Table 21.4 Sub-characteristics of usability as per proposed framework

Characteristic	Sub-characteristic
Usability	Learnability
	Simplicity
	Attractiveness
	Operability
	Error and recovery
	Satisfaction

21.6 Proposed Framework

The purpose of this paper is to propose a fuzzy-based evaluation model. In order to accomplish this, we will be using ISO 9126 as a baseline model. The three sub-characteristics that will be taken from ISO 9126 are Learn ability, Attractiveness, and Operatability. The other three attributes that will be taken into consideration are simplicity, error and recovery, and user satisfaction. As a result, our suggested model's definition may be broken down into six sub-characteristics. The fuzzy rate and fuzzy weight are two quantitative characteristics that may be used in order to measure the quality of a mobile game. The weight must be determined according to the priorities of the sub-characteristic, and the fuzzy rate is determined by the questionnaire. The experience of the beginner user is the only factor that determines the fuzzy rate. The triangle fuzzy function is used in order to take into consideration both of the parameters (Table 21.4).

21.6.1 Flow Graph of Proposed Model

Proposed model is described in the following steps and illustrated with the help of the diagram as shown in Fig. 21.1.

21.7 Design of Software Metrics for Usability Expectation: Case Study

21.7.1 Learnability

The ease with which a user may learn how to utilize a system for the first time is referred to as the learnability of the system. For beginner users, this is an essential sub-characteristic, since novice users are usually looking for models that are simple to understand. Naturally, a rookie user will always choose for natural learning as per

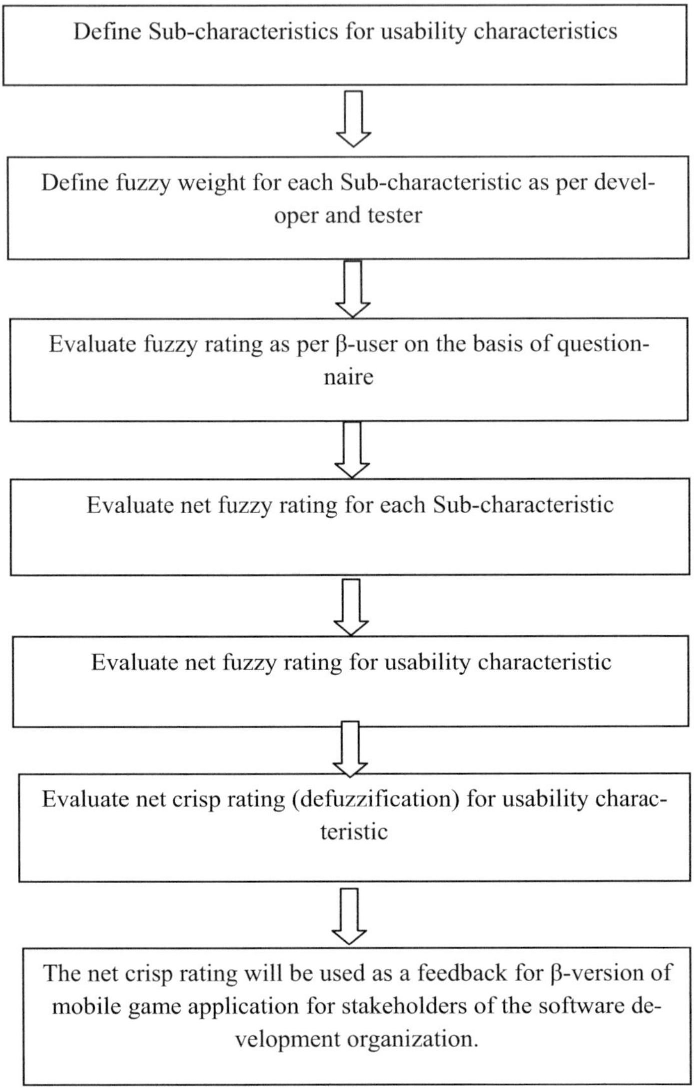

Fig. 21.1 Process flow diagram of the proposed model to find net crisp rating for mobile game application

his or her intuition, despite the fact that there are a variety of choices accessible in every mobile game in the shape of instructional text, demo, and so on, in order to comprehend the game (Table 21.5).

Metrics designed for learnability:

Calculation of average rating as per average fuzzy rate, and average fuzzy weight is calculated as:

Table 21.5 Metrics for learnability

Metrics	Description
Number of attempts	Game has optimal number of attempts to finish each task of each level
Probability for errors	Probability for error must be in lower side
Duration for searching a button	Time must be minimum for searching a button to perform specific function
Sufficient assistance	In built assist must be provided which helps users specially novice, during execution of game

Average Fuzzy rate (decided by novice users as per responses of questionnaires): High (0.50, 0.70, 0.90).

Fuzzy Weight (decided by software developer and tester): Very High (0.75, 1.0, 1.0).

Thus average rating calculated as:

$$\begin{aligned} R_{AVG(LEARNABILITY)} &= (0.50, 0.70, 0.90) \times (0.75, 1.0, 1.0) \\ &= (0.50 \times 0.75, 0.70 \times 1.0, 0.90 \times 1.0) \\ &= (0.38, 0.70, 0.9) \end{aligned}$$

21.7.2 *Simplicity*

The concept of simplicity refers to the degree to which a design is simple for anyone to understand (Table 21.6).

Matrices designed for simplicity:

Calculation of average rating as per average fuzzy rate and average fuzzy weight calculated as:

Average Fuzzy rate (decided by novice users as per responses of questionnaires): Medium (0.30, 0.50, 0.70).

Fuzzy Weight (decided by software developer and tester): High (0.5, 0.75, 1.0).

Thus average rating calculated as:

$$\begin{aligned} R_{AVG(SIMPLICITY)} &= (0.30, 0.50, 0.70) \times (0.5, 0.75, 1.0) \\ &= (0.30 \times 0.5, 0.50 \times 0.75, 0.70 \times 1.0) \\ &= (0.15,\ 0.38,\ 0.7) \end{aligned}$$

Table 21.6 Metrics for simplicity

Metrics	Description
Description of options	Menu options are descriptive in such a way that user easily understood
Characters are interactive	Characters attracts novice user
Navigation for destination	Proper navigation to reach destination
Modeling of symbols	Symbolic representation will be same for each task or level
Compatible with screen size	Mobile game interface compatible for each screen size
Proper orientation	Mobile game interface properly aligned as portrait or landscape
Compatibility of key board	Key board will be compatible as per the size of the screen
Time to search particular button	Avg. time will be minimized to search particular button for a first-time user
Zoom in or out	It is helpful and necessary to visualize a map based mobile game

21.7.3 Attractiveness

The degree to which the user is drawn to the system because of its features. For the purpose of this definition, attractiveness refers to the degree to which an application has been made appealing. This particular sub feature was designed with the inexperienced user in mind. The tradeoff between overall time spent and relaxation is what makes attractiveness so appealing (Table 21.7).

Matrices designed for attractiveness:

Calculation of average rating as per average fuzzy rate and average fuzzy weight calculated as:

Average Fuzzy rate (decided by novice users as per responses of questionnaires): Low (0.10, 0.30, 0.50).

Fuzzy Weight (decided by software developer and tester): High (0.5, 0.75, 1.0).

Thus average rating calculated as:

$$\begin{aligned} R_{\text{AVG(ATTRACTIVENESS)}} &= (0.10, 0.30, 0.50) \times (0.5, 0.75, 1.0) \\ &= (0.10 \times 0.5, 0.30 \times 0.75, 0.50 \times 1.0) \\ &= (0.05, 0.23, 0.5) \end{aligned}$$

Table 21.7 Metrics for attractiveness

Metrics	Description
Event based sound effect	A unique sound appears whenever an event occurs
Font of text and graphics	Font of text chosen as per the theme of game, while graphics suitable for latest graphics hardware
Night mode	Night mode available to play the game in the night

Table 21.8 Metrics for operability

Metrics	Description
Option for color blind person	Option available so that a color blind person plays the mobile game
Action &reaction	Each and every action properly responds as per the rule
Response	Proper response in terms of time and reaction whenever user operates with buttons (software or hardware)
Web browser version	Web browser version is available, if yes then fuzzy value will be VH else VL
Compatible for OS	Compatible for each mobile operating system, If yes then fuzzy value if VH else VL

21.7.4 Operability

The capability of the user to carry out certain activities in the order that they have specified. Operability is defined as the degree to which an application is easy to learn (Table 21.8).

Matrices designed for operability:

Calculation of average rating as per average fuzzy rate and average fuzzy weight calculated as:

Average Fuzzy rate (decided by novice users as per response of questionnaire): High (0.50, 0.70, 0.90).

Fuzzy Weight (decided by software developer and tester): Very High (0.75, 1.0, 1.0).

Thus, average rating calculated as:

$$
\begin{aligned}
\mathrm{R}_{\mathrm{AVG(OPERABILITY)}} &= (0.50, 0.70, 0.90) \times (0.75, 1.0, 1.0) \\
&= (0.50 \times 0.75, 0.70 \times 1.0, 0.90 \times 1.0) \\
&= (0.38, 0.70, 0.9)
\end{aligned}
$$

21.7.5 Error and Recovery

A measure of the frequency with which users make errors when using a system. During the execution of a job, mistakes may be described as both the quantity and the kind of errors that occur (Table 21.9).

Matrices designed for error and recovery:

Calculation of average rating as per average fuzzy rate and average fuzzy weight calculated as:

Average Fuzzy rate (decided by novice users as per response of questionnaire): High (0.50, 0.70, 0.90).

Table 21.9 Metrics for error and recovery

Metrics	Description
Recover from error	Defined as recoverable task divided by total no of errors
Error related with developer	If it is yes then very serious issue
Warning messages	Warning messages properly understood by first-time user
Interrupts during game	Offline game paused and restart during incoming voice call or video call, if yes then fuzzy value is VH, otherwise VL

Fuzzy Weight (decided by software developer and tester): High (0.5, 0.75, 1.0). Thus average rating calculated as:

$$\begin{aligned} R_{\text{AVG(ERROR\&RECOVERY)}} &= (0.50, 0.70, 0.90) \times (0.5, 0.75, 1.0) \\ &= (0.50 \times 0.5, 0.70 \times 0.75, 0.90 \times 1.0) \\ &= (0.25, 0.53, 0.9) \end{aligned}$$

21.7.6 *Satisfaction*

The user's overall level of satisfaction with the system. User's satisfaction evaluated in terms of feedback. Satisfaction defined as user's perceived level of expectation and need (Table 21.10).

Matrices designed for satisfaction:

Calculation of average rating as per average fuzzy rate and average fuzzy weight calculated as:

Average Fuzzy rate (decided by novice users as per response of questionnaire): High (0.50, 0.70, 0.90).

Table 21.10 Metrics for satisfaction

Metrics	Description
Execution of all functions	If answer is yes then fuzzy value is VH otherwise fuzzy value is VL
Experience as per expectation	Overall experience of the game as per expectation regarding its peers
Gaming experience	User satisfied with over all experience with the game
Overall user interface	User interface will be comfortable regarding long duration of the game
Overall functionality	Overall perception of playing the game
Overall challenge	Overall Challenge experience in the whole game

Fuzzy Weight (decided by software developer and tester): Very High (0.75, 1.0, 1.0). Figure 21.2 is mathematical reprentation for triangular fuzzy function and provide help for evaluation as crisp vale (Defuzzification).

Thus average rating calculated as:

$$\begin{aligned} \mathrm{R}_{\mathrm{AVG(SATISFACTION)}} &= (0.50, 0.70, 0.90) \times (0.75, 1.0, 1.0) \\ &= (0.50 \times 0.75, 0.70 \times 1.0, 0.90 \times 1.0) \\ &= (0.38, 0.70, 0.9) \end{aligned}$$

21.7.7 *Defuzzification*

Evaluation of overall quality of mobile game application (on applying centroid formula) (Fig. 21.2):

A line's equation that goes across (0.38, 0) and (0.7, 1):

$$\begin{aligned} \mu(\mathrm{z}) - 0 &= \frac{1-0}{0.7-0.38}(\mathrm{z} - 0.38) \\ \mu(\mathrm{z}) &= \frac{1}{0.32}(\mathrm{z} - 0.281) \\ \mu(\mathrm{z}) &= 3.125\mathrm{z} - 1.1875 \end{aligned}$$

A line's equation that goes across (0.9, 0) and (0.7, 1):

$$\begin{aligned} \mu(\mathrm{z}) - 0 &= \frac{1-0}{0.7-0.9}(\mathrm{z} - 0.9) \\ \mu(\mathrm{z}) &= \frac{1}{-0.2}(\mathrm{z} - 0.9) \end{aligned}$$

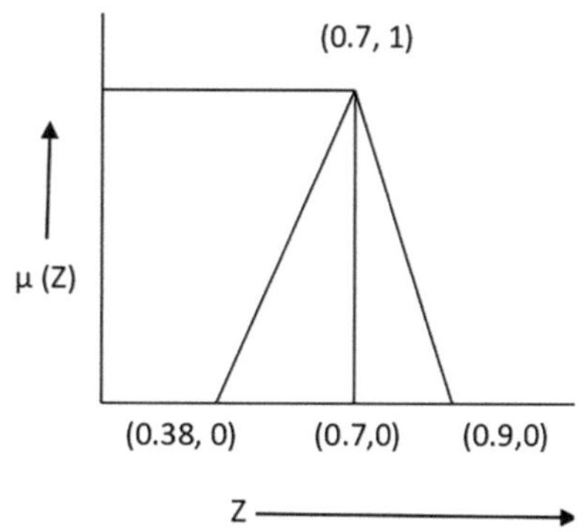

Fig. 21.2 Defuzzification (centroid method)

$$\mu(\mathrm{z}) = 4.5 - 5\mathrm{z}$$

$$Z* = \frac{\int_{0.38}^{0.7}(3.125z - 1.1875)zdz + \int_{0.7}^{0.9}(4.5 - 5z)zdz}{\int_{0.38}^{0.7}(3.125z - 1.1875)dz + \int_{0.7}^{0.9}(4.5 - 5z)dz}$$

$$Z* = \frac{0.0951 + 0.0766}{0.16 + 0.1}$$

$$Z* = \frac{0.171}{0.26}$$

$$Z* = 0.660769 \text{ or } 66.08\%$$

The numeric value is calculated as: 0.660769 or 66.08%. Thus, the software quality (usability) will be quantified.

21.7.8 Recommendations for Researchers

In any research endeavor, the analysis of the data that is collected is one of the most important components that must be considered. In spite of the fact that quality is a subjective topic, this analysis may be carried out with the help of numerical data. Furthermore, numerical data is the only kind of data that can be used for deductive reasoning. MAQM-MA is used in order to do this. The researchers have the choice of using this technique in order to quantify domain-specific mobile apps, evaluate and predict the numerical data, and therefore indicate the appropriate places in which alterations need to be implemented or implemented. When it comes to making judgments, researchers have the ability to employ numerical data in combination with any of the proper decision-making procedures, such as fuzzy TOPSIS, depending on the requirements that they have. If the data is in the form of quantitative information rather than qualitative information, then the objective of the analysis will be achievable. The researchers are able to make use of this method for the goal of defining alternative sets of weights in line with the various perspectives held by stakeholders of the mobile gaming company. This is a practical option for the researchers. After that, they will determine which prototype governs to have a more qualitative aspect by evaluating the overall quality of multiple alternatives of the prototype of the mobile game application when mobile application development is taking place. This will be done in order to determine which prototype rules best. For the purpose of determining which prototype governs to have a higher qualitative component, this will be done.

21.7.9 Recommendations for Mobile Application Business

It is very necessary to make use of quantitative data in order to ensure that decisions are made in a manner that is objective. As a consequence of this, it is helpful in determining which aspects need enhancement, evaluating the relevance of additions or updates, and boosting the efficiency of the process of development. With the assistance of this, the software development company will be able to construct success metrics that are not only evident but also measurable within the scope of their operations. Due to this, it is now possible to set specific objectives and monitor progress over the course of time. This makes it possible to do both of these things. When it comes to the process of more efficiently distributing the resources that are available, quantitative data may also be of aid. In the event that you evaluate the effects of the many modifications or additions that you have implemented, you will be able to direct your efforts toward the regions that will provide the biggest return on investment.

Surveys, user analytics, user testing, and feedback gathering tools are some of the approaches that organizations may use in order to turn qualitative qualities into quantitative data. This may be accomplished via the utilization of these methods. All of these different methodological approaches are responsible for evaluating fuzzy ratings in their own unique way. If the stakeholders and investors are able to come to a consensus over the significance of fuzzy weight, then it will be regarded substantial. As a consequence of this, the precision of the computation is dependent on the value determined by the fuzzy weight. Utilizing fuzzy AHP has the potential to raise the level of accuracy. The achievement of an accurate assessment will be made easier by this, which will give a base that is acceptable. As a result of this, it is highly recommended to all of the companies that are engaged in the development of mobile applications that they make use of this framework in order to obtain computation that is appropriate. This will guarantee that businesses who develop mobile apps are able to give a comprehensive knowledge of the performance of their app as well as the level of pleasure experienced by its users via the use of reliable quantitative data. Because of this, it will be possible to raise the amount of trust that investors and other stakeholders have in the organization being discussed.

21.8 Conclusion and Future Work

When it comes to ensuring that the user experience is optimal for all users, one of the most crucial steps is to conduct a quantitative evaluation of the usability of the system for beginner users. Using methods such as questionnaires, interviews, and several other metrics, it is feasible to ascertain whether or not a product or system is beneficial for users who are just starting out. The process of quantifying the usability element is particularly helpful in testing the β-version of mobile game applications. This ensures that the quality aspect is improved according to the end user, which in turn assists the mobile game industry in the final stages of the launch of mobile game applications. The principal component analysis (PCA) is a statistical approach that assists in the identification of significant sub-characteristics among all of the sub-characteristics. The mobile game industry may better design their goods and systems to fit the demands of beginner users if they have a better grasp of the usability challenges that novice users face. This will allow them to construct a game application that is of the highest possible quality as the end result. It is recommended that more study be carried out in the future on the most effective methods for assessing the usability of a product for first-time users. In addition, doing more study on the efficacy of various measures that are used to assess usability is something that should be done. In conclusion, further study has to be conducted in order to investigate the ways in which inexperienced users engage with various forms of technology, such as voice-based systems among others.

21.9 Future Work

One may consider the future scope of quantification in a number of different ways, depending on their own preferences. Whenever it comes to software development projects, the length of time and the budget are two of the most important factors to consider. In the not too distant future, it is feasible that software project management (SPM) will be included into the process of measuring the overall quality of mobile apps based on the developer, tester, and end-user. Furthermore, it is also possible that this technique will be offered as a standard. In the realm of software engineering, a new opportunity will reveal itself in the near future. It is feasible to think of the cost involved in quality (CIQ) as a new quality factor that may be seen as a function of both time and budget. This would imply that CIQ is equal to the function of time and budget.

This is something that is always up for argument when it comes to picking the ideal fuzzy membership function, and it is required to undertake a full mathematical analysis in order to justify a fuzzy membership function that is acceptable for the purpose of quantification. In addition, this is something that is always up for debate.

There is a possibility that the current research might be brought into the contemporary era via the use of artificial intelligence or machine learning. It is feasible to train and use cluster analysis, which can be utilized for a range of decision-making techniques, using the data that was created as a result of this study. There is a technique known as a fuzzy cognitive map that may be used in order to evaluate the impact that overall quality factors have on particular systems. As a consequence of this, modifications to a particular quality aspect will have an impact on another quality component. This will enable us to anticipate the behavior of the overall quality perception from the perspective of a developer, tester, and β-user. Automation has the ability to improve efficiency in terms of both time and money constraints, in addition to functioning as a fresh technique altogether. In order to get the greatest potential result, it is possible to include both human and automated operations into the process.

References

1. Boehm, B. W., Brown, J. R., & Lipow, M. L. Quantitative evaluation of software quality. In *Proceedings of the 2nd International Conference on Software Engineering*, San Francisco, CA, USA.
2. Dromey, R. G. (1995). A model for software product quality. *IEEE Transactions on Software Engineering, 21*(2), 146–162.
3. ISO/IEC 9126-1:2001. Software engineering-product quality—part 1: Quality model. In *Int'l Organization for Standardization*, Available at www.iso.org
4. Srivastava, P. R., Singh, A. P., Vageesh, K. V. (2010). *Assessment of software quality: A fuzzy multi—Criteria approach, evolution of computation and optimization algorithms in software engineering: Applications and techniques* (pp. 200–219). IGI Global USA, 2010, chapter - 11.
5. Dubey, S., Sharma, D. (2015) Software quality appraisal using multi-criteria decision approach. *ijieeb, 2*, 8–13.
6. Dadhich, R., & Mathur, B. (2012). Measuring reliability of an aspect oriented software using fuzzy logic approach. *IJEAT, 1*, 233–237.
7. Maryoly, O., Perez, M. A., & Rojas, T. (2003). Construction of a systemic quality model for evaluating software product. *Software Quality Journal, 11*(3), 219–242.
8. Slaughter, S. A., Harter, D. E., & Krishnan, M. S. (1998). Evaluating the cost of software quality. *Communications of the ACM, 41*(8), 67–73.
9. Srivastava, P. R., & Kumar, K. (2009). An approach towards software quality assessment. *Communications in Computer and Information Systems Series (CCIS Springer Verlag), 31*(6), 345–346.
10. Nitze, A.; Schmietendorf, A. (2015) A survey on mobile users' software quality perceptions and expectations. In *Software Testing, Verification and Validation Workshops (ICSTW), IEEE Eighth International Conference*, (pp. 1–2), 13–17 April 2015.
11. Idri, A., Mounane, K., Abran, A. (2013) On the use of software quality standard ISO/IEC 9126 in mobile environments. In *IEEE 20th Asia-pacific Software Engineering Conference.*
12. Horta, E., Ferreira, R. (2012) Quality factors in development best practices for mobile applications. In *proceedings of 12th international conference*, (pp. 632–645). Springer2, part IV.
13. Parsons, D., Ryu, H. *A framework for assessing the quality of mobile learning.*
14. Ghazizadeh, F. Z., & Vafadar, S. (2017). 'A quantitative evaluation of usability in mobile applications: An empirical study. In *Proc. Int. Symp. Comput. Sci. Softw. Eng. Conf. (CSSE)*, (pp. 1–6). Oct. 2017.
15. Barnett, L., Harvey, C., & Gatzidis, C. (2018). Novice user experiences in mobile games: An evaluation of usability. *Entertainment Computation, 27*, 82–88.
16. Cui, M., & Zhu, L. (2017). *Usability evaluation methods of user interface based on mobile games using fuzzy methods* (pp. 124–131). https://doi.org/10.1007/978-3-319-53838-9_10.
17. Darin, T., Andrade, R., & Sánchez, J. (2019). *Principles for evaluating usability in multimodal games for people who are blind*. https://doi.org/10.1007/978-3-030-23563-5_18.

Chapter 22
Comparative Analysis of AI Techniques for Customer Churn Prediction in Telecommunication

R. Elakkiya, P. Keerthana, and Alwin Joseph

Abstract Retention of customers is essential in any industry but a vital factor in telecommunications. It is required due to increased competition and the availability of competitors in the industry. Identifying customers unsatisfied with the current services and assisting them by creating new offers is an essential function of organizations. Acquiring a new customer costs more than retaining an old customer. Maintaining an existing customer is more important than customer acquisition. Customer churn/attrition is one of the most critical metrics for a growing business to evaluate. It measures the customer that leaves the company/service. Customer churn has drawn high business attention, particularly in the telecoms sector. The churn prediction models have been developed, which are heavily based on data mining principles and use machine learning and meta-heuristic algorithms. This paper examines some of the most significant churn prediction approaches created in recent years with the support of Machine Learning (ML) and Deep Learning (DL) techniques. The effective preprocessing methods employed by various researchers are studied and combined to create a preprocessing workflow for processing the data for the ML algorithms. These models aim to predict customer preference in the telecom sector.

Keywords Telecommunication · Customer retention · Deep learning · Customer Churn

R. Elakkiya (✉) · P. Keerthana · A. Joseph
CHRIST University, Pune, India
e-mail: elakkiya.r@msds.christuniversity.in

P. Keerthana
e-mail: keerthana.p@msds.christuniversity.com

A. Joseph
e-mail: alwin.joseph@christuniversity.in

P. K. Kapur et al. (eds.), *Data-centric Approaches to Industrial Decisions*, Asset Analytics, https://doi.org/10.1007/978-981-96-7556-2_22

22.1 Introduction

Churn is when a customer consciously decides to terminate its relationship with a service provider or vice versa. Customer Churn Prediction (CCP) studies identify customers at risk of churn. Customer retention is one of the main concerns of companies in the telecommunications industry. Between the numerous service providers, the telecommunications sector faces intense competition. As a result, they continue to roll out new promotions daily, forcing clients to migrate to their services from rival mobile service providers. Customers often seek better services at a lower cost, leading them to switch between operators as they please. Telecom companies frequently grapple with customer churn as they strive to predict when customers might leave. Developing an efficient churn prediction model becomes imperative to identify customers likely to quit services at a particular moment. Churn prediction holds significant importance in the telecommunications sector as it aids in retaining valuable consumers and enhancing customer relationship management (CRM). The impact on the industry's profits underscores the recognition of customer churn as a pressing issue. Consequently, telecommunications companies focus on reducing customer churn by implementing effective strategies. To counteract customer churn, the company has tried employing customer churn predictions. Customer churn typically includes higher costs, low service quality, limited features, and other factors. To foster customer growth and loyalty, the telecom industry has introduced enticing offers such as internet services and free calls to retain customers [1].

Customer churn may occur due to the necessity for the company to engage and maintain effective communication with its clients, appropriately address complaints, and manage negative feedback on various platforms such as social media and press. Understanding the reasons behind customer departures is crucial, and analysing relevant data can aid in predicting and comprehending customer churn patterns, ultimately helping the company proactively prevent its occurrence [2].

Churners are persons who move to other companies for various reasons. To reduce customer churn, the company should predict customers' behaviour correctly and establish connections between customer attrition and keep factors under their control [3]. The customers search for better services at a lower cost so they can switch between the operators at any time. Attrition growth in the telecom industry has resulted in a rise in attention to Customer Relationship Management. CRM supports marketing by selecting target consumers and creating cost-effective relationships with them [4].

The churn prediction modelling process is strongly dependent on the data mining process. Data mining is used to analyse data from various databases [5]. Data mining techniques can help build prediction models to discover future trends and patterns, allowing organizations to make intelligent decisions based on knowledge from data [6]. Recently, data mining techniques have emerged to tackle the challenging problems of customer churn in the telecommunication service field [7].

The motivation behind our study is to evaluate the suitability of state-of-the-art deep learning methods for the problem of churning. In general, many deep learning techniques predict customer churn with better accuracy.

22.2 Related Works

This section analyses relevant customer attrition forecasting work using machine learning models in other domains such as telecom, retail business, and banking. Researchers used an Enhanced Deep Forward Neural Network (EDFNN)-based model to predict banking sector industry churners. The Tukey outlier detection technique is used to overcome outliers. During the evacuation of expectancies, the information gathered is arbitrarily partitioned in an 80:20 proportion, with 80% showing the training set and 20% demonstrating the test set. The Deep Forward Neural Network architecture with five hidden layers and hidden layers with six internal nodes offers good generalization performance. The Adam optimizer is mainly used in this model because it is well suited for sparse or noisy data, and the log loss-based binary cross entropy is used to quantify the model's performance [8]. The proposed methodology's accuracy is compared with other existing approaches. The most accurate predictor to study customer attrition rates in the banking industry is the EDFNN Model.

Using data mining techniques to forecast the turnover rate in the labour market is suggested. This framework applies this neural network by considering three layers [9]. The activation function for each layer is either "ReLU'" or sigmoid. The suggested study uses the K-fold iterative cross-validation approach to improve the prediction from the provided dataset. The experimental model demonstrates that for forecasting the employment sector's churn rate, the suggested data mining approaches outperformed the stated baseline classifiers (such as Naive Bayes, SVM, KNN, Decision Trees, Adaboost, and Random Forest Classifier) by a large margin. Businesses have recently gathered large-scale customer turnover data with time series properties for research. Various churn prediction models based on time series properties have also been created utilizing the data. RNN has effectively handled the time series dataset as deep learning techniques have advanced.

The RNN model is commonly applied to the telecom industry to predict customer turnover. Using RNN with LSTM units and introducing product operation before the recurrent layer in the RNN model, sequential patterns from customer churn data that change over time are learned [10]. It is compared to various baseline models to show the accuracy of the proposed framework in predicting churners. The number of cells in the recurrent layers of the LSTM in the PENN and LST MNN and the depth of the hidden layer are two more tests. When performance indicators like the Maximum Profit measure (MPC) and Expected Maximum Profit measure (EMPC) are added, churners can be evaluated in a way that accounts for both costs and benefits and can estimate which percentage of the total is attributable to each of the customer to retain.

Predictions about whether a customer will leave a company/service are evaluated using CART and ANN algorithms. Relief-F's feature selection approach was used in this study to select relevant features from a large churn telecom dataset. The relevant features were classified using ANN and CART. The experimental result established a comparative study with existing works by comparing Relief-F-ANN and Relief-F-CART methods against other works that do not use this methodology, and the deep learning classifier ANN with the Relief-F feature selection method is proved to have a superior accuracy of 93.88% compared to the 91.60% CART classifier [11].

The neural network model is implemented in the Mobile telephone churn prediction dataset to predict customer attrition [12]. Experimentation with oversampling and under sampling by applying SMOTE and resample techniques on the dataset helped to fine-tune the data. Comparatively, comparisons with multiple activation functions like tanh, ReLU, and sigmoid function and optimizers such as Adam, Nadam, SGD, and RMSprop have experimented with deep learning algorithms. The number of epochs for the proposed models is determined by the early stopping technique to avoid overfitting and to attain a better-fit model. The performance of each of these classification methods for predicting customer churn is analysed and established using accuracy metrics. It is observed that the Deep Neural Network gave better accuracy of 96%.

22.3 Methodology

The suggested method is built on supervised categorization machine learning and deep learning approaches to forecasting whether or not a client would leave a telecom company. Additionally, four DL techniques—ANN, Early Stopping, CNN, and RNN—and the four most popular ML techniques—LR, KNN, Gradient Boosting, and Random Forest—were implemented in CCP to contrast our suggested model.

22.3.1 Dataset Description

The dataset utilized in this study, called Cell2Cell, has 58 characteristics and 71,047 occurrences. The Cell2Cell dataset has been utilized in several studies. The information is accessible on the "Centre for Customer Relationship Management Duke University's" website and on Kaggle [2, 13]. The IBM Telecom churn dataset is the additional data set utilized in the study. We were able to determine a logical link between the customer's behaviours and the turnover thanks to the dataset, which contains many customer-related information from a telecoms firm. The dataset, accessible on Kaggle, comprises 33 characteristics and 7043 rows. Numerous academics have utilized the IBM Telco data set to forecast client turnover [2, 14]. The dependent property (churn), which can be either (Yes) or (No), is included in both datasets and indicates whether the client has churned. Moreover, both datasets have segregated the characteristics into continuous attributes categorized under the numeric type and categorical attributes categorized under the object type.

22.3.2 Data Preprocessing Techniques

Data preparation has a critical function to play as one of the key phases of information discovery activities. Cleaning and preparing the data to be appropriate for the model is known as data preprocessing, improving a model's accuracy and effectiveness [15]. It necessitates several stages, including data reduction and data transformation. The accuracy and effectiveness of learning algorithms will be degraded if raw input is transformed into poor-quality data. Data preparation allows precise data preparation stages and the selection of appropriate learning algorithms to accurately assess the obtained data [11]. Missing values, non-numeric characteristics, inconsistent component scales, etc., are just a few problems with telecom datasets. As a result, it is crucial to preprocess the data before applying a learning model. The datasets utilized, Cell2cell and IBM Telco, have particular and typical properties that are listed in Table 22.1. Datasets used in this study were preprocessed using the same techniques. The preparation phase comprises managing the missing data and "label encoding" and "one hot encoding" the categorical variables (Fig. 22.1).

Furthermore, to address the issue of high variance values, the normalization technique was employed by scaling the data using "standard scaler" and "Min–Max Scaler" methods. Additionally, irrelevant features that do not significantly impact the dependent variable were removed. Subsequently, some Exploratory Data Analysis (EDA) techniques were applied to gain a better understanding of the provided data.

22.3.2.1 Handling the Missing Data

For a variable in the critical observation, missing data (or missing values) refers to the value of the data that is not saved. It is crucial to understand the proportion and location of missing values as both data sets them. The overall charges for the first data set, IBM Telco, are less than 30% [16]. The fourteen numeric kinds having missing

Table 22.1 Characteristics of Datasets

Data sources	IBM	Cell2Cell
Samples	7043	51,047
Features	32	58
Missing value	Yes	Yes
Churn	26.5%	28.8%
Not churn	73.5%	71.2%

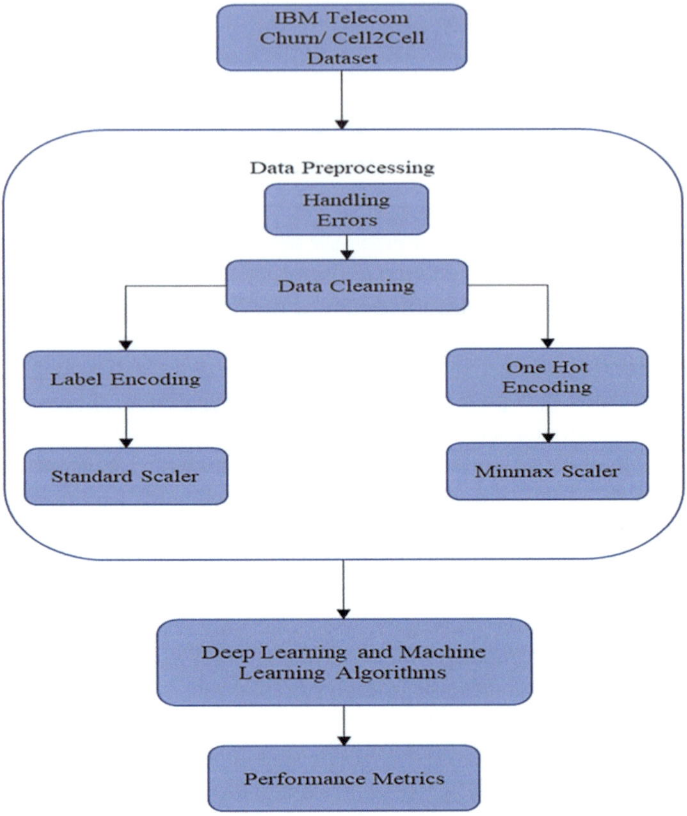

Fig. 22.1 Flowchart of the pipeline included in the dataset

values of varying proportions—yet still less than 30%—in the Cell2Cell dataset are called features. It was mentioned that it is preferable to eliminate missing data when they exceed the 30% threshold; otherwise, the residual mean should be used for numerical values and the most common for categorical values. As a result, both sets of missing data are handled using the simple imputer approach using the module. With multiple means, medians, most common values, and constant values filled in for the missing values, simple imputer assigns or replaces incomplete continuous and categorical data.

22.3.2.2 Data Transformation

Data transformation is a technical process that converts data from one format, standard, or structure to another without changing the substance of the dataset. Usually, the preparation is done to enhance the data's quality [17]. It can considerably raise the effectiveness of churn prediction overall. The IBM Telco data set has been transformed using Label Encoding, One Hot Encoding, and Normalization techniques. The Cell2cell data set has also been transformed using label encoding, one hot encoding, normalization, and PCA's dimensionality reduction methodology [2].

22.3.2.3 Label Encoding

Is a crucial approach for nominal qualities to translate categories as continuous numbers. With categorical variables like (yes or no) and (male or female) that only have two distinct values, the LE approach is more useful [18]. The LE approach was used to encode these attributes, replacing the words "yes" and "no" with one and "male" and "female" with zero, respectively. The Cell2cell data collection includes fifteen characteristics with a yes/no binary encoding, similar to the LE approach used in the IBM Telecom Dataset.

22.3.2.4 One Hot Encoding (OHE) (oR DUMMY)

Is a crucial step in transforming categorical information for machine learning and deep learning algorithms, enhancing a model's classification and prediction accuracy. A binary vector represents each integer value. A one represents the index, and all values are zero. Consequently, One Hot Encoding should not be utilized for high cardinality columns [19].

22.3.2.5 Data Scaling

Both datasets were standardized to enhance consumer prediction effectiveness with machine learning (ML) approaches. Standard Scaler and Min-Max Scaler are two kinds of scaling algorithms for transforming category variables into numerical variables, and they are two of the most popular normalization techniques [20].

22.3.2.6 Dimensionality Reduction

Also known as (data compression) it uses the mechanisms of encoding to reduce the size of the dataset. Here Principal Component Analysis (PCA) is used to reduce the dimensionality of the dataset [21]. And PCA was used in the Cell2Cell dataset to decrease the size of the dataset. And PCA technique is not used in the IBM dataset because using PCA in structured data gives biased results.

22.3.2.7 Evaluation Metrics

Prediction metrics such as Accuracy, Recall, F1 score, and precision are then calculated to evaluate the model in CCP. True Positive, False Positive, True Negative, and False Negative can be calculated from the confusion matrix [22].

True Positive: The customer who will churn is classified as positive.

True Negative: The customer who will not churn is classified as non-churners (negative).

False Positive: The customer who will not churn is classified as a churner (positive).

False Negative: The customer who will not churn is classified as unfavourable. Accuracy, Recall, and precision can be interpreted from the above measures. F1 score can be construed from Precision and Recall.

22.4 Results and Discussion

These datasets have been subjected to multiple classification algorithms. Some significant baseline model classifiers, including Logistic Regression, Random Forest, KNN, and Gradient Boosting, have been subjected to reference to compare the telecom industry data. After the model has been constructed and produced, the dataset is set up to fit the suggested models. The dataset has tested two distinct subsets of data preparation methods on two different datasets. The pipeline indicated above performs data preparation [23]. These tests are being conducted for this purpose since it has been found that the methods employed commonly in many churn predictions produce uncertainty in performance measures.

22.4.1 Preprocessing Workflow 1

The overall churn rate in IBM dataset is 73.5%- not churn and 26.5%- churn implies that the dataset is imbalanced. So, the skewed dataset is handled using the Random Sampling Technique (ROS). After handling the errors and missing values, the dataset was used for training and validation with a 60:40 split. The first subset of data preprocessing-1 includes label encoding the variables and standard scaling the attributes. Since the IBM Telecom dataset contains only essential features, applying label encoding gives a reliable machine-readable form of data. A standard scaler would standardize the dataset by rescaling the distribution of observed values so that the average value is 0 and the standard deviation is 1. For the first set experiment, the result shows that the classification from the learned CNN model gives the best values in classifying the churn problems next to other baseline models. CNN classified 1686 customers will not churn, and 729 customers will churn from the company compared with different proposed algorithms. The evaluation performance

Table 22.2 The evaluation performance with preprocessing-1 in the IBM dataset

Model	Accuracy	Precision	Recall	F1 score
Logistic regression	82	60	1.0	75
Random forest	85	64	98	77
Gradient boosting	83	63	97	76
ANN	86	61	98	76
CNN	84	66	97	79
RNN	83	62	93	75

Table 22.3 The evaluation performance with preprocessing-1 in the Cell2Cell dataset

Model	Accuracy	Precision	Recall	F1 score
Logistic regression	71	72	98	83
Random forest	72	72	1.0	84
Gradient boosting	72	73	98	84
ANN	72	72	1.0	83
CNN	72	72	1.0	80
RNN	73	73	98	83

of our model for IBM Telecom is shown in Table 22.2. The CNN model outperforms other models and achieves the best accuracy using the first subset of data preprocessing techniques. Cell2cell dataset, we include the preprocessing methods. The result shows that the model gradient Recurrent Neural Network (RNN) gives the best data preprocessing workflow-one technique accuracy in both ml and dl algorithms. And the other machine learning and deep learning algorithms almost give near predictions. The cell2cell dataset of evaluation performance of the algorithms using the preprocessing one method is shown in the below-mentioned Table 22.3.

22.4.2 *Preprocessing Workflow 2*

The second experiment includes One Hot Encoding and Min–Max Scaler in the preprocessing. As a result, the characteristics in the IBM dataset are increased to 36 after the train test split. Min–Max Scaler scales each feature by rescaling the variables from 1 to 0. The ANN model obtains the best classification results in this model with an accuracy of 91%. The model was able to classify 1893 customers who will not churn. The obtained classification results of ANN in this experiment outperform other algorithms. The evaluation performance of the algorithms using the second preprocessing technique is shown in Table 22.4. Cell2cell data preprocessing techniques include another method, i.e., Principal Component Analysis (PCA) to reduce the dimension of the dataset. In this result, Artificial Neural Network (ANN) gives the best classification model with an accuracy of 72%, and the other algorithms provide almost the exact predictions. The cell2cell dataset of evaluation performance of the algorithms using the preprocessing two techniques is shown in Table 22.5.

The result shows that the Preprocessing-2 method outperforms the preprocessing-1 method in the IBM telecom dataset. Preprocessing is a vital pipeline to build the machine learning and deep learning model. This method includes one hot encoding, a standard scaler, and other techniques, such as a preprocessing pipeline. One Hot Encoding is a usual way of converting categorical attributes for a machine learning model. It adds additional features for the given dataset during the preprocessing stages, assigning a new integer value for each unique integer value. IBM dataset consists mainly of categorical variables such as gender, churn value, and not much of ordinal values. It converted the categorical data into more exact integer values for machine convenience. In both methods, the ANN is applied with the early stopping technique, as deep learning models work well with data with more information. The early stopping technique chooses the perfect epoch where the model gives the best results. In contrast, data with another preprocessing process achieved much less accuracy, 86%, but outperformed other models.

The techniques for preprocessing and various algorithms work well on the Cell2Cell dataset. First, Recurrent Neural Network (RNN) exceeds other algorithms in preprocessing-1 after completing preprocessing pipelines, which include label encoding and standard scaler and implementing machine learning and deep learning models. There are various methods available for rescaling data, including standardization and binarization. However, normalization has been found to yield the most

Table 22.4 The evaluation performance with preprocessing-2 in the IBM dataset

Model	Accuracy	Precision	Recall	F1 score
Logistic regression	90	76	92	83
Random forest	91	78	92	85
Gradient boosting	91	80	91	85
ANN	92	79	90	84
CNN	89	75	91	82
RNN	90	76	89	82

Table 22.5 The evaluation performance with preprocessing-2 in the Cell2Cell dataset

Model	Accuracy	Precision	Recall	F1 score
Logistic regression	71	71	1.0	83
Random forest	71	71	1.0	83
Gradient boosting	72	71	99	84
ANN	74	72	99	83
CNN	71	70	98	80
RNN	72	71	1.0	81

favourable performance results. In addition to preprocessing-2, one hot encoding techniques are applied. A novel preprocessing method called Principal Component Analysis (PCA) is introduced in conjunction with other procedures for the Cell2Cell dataset. The purpose of PCA is to reduce the dataset's dimensionality effectively. It has proven to be particularly efficient in reducing the size of the Cell2Cell dataset, resulting in improved outcomes. Following this preprocessing pipeline, the ANN model is implemented using the early stopping strategy. Deep learning models tend to perform better when they have access to data with more comprehensive information.

22.5 Conclusion

Deep learning techniques play a vital role for businesses dealing with massive amounts of data, which traditional machine learning methods struggle to handle. In the current landscape, customer churn prediction models hold significant importance, especially in the telecom sector. Recent research recognizes the value of employing deep learning algorithms to develop predictive models for customer turnover. Our study gains importance as it focuses on creating a deep learning model that surpasses existing approaches in accurately forecasting customer churn while effectively handling vast volumes of data.

To develop CCP models for this study, telecom churn prediction datasets from IBM and Cell2Cell were utilized. This study seeks to identify the most reliable model for telecom churn prediction and the critical causes of customer churn. The samples are normalized because the original data collection includes many data values. As a result, performance is assessed using machine learning algorithms based on precision, recall, and accuracy. Additionally, we can enhance our results by applying massive data sets and deep learning-based techniques. These approaches' deep designs, however, require more significant computational resources. This study can be expanded to improve customer churn prediction for the telecom industry utilizing deep learning-based methods. Future initiatives will focus on enhancing the speed and effectiveness of this system.

References

1. Mahalekshmi, A., & Chellam, G. H. (2022). Analysis of customer churn prediction using machine learning and deep learning algorithms. *International Journal of Health Sciences, 6*, 11684–11693.
2. Wael Fujo, S., Subramanian, S., & Ahmad Khder, M. (2022). Customer churn prediction in telecommunication industry using deep learning. *Information Sciences Letters, 11*, 24.
3. Kumar, S., & Chandrakala, D. (2016). A survey on customer churn prediction using machine learning techniques. *International Journal of Computer Applications, 975*, 8887.
4. Sabbeh, S. F. (2018) Machine-learning techniques for customer retention: A comparative study. *International Journal of Advanced Computer Science and Applications.*

5. Maheswari, K., & Priya, P. P. A. (2017). Predicting customer behavior in online shopping using SVM classifier. In *2017 IEEE international conference on intelligent techniques in control, optimization and signal processing (INCOS)*.
6. Brandusoiu & Toderean, G. (2013). Churn prediction in the telecommunications sector using support vector machines. *Margin, 1*, x1.
7. Huang, B., Kechadi, M. T., & Buckley, B. (2012). Customer churn prediction in telecommunications. *Expert Systems with Applications, 39*, 1414–1425.
8. Mundada, M. R., et al. (2019). Enhanced deep feed forward neural network model for the customer attrition analysis in banking sector. *International Journal of Intelligent Systems and Applications, 11*, 10.
9. Dutta, S., Bandyopadhyay, S. K., & Kumar Bandyopadhyay, S. (2020). Employee attrition prediction using neural network cross validation method. *International Journal of Commerce and Management Research, 6*, 80–85.
10. Abdulsalam, S. O., Arowolo, M. O., Saheed, Y. K., & Afolayan, J. O. (2022). Customer churn prediction in telecommunication industry using classification and regression trees and artificial neural network algorithms. *Indonesian Journal of Electrical Engineering and Informatics (IJEEI), 10*, 431–440.
11. Singh, Y., Pandit, Y., Joshi, N. & Avhad, V. *Prediction of customer churn using machine learning*.
12. Ahmed, M., Afzal, H., Majeed, A., & Khan, B. (2017). A survey of evolution in predictive models and impacting factors in customer churn. *Advances in Data Science and Adaptive Analysis, 9*, 1750007.
13. Liu, R., Ali, S., Bilal, S. F., Sakhawat, Z., Imran, A., Almuhaimeed, A., Alzahrani, A., & Sun, G. (2022). An intelligent hybrid scheme for customer churn prediction integrating clustering and classification algorithms. *Applied Sciences, 12*, 9355.
14. Beeharry, Y., & Tsokizep Fokone, R. (2022). Hybrid approach using machine learning algorithms for customers' churn prediction in the telecommunications industry. *Concurrency and Computation: Practice and Experience, 34*, e6627.
15. Almufadi, N., & Qamar, A. M. (2022). Deep convolutional neural network based churn prediction for telecommunication industry. *Computer System Science Engineering, 43*, 1255–1270.
16. Gabhane, M. D., & Aslam Suriya, D. S. B. (2022). Churn prediction in telecommunication business using CNN and ANN. *Journal of Positive School Psychology*.
17. Hu, J., Zhuang, Y., Yang, J., Lei, L., Huang, M., Zhu, R., & Dong, S. (2018). pRNN: A recurrent neural network based approach for customer churn prediction in telecommunication sector. In *2018 IEEE International Conference on Big Data (Big Data)*.
18. Agrawal, S., Das, A., Gaikwad, A., & Dhage, S. (2018). Customer churn prediction modelling based on behavioural patterns analysis using deep learning. In *2018 International conference on smart computing and electronic enterprise (ICSCEE)*.
19. Singh, M., Singh, S., Seen, N., Kaushal, S., & Kumar, H. (2018). Comparison of learning techniques for prediction of customer churn in telecommunication. In *2018 28th International Telecommunication Networks and Applications Conference (ITNAC)*.
20. Soeini, R. A., & Rodpysh, K. V. (2012). Applying data mining to insurance customer churn management. *International Proceedings of Computer Science and Information Technology, 30*, 82–92.
21. Sato, T., Huang, B. Q., Huang, Y., Kechadi, M.-T., & Buckley, B. (2010). Using PCA to predict customer churn in telecommunication dataset. In *International Conference on Advanced Data Mining and Applications*.
22. Faritha Banu, J., Neelakandan, S., Geetha, B. T., Selvalakshmi, V., Umadevi, A., Martinson, E. O., et al. (2022) Artificial intelligence based customer churn prediction model for business markets. *Computational Intelligence and Neuroscience*.
23. Muneer, A., Ali, R. F., Alghamdi, A., Taib, S. M., Almaghthawi, A., & Abdullah Ghaleb, E. A. (2022). Predicting customers churning in banking industry: A machine learning approach. *Indonesian Journal of Electrical Engineering and Computer Science, 26*, 539–549.

Chapter 23
Aspect Level Sentiment Analysis on Twitter Tweets for Sarcasm Detection

Simran Garg, Devang Chaturvedi, Tanya Jain, Anju Mishra, and Archana Singh

Abstract Sentiment analysis is contextual mining of text which identifies and extracts subjective information in the source material and helps a business to understand the social sentiment of their brand, product, or service while monitoring online conversations. Analysis of social media streams is usually restricted to just basic sentiment analysis and count-based metrics. Automatic systems that rely on machine learning techniques to learn from data Hybrid systems. Uses of sentiment analysis include Surveys, text analytics, Forecast sales, and much more. Sentiment analysis is an automated process capable of understanding the feelings or opinions that underlie a text. One fundamental problem in sentiment analysis is the categorization of sentiment polarity. Compared to previous approaches in sentiment topics, additional findings showed that adding the semantic feature produces better Recall to compute the score) in negative sentiment classification. Approaches of Sentiment Analysis depend on how you wish to interpret client feedback and inquiries, you can define and customize your categories to match your sentiment analysis needs. The advantages of sentiment analysis are discussed in the paper. Sentiment analysis will enable you to have all kinds of market research and competitive analysis. While some NLP models are more emotionally intelligent than others, sentiment classification systems generally use one of the three algorithms: Rule-Based Systems, Automated Systems, or Hybrid Systems. We can clearly see that sentiment analysis is becoming more popular as e-commerce, SaaS solutions, and digital technologies advance.

S. Garg (✉) · D. Chaturvedi · T. Jain · A. Mishra · A. Singh
Amity University Uttar Pradesh, Noida, India
e-mail: simrangarg36@gmail.com

D. Chaturvedi
e-mail: devu8chaturvedi@gmail.com

T. Jain
e-mail: tanya10032001@gmail.com

A. Mishra
e-mail: amishra1@amity.edu

A. Singh
e-mail: asingh27@amity.edu

P. K. Kapur et al. (eds.), *Data-centric Approaches to Industrial Decisions*, Asset Analytics, https://doi.org/10.1007/978-981-96-7556-2_23

Keywords Sentiment analysis · Twitter tweets · Sarcasm detector · Naïve Bayes · Positive-Neutral

23.1 Introduction

The emotional undertones of a text are dissected by means of sentiment analysis. Typical applications include customer evaluations, survey responses, and market research. This has applications in a number of fields, including keeping tabs on what people are saying online, managing your reputation, and providing excellent customer service [1]. For instance, if hundreds of evaluations are analyzed, insights about price and features may be gleaned. Public opinion is a major factor in determining consumers' propensity to engage with brands and in shaping consumers' overall impressions of such brands [2]. According to a poll conducted by Podium, 93% of consumers admit that internet reviews have an impact on their final purchase choices. Some people may be put off trying your product or service after reading critical feedback. There will be no investigation on the sincerity of the comments. However, conventional social media analysis is limited to simple counts and mood ratings [1]. To do so is equivalent to only touching the surface, preventing one from gaining access to the wealth of undiscovered, high-value ideas. What, therefore, should a company do to reach that easy target audience? The development of deep learning has greatly enhanced the capacity of algorithms to analyze text. Applying cutting-edge AI methods creatively may be a useful instrument for comprehensive study [1, 2]. A variety of algorithms and techniques from the field of Natural Language Processing (NLP) are used in sentiment analysis.

- Rules-Based Systems: using a set of manually crafted rules.
- Automatic systems: rely on machine learning techniques to learn from data
- Hybrid systems: combine both rule-based and automatic approaches (Fig. 23.1).

Fig. 23.1 Sentiment analysis [3]

23.1.1 History

Many different granularities of the Natural Language Processing job of sentiment analysis have been implemented [2]. It was originally classified at the document level, but has since been broken down into sentence- and, most recently, phrase-level tasks (Wilson et al., 2005; Agarwal et al., 2009). Twitter data, in which individuals share instantaneous thoughts and views on everything, and everything presents novel difficulties. Go et al. (2009), Bermingham and Smeaton (2010), and Pak and Paroubek (2013) are among the earliest and most recent findings on sentiment analysis of Twitter data, respectively (2010). Sentiment data was collected remotely using distant learning by Go et al. (2009) [2]. Tweets are used to express both happy and negative feelings. They use Naive Bayes, Max Ent, and Support Vector Machines (SVM) to create their models, and they claim SVM is the most effective classifier. They test the waters of a Unigram-Bigram model combined with POS characteristics. They mention how the unigram model is the most effective of all the ones they have tried. We can rule out the use of bigrams and POS elements in particular. Similar data collection techniques used by Pak and Paroubek (2010)'s remote learning paradigm are described. However, they are engaged in a job of categorization that differs between the two: subjective and objective. Similar to Go et al., they collect tweets that finish in emoticons to get a sense of the tweeters' moods (2009) [4]. The Twitter feeds of well-known publications like the New York Times, the Washington Post, and others are trolled for their unbiased reporting. They found that both POS and bigrams were beneficial (opposite of the findings reported by Go et al., 2009). But unigram models form the backbone of each of these methods. Furthermore, they employ biased search query data for both training and testing. In contrast, the features we provide here perform noticeably better than a unigram control. Moreover, we look at a different data representation method and discover that it excels where unigram models fall short. Use of unbiased, manually [1] annotated data to explain results is another novel aspect of this work. Our data will be collected at random from a stream of tweets, as opposed to systematically answering specific questions. Because we have so much manually annotated data, we can undertake cross-validation tests to discover why the classifier's performance varies over folds [2]. There has been significant improvement in the categorization of Twitter users' moods thanks to the work of Barbosa and Feng (2010). As noise labels, they use polarity predictions from three different websites, in addition to a thousand personally labeled tweets for fine-tuning and another thousand for testing. Unfortunately, they do not detail the sources of the data used in their investigations. They propose using syntactic features of tweets including retweet, hash tags, links, punctuation, and exclamation marks in addition to other characteristics like preceding polarity of words and POS of words. To surpass them, [4] we use actual valued prior polarity and incorporate it into POS. Based on our findings, features that take into account both the previous polarity of words and their parts of speech seem to improve our classifiers' performance the most. Classifying the intensity of emotions is a major challenge for sentiment analysis. The challenge here is to determine if a given piece of text is meant to be read in a positive or negative light (or neutral). Sentiment polarity may be classified on three different levels, depending on how far back in the text one wants to go: the document level, the sentence level, and the entity and aspect level. At the document level, we care about whether or not the document as a whole reflects positive or negative emotion; at the sentence level [4], we classify the sentiment of individual sentences. Using Twitter data, Gann et al. selected 6799 tokens, each of which is either a good token (with a high TSI) or a negative token (with a low TSI), depending on the overall tone of its associated tweets. Moreover, it was shown that the popular Twitter "geo-tagged" feature may be used to determine the polarity of US political candidates via the use of sentiment analysis algorithms for foreseeing the outcomes of events like the US presidential elections. Additional results show that negative sentiment classification benefits from the inclusion of the semantic feature, as measured by Recall (retrieved documents used to calculate the score).

23.1.2 Approaches of Sentiment Analysis

You may construct and tweak your categories to meet your sentiment analysis requirements, depending on how you want to make sense of customer comments and questions [5]. Meanwhile, some of the most widespread approaches to sentiment analysis are as follows:

1. Fine-grained Sentiment Analysis: provides a more nuanced polarity by dividing it into smaller groups, typically ranging from extremely positive to extremely negative [5]. It is like giving your opinion of something a rating out of five stars.
2. Aspect-based Sentiment Analysis (ABSA): It is most helpful when it is connected to some quality or characteristic mentioned in the text. ABSA is the method used to unearth these characteristics and the feelings associated with them. At Thematic, we call these elements themes.
3. Emotion detection: [6] Emotion detection is more nuanced than simply looking for happy or sad faces. There is a wide range of human emotions, including joy, annoyance, surprise, anger, and sadness.
4. Intent-based: Intent-based analysis can tell the difference between objective data and subjective commentary [5]. For instance, if you express dissatisfaction with the process of replacing a battery in an online review, this may prompt customer support to get in touch with you to find a solution.

23.1.2.1 Advantages of Sentiment Analysis

Following are the few advantages of sentiment analysis (Fig. 23.2).

- Quantitative analysis of qualitative data: [5] Complex question formats such as the semantic differential scale, side-by-side matrix, flex matrix, etc., cannot always be relied upon to accurately capture respondents' feelings and sentiments.

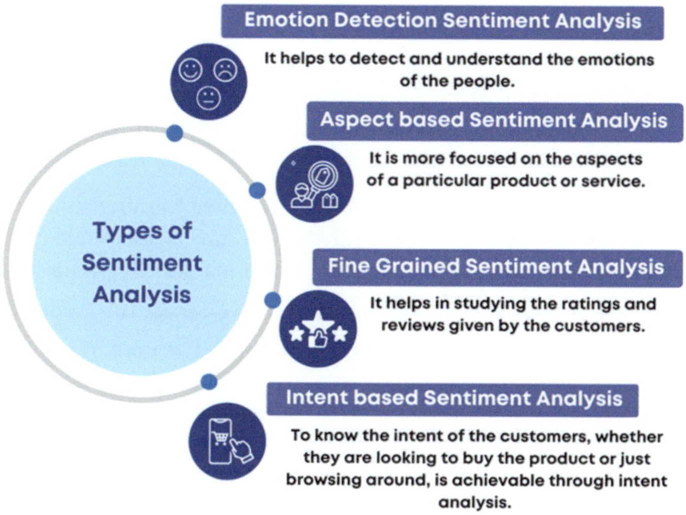

Fig. 23.2 Types of sentiment analysis [7]

- Customer experience measurement: Analysing the tone of client feedback is a great way to learn about their overall experience. As an example, airlines may utilize a sentiment analysis technology to compile passenger comments and identify areas for improvement.
- Competitor analysis: Sentiment research may also be used to learn how the public views the performance of your rivals [5]. In order to keep ahead of the competition, it is essential to do market research to help you review your priorities.
- Identification of trends: You can predict the likelihood of a good or bad outcome by doing a sentiment analysis.
- Time-effective: To better understand [5] the data, sentiment analysis uses bubble charts rather than traditional numerical reports. Time spent on statistics data may be greatly reduced by switching instead to visually appealing charts.

23.2 Literature Review

In recent years, sentiment analysis has gained significant attention due to the growing importance of understanding customer feedback and online reviews. This literature review summarizes the key concepts and recent developments in sentiment analysis.

23.2.1 Classification Techniques

Lexicon-Based Methods - Manually generated dictionaries and corpora are examples of lexicon-based, or knowledge-based, methodologies [6]. The lexical approach often does not detect irony, sarcasm, misspellings, or grammatical errors.

1. **Automated/Machine Learning Methods -** Machine learning algorithms, which classify sentiment based on statistical models, are at the heart of automated sentiment analysis approaches. In order to use machine learning methods, the phrases must be converted to vector space.
   1.1. **Naive Bayes -** This method employs Bayes' Theorem to provide a supervised, probabilistic classification strategy, [6] which may then be utilized to extract relevant features. This method accounts for typos and grammatical errors by assuming that each token or feature exists in isolation.
   1.2. **Support Vector Machine -** It is a method for categorizing data that does not rely on probability, and it works by finding the hyperplane that minimizes the distance between pairs of vectors.
2. **Hybrid Approaches -** There are benefits and drawbacks to using both automated and lexicon-based approaches [4]. Companies can therefore implement hybrid approaches, combining automated and lexicon-based methods, to make up for the shortcomings of each. The two methods can be used together in parallel or in separate phases of analysis (Fig. 23.3).

23.3 Factors Affecting Sentiment Analysis

Going through the literature thoroughly we find out certain critical factor that affect the sentiment analysis.

1. Tone - It is challenging enough to decipher someone is tone when they are speaking, and it is even more challenging [4] when it is written down. But a competent sentiment analysis program will have the fundamental ability to differentiate between subjective and objective statements and to identify the appropriate tone in each.
2. Polarity - Polarity scores for words like "love" and "hate" tend to be very positive (+1) and very negative (-1) respectively. These are simple and straightforward [6]. But in order to provide a complete picture of a statement, sentiment analysis techniques can quickly identify phrases and words that fall in the middle.
3. Sarcasm - Slang and internet memes are rife with examples of irony and sarcasm. Backhanded compliments are a common way [9] to communicate unpleasant emotions, and they may muddy the waters for sentiment analysis systems trying to figure out what was really meant.

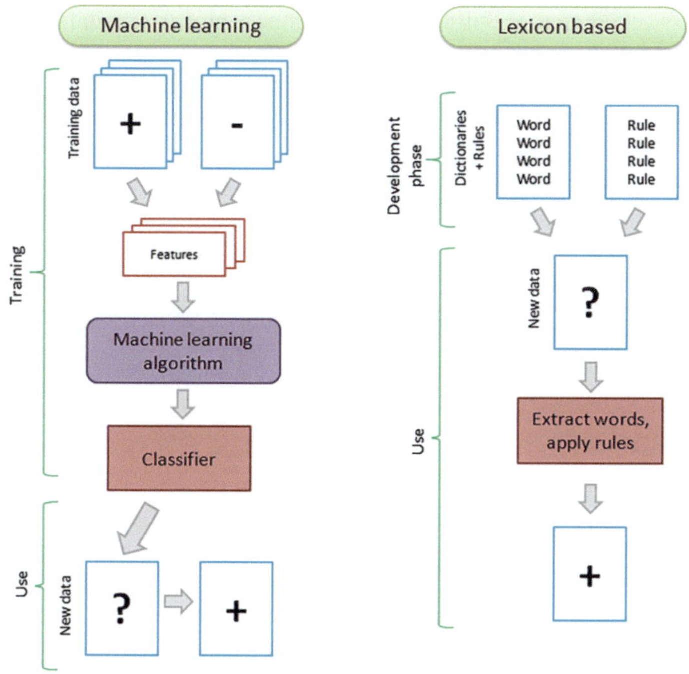

Fig. 23.3 Visualized comparison of automated and lexicon-based sentiment analysis methods [8]

4. Emojis - Twitter and other text-based social media platforms suffer from an overabundance of emojis. When training NLP tasks, a [4] language's unique linguistic features are prioritized. In general, machine learning algorithms do not grasp the concept of figurative language.

23.3.1 Application of Sentiment Analysis

1. Brand Monitoring - A company's name is more than the goods it produces. Marketing a business online, conducting social campaigns, [10] creating content to advertise, and providing customer service are all factors.
2. Customer Service - Firms in the customer service industry often employ sentiment analysis to automatically [10] sort incoming calls from users into "urgent" and "non-urgent" categories. Emails or phone calls may be sorted depending on the tone of the sender to detect irate clients in advance.
3. Finance and Stock Monitoring - There is always some degree of danger when investing in stocks and the market, but that risk may be mitigated with proper preparation [10]. If, for instance, you are trying to decide between two competing automakers to invest in but cannot make up your mind, reading customer reviews of each manufacturer's newest offerings will help you make an informed decision.

4. Business Intelligence Build-up - The importance of digital marketing in the commercial world cannot be overstated. The responses and opinions of consumers to a product are often shared on social media [10]. The answers your rivals provide may likewise be studied. The results of the survey might help you provide better service to your customers.

23.4 Proposed System

The proposed system is explained below in different part.

23.4.1 Dataset

Twitter datasets gathered using hashtag-based supervision have been widely used in previous Sarcasm Detection research, [11] despite the fact that these datasets are linguistically and labeling noisy. The News Headlines dataset for Sarcasm Detection is an amalgamation of data from two news websites and was created to help overcome the noise limitations of Twitter datasets. We selected all of the headlines from TheOnion's (sarcastic) [11] News in Brief and News in Photos sections for our parody of current events. Without irony, we use only genuine headlines from the Huffington Post. Compared to the previous Twitter datasets, this one has the following advantages:

- Spelling and grammar mistakes are avoided when journalists write headlines because they are expected to follow strict style guidelines. Sparsity is reduced, and the likelihood of discovering pre-trained embedding is increased as a result.
- Because TheOnion's [11] primary goal is to publish satirical news, we also acquire high-quality labels with significantly less noise than Twitter datasets.
- Unlike tweets, which frequently respond to other tweets, the news headlines we gathered stand on their own. This could be helpful in determining the content's irony.

Content: Each record has three characteristics:

- is_arrogant:1 in the event that the track is sarcastic; otherwise,
- headline:the news article's headline
- article_link:provide a link to the original news story useful for acquiring additional data
- [11] This GitHub repository includes summary data statistics, Python code for reading the data, and some initial exploratory analysis. This GitHub repository contains a hybrid NN architecture that was trained using this dataset (Fig. 23.4).

```
import json

def parse_data(file):
    for l in open(file,'r'):
        yield json.loads(l)

data =
list(parse_data('./Sarcasm_Headlines_Datas
et.json'))
```

Fig. 23.4 Reading the data: following code snippet could be used to read the data

23.5 Methodology

23.5.1 Creating Raw Tweets Database: [12]

Data collection is the first step in data visualization. We used the Kaggle platform to look for the data, and then we used Google Colaboratory using these methods to extract the data.

1. Go to your kaggle account, Scroll to API section and Click Expire API Token to remove previous tokens
2. Click on Create New API Token - It will download kaggle. json file on your machine.
3. Go to your Google Colab project file and run the following commands:

23.5.2 Preprocessing of Tweets

The term "data preprocessing" is used to describe any operation carried out on raw data in order to get it ready for further processing [12]. It has always been a crucial first stage in the data mining process. Preprocessing may be broken down into two broad categories: data purification and feature engineering. Several methods are included in each, which are described in more depth below.

- Data Cleansing - Methods such as these are available for sanitizing jumbled data. Find and organize the missing information [13]. The absence of some data fields in a given dataset may have been deliberate or unintentional.
- Reduce Noisy Data: Noisy real-world data might skew the results of an analytical or AI model [13]. For instance, a temperature sensor that normally gives an accurate reading of 75 degrees Fahrenheit can provide an inaccurate reading of 250 degrees.
- Data Reduction: When [13] developing a new machine learning or analytics model, data scientists must frequently combine data from multiple sources (Fig. 23.5).

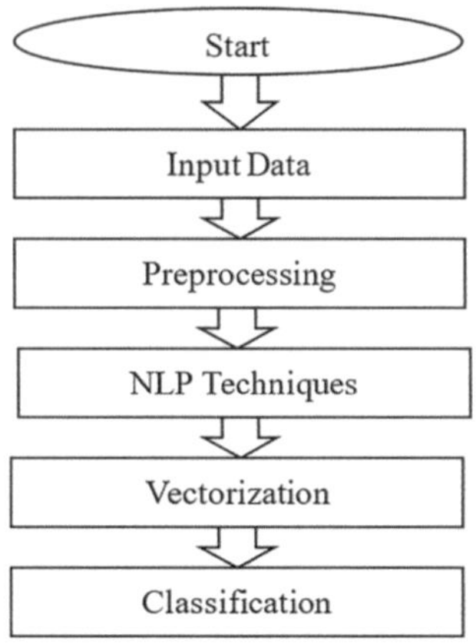

Fig. 23.5 Flow diagram of the proposed model

23.5.3 *Categorizing of Data*

Each [11] entity must be assigned to exactly one class within a set of classes that are designed to be mutually exclusive and no overlapping; this is the essence of classification as a process.

23.5.4 *Analysis Sentiment of Dataset*

(See Fig. 23.6).

23.5.5 *Classification into Sentiments*

It is possible to automatically categorize the positive, negative, and neutral feelings expressed by consumers inside text using a method called sentiment classification [14]. Sentiment categorization uses natural language processing to assess client feedback about your brand, goods, and services [14]. Natural language processing (NLP) that combines linguistics and computer science to translate human-written text into a format that computers can comprehend. Sentiment classification systems typically utilize one of three techniques, [12] while some natural language processing models are more emotionally sophisticated than others: Rule-Based Systems, Automated Systems (Based on Machine Learning) and Hybrid Systems.

```
labels = ['Sarcasm', 'Not Sarcasm']
values = [df['is_sarcastic'].sum(), len(df) - df['is_sarcastic'].sum()]

trace0 = go.Bar(x = [labels[1]], y = [values[1]], name = 'No Sarcasm')
trace1 = go.Bar(x = [labels[0]], y = [values[0]], marker= {'color': '#00cc66'} ,


data = [trace0, trace1]

layout = go.Layout(title = 'Number of Sarcastic Articles',
                   width = 800,
                   height = 500,
                  yaxis= dict(title = 'Number of articles'),)

fig = go.Figure(data, layout)

pyo.offline.iplot(fig)
```

Fig. 23.6 Code snippet [11]

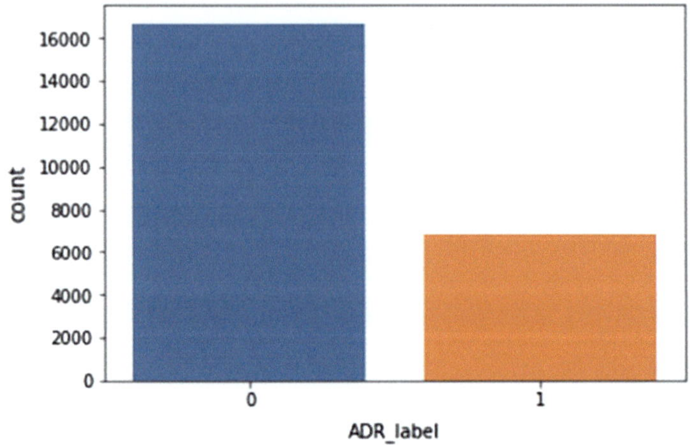

Fig. 23.7 ADR_Label [11]

23.6 Result and Future Scope

As the use of digital technologies like e-commerce, SaaS, and other digital services continues to grow in popularity, it is easy to understand why sentiment analysis is gaining ground. We have covered its operation and looked at its most prevalent business uses [13]. We have also covered the problems and restrictions that currently exist in the analysis. Sentiment defined by humans, lies in the fact that the latter follow a preset set of rules. Examples of using sentiment analysis were shown. Understanding customers' remarks, critiques, and praises using sentiment analysis in customer feedback surveys [1, 2]. People's reactions to breaking news and current events are analyzed using social media data. And public relations firms manage their customers' online reputations using sentiment analysis software. More besides an automated procedure, sentiment analysis may decipher the underlying emotions or perspectives of a document. This is one of the most intriguing research areas in Natural Language Processing (NLP), a branch of artificial intelligence that tries to teach computers how to read, write, and converse effectively in human language [12, 15]. The introduction of the semantic feature enhances Recall (retrieved documents used to compute the score), which is evident when comparing this approach to others that have been employed to research this topic [4]. Previous attempts at sarcasm detection mainly relied on Twitter datasets obtained using hashtag-based supervision, despite the fact that these datasets are noisy in terms of labels and language. Since many tweets are replies to other tweets, it is essential to have access to contextual tweets because this will make it easier to spot irony in these replies. Furthermore, when new social media platforms and capabilities emerge, the insights gathered through these technologies become increasingly nuanced (Figs. 23.7, 23.8).

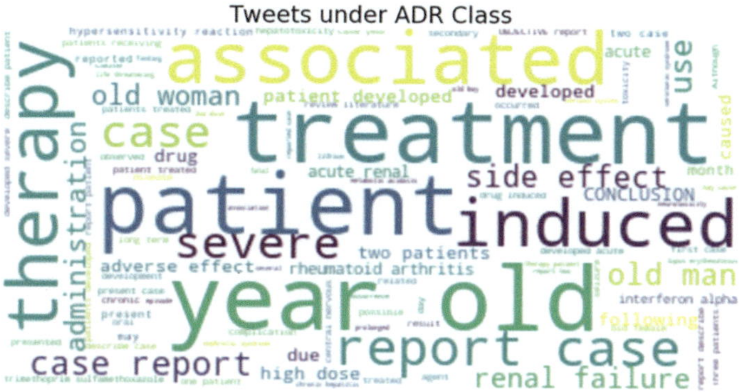

Fig. 23.8 Tweets under ADR class [11]

References

1. Sakhawat Hossain, G. M., Sakhawat Hossain, G. M., Mynoddin, Md., Mynoddin, Md. (2022). *Deep learning approach for public sentiment analysis in COVID-19 pandemic.* (pp. 1–6). Preprints
2. Vishal, A. K., & Sonawane, S. S. (2016). Sentiment analysis of twitter data: A survey of techniques. *International Journal of Computer Applications, 0975–8887*, 1–8.
3. Blog: < https://phedone.com/en/blog/sentiment-analysis-can-we-believe-in-it/>
4. Ankita, S., Udayan, G. (2020) Sentimental analysis of Twitter data with respect to general elections in India. In *International conference on smart sustainable intelligent computing and applications under ICITETM*, (pp. 3–10).
5. Anusuya, D., Arkadeb, S., Sourish, S., Pranit, B. (2018). *Sentiment analysis of product-based reviews using machine learning approaches.* RCC Institute of Information Technology (pp. 2–12)
6. Batra, S., & Rao, D. (2010). *Entity based sentiment analysis on Twitter.* Stanford University, (pp. 2–10)
7. Project research: < https://www.analyticsvidhya.com/blog/2022/04/a-comprehensive-overview-of-sentiment-analysis/>
8. Maite T. (2016). Sentiment analysis: An overview from linguistics. In *2016 Research Gate*, (pp. 2–6)
9. Efthymios, K., & Johanna, M. (2012). *IJCSI International Journal of Computer Science Issues, 9*(4), 6–9
10. Afroze Ibrahim, B. (2011). Twitter sentiment analysis: The good the bad and the OMG! In: *Proceedings of the Fifth International AAAI Conference on Weblogs and Social Media.* (pp. 2–7)
11. Kaggle Dataset: < https://www.kaggle.com/datasets/rmisra/news-headlines-dataset-for-sarcasm-detection >
12. Kavya, S., Narasinga, R. (2019). Sentiment analysis using Naïve Bayes classifier. *International Journal of Innovative Technology and Exploring Engineering (IJITEE) 8*(8).
13. Aliza, S., Chayanit, N., Shuib, B. (2014). Twitter sentiment analysis. In *2014 International Conference on Information Technology and Multimedia (ICIMU)*, (pp. 2–6). November 18–20.
14. Shiv Naresh, S., & Saritha K. Emotion detection from text department of CSE and IT, Maulana Azad National Institute of Technology, Bhopal, Madhya Pradesh, India.
15. Mohsen, F., Erik, R. (2016). Fundamentals of sentiment analysis and its applications. *Research Gate* (pp. 2–8).

Chapter 24
Online Fraud on Social Media Sites

Ravneet Singh Bhandari and Vivek Kumar

Abstract The enormous usage of online social networking applications has changed the way in which people connect with each other. However, the rise in utilization of these online applications has also discovered new ways of fraud. In this paper, the authors exhibited a concise examination of interaction without online fraud and interaction with online fraud on social networking sites. Furthermore, the authors also investigate some of the components that can influence the motive of accomplishing fraud on social networking sites and proposed a deception model to show diverse online frauds. The authors additionally examine the complexity of deploying and accomplishing these fraudulent techniques. The social media space continuously advances and keeps on being stretched out with various innovations that fraudsters can utilize; therefore, a structured model is required for online fraud. Also, the authors featured a few drawbacks of social networking architecture that developers must take into consideration in order to protect social networking users from online fraud dealings.

Keywords Social media sites · Online fraud · Internet · Identity fraud · Content fraud

24.1 Introduction

The rapid expansion of internet has changed the ways in which content is produced and exchanged over the internet, leading to a fast development internet-based applications. Internet-based social networking sites empowered the creation and trading of user-generated content on these applications [1]. This development has increased the number of internet-based applications and also the number of users joining such sites. Between 2010 and 2018, there had been a 60 per cent expansion in the number of individuals utilizing social networking sites. For example, Facebook usage expanded

R. S. Bhandari (✉) · V. Kumar
Amity International Business School, Amity University, Noida, India
e-mail: ravneetsinghbhandari@gmail.com

P. K. Kapur et al. (eds.), *Data-centric Approaches to Industrial Decisions*, Asset Analytics, https://doi.org/10.1007/978-981-96-7556-2_24

by 50 per cent in the period of 2013–2018 [2]. A total of 2.23 billion users interface through Facebook with their accounts [3]. However, with the ease of registering an account on the social media platforms, it takes very less effort for people to deceive others. Previous research on online fraud has discovered that individuals who committed fraud (offline) in the past endeavoured to commit deception on social networking sites [4]. All through in past, fraud activities have been increased alongside technological advancements to deceive people [5]. Social networking sites provide new environment and settings to the potential fraudsters. There are various cases of individuals who have been deceived by social networking sites, which caused financial and psychological distressing consequences to the users [6].

In this paper, the authors considered online fraud as a deliberate attempt with the ultimate goal to misdirect others while the victims are not aware of such a fraudulent action and the objective of the deceiver is to get monetary or non-monetary benefits from the cheated ones [7]. This perspective on online fraud turns out to be appropriate especially while analysing social networking sites where the boundaries ensuring one's protection and deceiving others are unambiguous. Moreover, the authors also analysed that these fraud convictions are exchanged through verbal and non-verbal communication and the consequences of online fraud is quantifiable and identifiable through financial analysis (monetary loss) and physiological analysis (e.g., stress) [8]. Past researches had contended that structuring and implementing of protection policies, for example, defining what personnel data to be shared on the social networking site, might be a success for saving users from online fraud [9]. However, individuals who are aware of online fraud are more careful while using social networking sites than individuals who do not have any knowledge of a security challenge while sharing personal data, despite the fact that there are so many frauds happening on social networking sites making themselves more defenceless against assaults [10], to worsen the situation, social networking sites, as a part of Internet-based applications, are incorporated with multiple virtual conditions that are not effectively regulated by the administrators [7].

This paper tried to explain online fraud and investigates its utilization in social networking sites in specific. The authors based this research on the inspirations for fraudulent actions in social networking sites and investigated different fraudulent strategies that have been utilized as now and their effect on social networking users [11]. Finally, the authors talked about the set of challenges that need to be addressed in future about online fraud in social networking sites [12]. While identifying and anticipating online fraud, there are two essential perspectives that are basic with the subject, understanding the social networking online fraud and the characterizing strategies utilized to reduce the rate of fraud at the beginning for the battle [13].

24.2 Social Media

Social networking sites can be classified based on user-generated content on social media for self-representation/self-disclosure [14]. User-generated content can be impacted by the structure and nature of the medium in which the correspondence takes place, while media content depicts the kind of data (i.e. pictures, videos, reviews) that can be transmitted in the media at a given point in time. Self-representation describes the control that users have in disclosing themselves on a social media platform which involves uncovering one's personal data whether knowingly or unknowingly [15]. Considering the above-mentioned attributes, a matrix was created by [16], which included various type of social media: blogs, collaborative projects (e.g., Wikipedia), social networking sites (e.g., Facebook), business networking sites (e.g., LinkedIn), and video content communities (e.g., YouTube). The authors extended the order of social media sites (as displayed in Table 24.1) which additionally incorporates micro blogging (e.g., Twitter) image content communities (e.g., Instagram) and review sites (e.g., Quora). The authors placed micro blogging, i.e., Twitter between collaborative projects, i.e., Wikipedia and social networking sites, i.e., Facebook and Business networking sites, i.e., LinkedIn above Blogging considering the high degree of self-disclosure due to earning opportunities, i.e., Affiliate marketing [17].

As shown in Table 24.1, social media sites, which provide users a great extent of flexibility for introducing themselves, are in the second row, while social media sites that power users to adjust to specific components or have no alternative for uncovering data about their personalities are in the first row [18]. However, with an increment in user-generated content on social media, the authors discovered that high contribution of users in terms of content for content-based communities than less content-based communities resulted in more content-based interchanges for virtual communications and virtual socialization [19]. The contrasts between these sorts of social media sites also influence how online fraudulent activities are actualized and performed.

In most social media networks, most interchanges are content based and are performed non-concurrently. In such conditions, fraudulent individuals have a favourable position for manipulating content which is an easiest method to trick

Table 24.1 Social media classifications

		User-generated content			
		Low			High
Self-representation/Self-disclosure	Low	Collaborative projects	Video content communities	Micro blogging	Image content communities
	High	Blogs	Social networking sites	Business networking sites	Review forums

Source Author's compilation based on [16]'s matrix

others [20]. Webster and Drew [21] pointed out that conducting a fraud offline is much more difficult when compared to online. For instance, in reality, if an individual needs to turn from male to female, he may dress like a female or get hair like a female but perhaps he may be caught with these artificial signs, and therefore it would be significantly harder to counterfeit offline. Social media anyway provide a domain in which personal evaluation signals are not cross-checked, which make committing fraud simpler to accomplish. For example, sexual orientation exchanging on the web requires usually just a name change [22].

24.3 Online Fraud

The process of committing fraud on the web is impacted by various components associated with the fraudster, the social networking administration, the fraudulent demonstration, and the potential victim. These elements would decide how simple or complicated for a fraudster to commit an online fraud act [23]. In case of high complications in accomplishing social media fraud may prevent potential victims, while low trouble might be viewed as a simple chance to fraud others [24]. Figure 24.1 demonstrates the different entities (fraudster, social media, and potential victim) associated in social media fraud.

24.3.1 The Fraudster

Numerous components related to the fraudster will motivate in accomplishing online fraud. These elements incorporate psychological behaviour, desires, objectives, connection to the victim, and the victim's level of suspicion. Psychological behaviour favours fraudulent actions as an instrument for targeting online fraud. For instance, psychology of an individual defines the value system of the activities online as well as offline and if the values of the user claim that committing any fraud online or offline is acceptable for them, then they indulge in such activities with pride [25]. Also, people have been utilizing fraudulent activities with the intent of monetary gains, but not all frauds are conducted with an objective of monetary gains,in some cases, revenge, envy, etcetera are the prime motives behind committing fraud [26].

Desire is a factor that decides the probability of achievement in fraud. More formal messages have a higher probability of success. Objectives decide the motive of fraud. They specifically impact the method and type of fraudulent act [27]. De Salve et al. [28] depicted three distinct inspirations for online fraud on social media sites: (i) an instrument where one can manipulate his online profile to get more lucrative offers, for example, lying on one's resume on a social medium to get better job position offers, (ii) Social status, for example, pretending to have social relations in the form of online social networks, and (iii) Anonymity, for example, saving one's actual identity from despicable occasions on their online profile.

Fig. 24.1 Entities involved in a social media fraud

The above-mentioned motivational factors thus decide the online fraud process (i.e., the level of complexity in accomplishing fraud). For instance, a fraudster who wanted to fake his or her profile requires more effort offline to succeed because it is possible to verify offline in case of any suspicion but not online where user can manipulate his personality just by uploading fake information (sexual orientation, age, and etc.) which doesn't include any formal verification [29]. The process of online fraud is also affected by the fraudulent connection to the victim. Familiarity with the victim makes it easy to get trust and decreases the trouble in accomplishing fraud. Numerous social media users expect that with innovation there is upgraded security resulting in less suspicion towards others,likewise lessening their level of doubt results expanding their odds of being a fraud [30]. The ethical cost is also considered as an important factor in accomplishing fraud. Ethics can intensely impact what fraudster considers unacceptable concerning sharing data or even presenting fake data [31]. Conversely, on account of online situations, connection and anonymity add to lose morals and thus the ethical cost is down for fraudsters.

24.3.2 Social Media

Social media incorporates growing connections, but these connections can be used to fraud recipients. For example, the Interpersonal Deception Theory (IDT) states that the communication on social media between a sender and a receiver is a session where the information shared is not verified by any regulatory body which perhaps accelerates the fraudulent activities. Burns and Roberts [32] have proposed that if fraudulent dealing is common in a specific social media (e.g., Facebook page) then at that point the probability of reducing the fraud is increased. Thus, the prevalence of online fraud would be more active if the social media administrators analyse all the activities on the respective platforms and take corrective measures in case of any suspicious activity. Social media administrations which experience high volumes of frauds will restrict networks that are more suspicious. This will increase the number of failed fraud endeavours [2]. However, by expanding the restrictions may prevent fraudsters from the network. The statistics, however, recommends that networks with low cases of fraud will probably be more vulnerable to assaults since suspicion will stay low for potential victims. Deciding the predominance of fraud in a network remains a complex assignment [33].

Thus, the fundamental programming of social media sites should increase the level of suspicion: the level of suspicion for fraud expands the odds of accomplishment for the fraudster. Security configurations can make a few assumptions being made by users about the level of security it offers. A few aspects of the configurations can make users concern of potential dangers of being fraud. For instance, social media administrators provide verification of the user on their respective profiles (i.e., phone verification, where documents of the user are already submitted), In addition, verification and trust instruments for a framework will decide the level of trust between the sender and receiver. Verification components can either decrease the possibility of fraud or increment the suspicion of deceivers [34].

High suspicion will expand the trouble for fraudsters particularly when the regulatory bodies are on high alert. Verification components are considered to be powerful in specific situations where it is contended that the requirement for trust can be totally decreased. In social media, verification systems are significantly harder to execute and in that capacity of being caught might appear to be lower than those in offline scenarios; thus the rate of fraud is lower compared to social media [35].

At last, media popularity is additionally a factor that decides the trouble in accomplishing fraud. In this specific situation [36], in their investigation of fraud on the internet found that fraudsters experienced more pressure when communicating with their victims through oral as opposed to writing or content-based (i.e., image, video, audio) communication.

24.3.3 Fraudulent Actions

Privacy policy of the social media site and the number of assaults also effects fraudulent action on the social media site. (Reference) stated, privacy policy should be stringent and in case of any unacceptable activities, legitimate actions should be taken. In addition, the quality of the privacy policy of the social media is to be well marketed so the people committing online frauds must be made aware of the consequences in case they are caught as it build a sense of fear amongst them [37]. For scenarios where online fraud can cause major setback for the nation, the cost of prevention technique of the online fraud may exceed and even the punishments are also harsh. The social space where fraudsters are connected to the number of social media users to target who are required to be deceived also influence the level of trouble in actualizing the fraud strategy. A government official or politician or even the celebrity need to maintain privacy in his/her profile as the greater part of fraudsters will confront to get the critical information. The sort of information is also another essential factor. Complex kinds of fraudulent acts (e.g., identity theft and instrumental) that are guided by different objectives are harder to accomplish where privacy settings are stringent [38].

24.3.4 The Potential Victim

In offline fraud, the victim's capacity to identify fraud might be higher which decides the trouble in accomplishing fraud, whereas in online fraud by all accounts significantly harder to identify by users [39]. For instance, in an investigation of online misrepresentation utilizing page-jacking methods even experienced technicians unable to recognize the fraud and very few who detected fraud demonstrated that it is always very difficult to identify online fraud. Hence, in social media, the potential victim's capacity to identify online fraud depends to some degree on his Information Communication Technology (ICT) proficiency [40]. Users with high ICT proficiency are difficult to deceive online and can have a significant advantage over casual social media users.

24.4 Fraudulent Techniques in the Social Media Environment

There are different methods mentioned in the past researches that can be utilized to fraud others on social media sites, i.e., feign identity, imitation (e.g., mirroring a site), fakery (e.g., developing fake social media presence), data evasion, commercial use of personal data, site page re-direction (e.g., deluding somebody to an inappropriate page), and disguised (e.g., concealing data from one's profile). We utilize the social media model proposed by [41] to counter fraud procedures.

24.4.1 Fraudulent Model

The model (shown in Fig. 24.2) comprises a sender (S), the content(C), the social media channel through which correspondence happens (SM), and the receiver (R). The model shows two different scenarios. Scenario 1: Normal interaction model where no fraud has happened. The sender sends his/her content using any selected social media channel to the receiver. This is likewise in accordance with [42], meaning of social media communication which contends that all the channel participants must be true to their respective identities. On the other hand, Scenario 2: by manipulating any of the S, C, SM, and R components or blends of these leads to fraudulent action. The authors propose in this paper a review of social media sites and distinguishes factors (S, C, SM, and R) and how fraud composes on social media. These were recognized from firmly related past researches on the theme and the authors tried to explain the concept further in the following segments.

24.4.2 Sender Fraud

Sender fraud is accomplished by manipulating the sender's (S) individual data. 'Fake identity' is a typical illustration which frequently results in sender duplicity. This fraud falls under the category of individual fraud in online fraud. For such a scenario, the fraudster may access the individual's data, for example, house address, date of birth, and cell number and utilize it to commit fraud with them as well as their peers. The inability to verify the sender's authenticity will prompt abundant frauds. Social

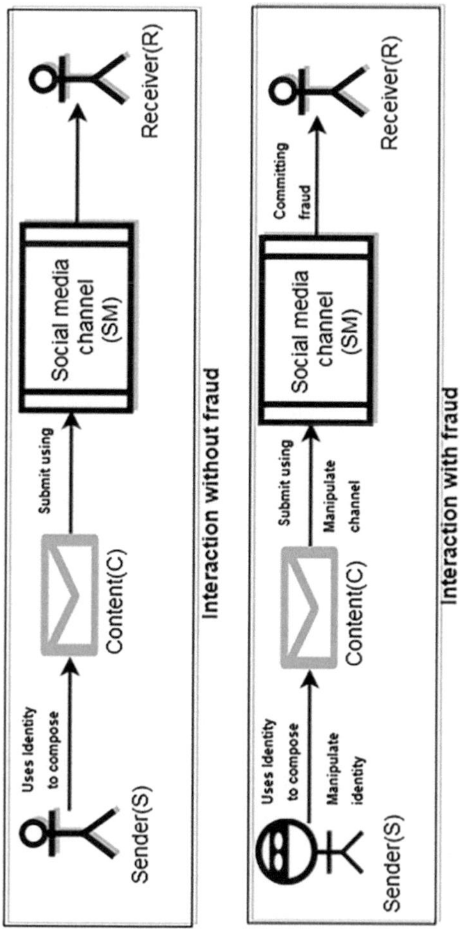

Fig. 24.2 Communication model with/without fraud

media sites have in-manufactured authentication mechanisms which are ineffective to authenticate the sender's identity. Perhaps social networking sites, i.e., Facebook, and micro blogging sites, i.e., Twitter, can prompt stolen identities on the web as there are no control systems to confirm the authenticity of the data submitted by the user while registering data. However, the harm caused by such online frauds with these kinds of social media data verification mechanisms is additionally prone to high and long haul damages to the receiver of such frauds. Senders on social media, with the character of fraudulent conspiracies push manipulation on different social media channels which may affect the brand value of the social media channel. In the case of social media sites, the efforts required at the senders' end to commit an online fraud is subjectively based on the authentication mechanism of the social media locale. The consequences, however, appear to be substantially callous in a social medium locale where family and peers are involved in the network since access to a user's personal organization can prompt emotional stress. The objective of the fraudulent is in the conspiracy to commit online fraud, perhaps not keen to get the full data of the user but rather interested in collecting information which helps in committing the fraud. With no privacy instruments set up for recognizing a source for registered fraud users who have a record of online frauds on social media, might be more harmful for the registered users who use social media for normal usage.

Social media sites (e.g., Wikipedia or YouTube) with low self-introduction and user-generated content are probably going to be more secured for online fraud, this could be somewhat credited to their content capacity. Collaborative projects and video content community data base which have to be verified by the site administrative of the respective sites. A sender, who needs to share any content on the low self-introduction sites needs to fulfil the benchmarks, set by the site administrative but the administrative cost is probably going to be high and the achievement of the online fraud would be low. Sender fraud includes manipulating data related to an identity. These online fraud assaults can be arranged as identity disguise where part of the data for an original identity is kept hidden or changed and in few cases identity imitation where another fake identity is formed to fraud people. For instance, individuals on social media sites may endeavour to counterfeit a few of the data in their profiles keeping in mind the end goal to get trust or misrepresent themselves in an alternate way. In social media locales, senders may use identity misrepresentation to gain others' confidence of various monetary and non-monetary benefits.

24.4.3 Content Fraud

Manipulating content is another common method for online fraud. In social media, this can be accomplished by distorting content. In social media sites, the entire structure is based on content such as blogs, images, videos, etcetera, which are profoundly vulnerable to such misleading. Technological advancement and its ease of availability today enable users to control and manipulate content to a phenomenal degree. Tampering text files is a common method to manipulate content, for example, faking identity and tampering documents to represent himself from a financial institution (i.e., Bank) and cheating others on a financial transaction. Such a technique may assist a fraudster with elevating their societal position and gaining the trust of a casualty so as to acquire additional data. Additionally, pictures and videos can be manipulated for causing frauds that on occasion, is vigorously in view of efforts takes into consideration a minimal effort for online fraud perhaps results in big online frauds. There are various factors responsible for such frauds, such as a low technical education of recipients (e.g., basically assessing content), the absence of suspicious nature and even responsibility. Social media locales that offer profile administration for their users, for example, social networking sites, i.e., Facebook and image content communities, i.e., Instagram are also vulnerable in situations where the inception of new connections is promoted. An equipped fraudster with content may have a considerable favourable position on these sorts of social media locales.

Conversely, collaborative project, i.e., Wikipedia, is less likely to be influenced by this sort of misdirection (i.e., content manipulation). The trouble in accomplishing online fraud may appear to be low; however, the odds of progress for such sites are also low. This is on account of the content management of these sorts of social media locales where site admin correspondence is upheld and empowers individuals to review the content. Cases of content fraud can be seen on Facebook where no such high standards of content authentication measures are applied. Moreover, authentication instruments, for example, the necessity for content legitimacy (following content back to its source) should be incorporated with the framework to guarantee that content manipulation turns out to be more noticeable.

Another common case of content-based fraud in these social media sites is with open source internet usage by different users where it is substantially easier to include pernicious content and prevail in online fraud in light of the fact that numerous people can access the network if it is being properly signed out. Business networking sites, i.e., LinkedIn also have a high likelihood for online fraud due to the firmly commercial components, for example, sharing highly professional information on the portal.

24.4.4 Social Media Channel Fraud

Controlling the social media channel requires a higher specialized ability level which increments the difficulty of online fraud. Manipulation of the channel incorporates tampering in as well as out messages and re-directing of movement of messages. Jamming correspondences have been utilized in review sites to manipulate channel. Hayat et al. [43] contended users' reviews in sites can be hacked with a specific end goal to remove the negative reviews. Once the hacker has access to the user' social media channel, he/she can throw the user out and continue with an online fraud. In few cases, the objective of a fraudster may not be to acquire data but rather to harm a victim's reputation. It is necessary to call attention to such cases where there is data theft and which ultimately leads to financial frauds.

Advantage that the fraudster may have on channel fraud, as there are so many parties involved in a social media channel thus victims don't even understand by whom they are being fraud. In real world, physical behaviour, for example, body movements or face expressions can uncover fraudster, but through social media, a fraudster may manipulate their behaviour in video or sound to cover their misleading, viably expanding the odds of accomplishment of the online fraud. Victims will have a troublesome in differentiating between a questionable/fraudulent association and the genuine association. Since channel fraud frequently includes technological knowledge, in few cases, social media administrations might be helpless to an assault; particularly those utilize non effective structure of online security. Social media administrations that have a higher degree of their user data will be more inclined to assaults while those that depend on server data will most likely be more secure.

Administrations with high user content tend to depend greatly on customer programming just like the case with review sites and business networking sites. Fraudulent actions by misusing social media channels are basic in such services. Server-based applications, for example, social networking sites (i.e., Facebook) or content-based sites are more inclined to channel fraud dealing since misuses depend on vulnerabilities of security settings of the media channel which are so for not solidified and made secured enough.

24.4.5 Hybrid Fraud Methods

Hybrid fraud methods include the control of different as well as all components (S, C, SM) in the communication model depicted above and can be more viable in propelling fraudulent assaults. The connections between S, C, and SM, as portrayed by [13], convey a reliable view for social media communication. Any one component of the social media communication model demonstrates a marginally unexpected conduct,this may be because of poor co-relation between two or more components (e.g., S and C). For instance, a social media message sent by one's user may lose its validity if the source data of the message does not coordinate with receiver.

Various hybrid fraud methods that control the content as well as sender's identity have been accounted in the past researches. These incorporate cases, for example, identity forgery data manipulation, personality falsification, web forgery, email misrepresentation. These frauds are very common in social media locales for example, social networking sites, i.e., Facebook, content-based sites, i.e., Instagram and YouTube, micro blogging, i.e., Twitter which feature user individual profile for one to one communication or one to many communication.

These online frauds on social media sites have been succeeded on various cases and to be successful as well as their outcomes can prompt tragic results including the bankruptcy, emotional breakdown, and depression in few extreme cases results to death. The social media administration needs to set high standards of registration for individuals who need to start new connections on social media and the absence of authentication checking for new users creates possibility of online fraud. On the social media sites, fraud can also have financial consequences, identity forgery used to get access to financial accounts of the user. Data manipulation includes controlling the sender's data and the content. Data manipulation is common for social media due to the security pattern of social media. Indeed, even after an internal audit mechanism on social media locales they are not compelling in recognizing fraudulent actions thus vulnerabilities as of now may be always present in social media locales.

24.5 Discussion

Online fraud in social media locales is filled with challenges, which include: (a) the absence of a standard uniformed theory for online frauds, (b) the absence of computationally productive techniques for online fraud detection for widespread social media channels, and (c) the absence of effective strategy for online fraud prevention by social media administrates.

24.5.1 No Standard Uniformed Theory for Online Frauds

Presently, there are only a few theories for fraud on social media sites as compared to offline fraud theories, although there have been various researches conducted and proposed for identifying double dealing including Management Obfuscation Hypothesis (MOH), Information Manipulation Theory (IMT), Identity Deception Theory (IDT), Four Factor Theory (FFT), and Leakage Theory (LT).

Above-mentioned theories mainly focus on frauds caused in an offline environment, the principal limitation with these strategies is that they depend on verbal and non-verbal prompts that may not apply to the social media. For instance, the existence of non-verbal elements in some social media locales requires revaluating as to what implications it can cause to users since they are not prone to exist online in the structure of social media as they exist in the physical offline world. A uniformed theory is required for online frauds as well. Researches towards a uniformed theory have been made but no such generalized theory has being developed. Current researches demonstrate that there are alternatives accessible for tending to issues of online fraud on the web. However, these previously mentioned theories can't specifically address a wide range of online fraud for various types of social media locales on the grounds that: (a) there is a huge variety amongst social media locales regarding outline and the sort and measure of data that is permitted to be traded amongst users and (b) it is hard to decide the setting in which security setting will be ideal for various frauds. Essentially, the field does not have a strong system that catches the interdependencies and connections amongst various detection techniques for the kind of online frauds on social media locales.

24.5.2 Absence of Online Fraud Detection Techniques for Widespread Social Media Channels

The procedures that are right now utilized for online fraud are very content based and a significant number of them can't be connected to the social media condition. Probably the most mainstream online fraud detection technique is managing verbal correspondence incorporating Content-Based Criteria Analysis (CBCA), Scientific Content Analysis (SCAN), and Real-time Monitoring (RTM)14. The pertinence and viability of these online fraud detection strategies to social media locales are misty. Techniques managing verbal signs, for example, video examination, might be computationally inefficient. Additionally, techniques that expect to identify sender fraud (identity misrepresentation) and utilize profile investigations to analyse user might be achievable for small datasets yet a correlation of all records with each other outcomes is a difficult task. In a few settings where the profile data is accessible and content correlation is workable for detecting online fraud on a profile, the algorithm keeps on changing for the social media locales. Basically, the potential for utilizing this recently accessible social information is there, be that as it may, computational effectiveness should be tended to for widespread social media locales. On a positive note, the social media administrates are carrying out on-going researches for developing more effective online fraud detection technologies.

24.5.3 No Effective Strategy for Online Fraud Prevention by Social Media Administrates

Social media administrates need to endeavour to address the issue frauds in social media locales. For instance, Wikipedia's approach requires data added to articles referred back to its source and has guaranteed that numerous unmerited contentions are presented to per users. Other social media administrations need to address the issue of identity confirmation. For instance, incomprehensibly, people who don't have an active profile today on a popular social media site (i.e. Facebook) will probably get in online fraud or identity misrepresentation alongside their real-world known ones. The issue is that users and different friends turn up worried when they find duplicate accounts particularly when an active original account of a user is already active. However, when a fraudster registers as a user that did not exist on a social media locale previously, others may probably accept the request to goodness that the user simply joined the social media locale. While trying to build their user base, social media administrations alongside their standard enlistment of access and usage highlights should inform about the possibilities of online frauds.

24.6 Conclusion

The social media space is continuously advancing and keeps on being stretched out with various innovations that fraudsters can utilize. While the physical distance that isolates the fraudster and the victim may appear to be wide in social media, therefore, the harm in social media sites is even more problematic for people, associations, and governments. Considering how online fraud works through social media locales and future advances remains a huge test. To address this test, social media administrates have to outline sites with different guidelines and standards that are conventional and cannot be deceived so easily. Human craving for technological development has brought about different social media locales that we don't yet completely comprehend and their vulnerabilities are at present being abused in different routes by different aggressors incorporating those included with fraudulent assaults. It is time we begin contemplating how to outline social media locales in regular usage conditions to defend and shield social media users from the unanticipated results of online frauds [44].

References

1. Humpherys, S., Moffitt, K., Burns, M., Burgoon, J., & Felix, W. (2011). Identification of fraudulent financial statements using linguistic credibility analysis. *Decision Support System, 50*(3), 585–594.
2. Peng, S. Y., Yu, S., Niu, J., & Jia, W. (2018). Influence Analysis in Social Networks: A Survey. *Journal of Network and Computer Applications, 106*(1), 17–32.
3. Mori, P., De Salve, A., & Ricci, L. (2018). A survey on privacy in decentralized online social networks. *Computer Science Review, 27*(1), 154–176.
4. Hu, H., Ahn, G., & Jorgensen, J. (2012). Multiparty access control for online social networks: Model and mechanisms. *Transactions on Knowledge and Data Engineering, 25*(7), 1614–1627.
5. Humphreys, L. (2007). Mobile social networks and social practice: A case study of dodgeball. *Journal of Computer-Mediated Communication, 13*(1), 341–360.
6. Guo, L., Zhang, C., & Fang, Y. (2015). A trust-based privacy-preserving friend recommendation scheme for online social networks. *Transactions on Dependable and Secure Computing, 12*(4), 413–427.
7. Samuel-Azran, T., Hayat, T. Z., & Lesser, O. (2017). Gendered discourse patterns on online social networks: A social network analysis perspective. *Computers in Human Behavior, 77*(1), 132–139.
8. Chen, C., He, B. Z., Su, Y. P., & Sun, H. M. (2014). A Defence scheme against identity theft attack based on multiple social networks. *Expert Systems with Applications, 41*(5), 2345–2352.
9. Wall, D. S. (2007). Policing cybercrimes: Situating the public police in networks of security within cyberspace. *Police Practice and Research: An International Journal, 8*(2), 183–205.
10. He, B. Z., Chen, C., Su, Y., & Sun, H. (2014). A defence scheme against identity theft attack based on multiple social networks. *Expert Systems with Applications, 41*(5), 2345–2352.
11. Holt, T., & Bossler, A. M. (2008). Examining the applicability of lifestyle-routine activities theory for cybercrime victimization. *Deviant Behavior, 30*(1), 1–25.
12. Whitty, M. B. (2012). The online romance scam: A serious cybercrime. Cyberpsychology. *Behavior and Social Networking, 15*(3), 181–183.
13. Campana, M. G., & Delmastro, F. (2017). Recommender systems for online and mobile social networks. *Online Social Networks and Media, 3*(1), 75–97.

14. Van Wilsem, J. (2011). Worlds tied together? Online and non-domestic routine activities and their impact on digital and traditional threat victimization. Digital and traditional threat victimization. *European Journal of Criminology, 8*(2), 115–127.
15. Davis, J. T. (2012). Examining perceptions of local law enforcement in the fight against crimes with a cyber component. *Policing: An International Journal of Police Strategies & Management, 35*(2), 272–284.
16. Kaplan, A. M., & Haenlein, M. (2010). Users of the world, unite! The challenges and opportunities of social media. *Business Horizon, 53*(1), 59–68.
17. Reyns, B. W. (2010). A situational crime prevention approach to cyberstalking victimization: Preventive tactics for Internet users and online place managers. *Crime Prevention and Community Safety, 12*(2), 99–118.
18. Klier, M., Heidemann, J., & Probst, F. (2012). Online social networks: A survey of a global phenomenon. *Computer Networks, 56*(18), 3866–3878.
19. Cross, C. (2016). Using financial intelligence to target online fraud victimization: Applying a tertiary prevention perspective. *Criminal Justice Studies, 29*(2), 124–145.
20. Leukfeldt, E. R. (2014). Phishing for suitable targets in the Netherlands: Routine activity theory and phishing victimization. *Cyberpsychology, Behavior and Social Networking, 17*(8), 551–555.
21. Webster, J., & Drew, J. M. (2017). Policing advance fee fraud (AFF): Experiences of fraud detectives using a victimfocused approach.,. *International Journal of Police Science and Management, 19*(1), 39–53.
22. Yar, M., & Leukfeldt, E. R. (2016). Applying routine activity theory to cybercrime: A theoretical and empirical analysis. *Deviant Behavior, 37*(3), 263–280.
23. Savage, D. Z., Yu, X., Chou, P., & Wang, Q. (2014). Anomaly detection in online social networks. *Social Networks, 39*(1), 62–70.
24. Tian, Y., Yuan, J., & Yu, S. (2016). SBPA: Social behavior based cross social network phishing attacks. In *Conference on Communications and Network Security (CNS)* (pp. 366–367). IEEE.
25. Alghamdi, B., Watson, J., & Xu, Y. (2016). Toward detecting malicious links in online social networks through user behavior. In *International Conference on Web Intelligence Workshops (WIW)* (pp. 5–8). IEEE.
26. Kelly, M., & Cross, C. (2016). The problem of "white noise": Examining current prevention approaches to onlinefraud. *Journal of Financial Crime, 23*(4), 806–814.
27. Ramalingam, D., & Chinnaiah, V. (2018). Fake profile detection techniques in large-scale online social networks: A comprehensive review. *Computers & Electrical Engineering, 65*(1), 165–177.
28. De Salve, A., Mori, P., & Ricci, L. (2018). A survey on privacy in decentralized online social networks. *Computer Science Review, 27*(1), 154–176.
29. AlRubaian, M., Al-Qurishi, M., Rahman, & Alamri, A. (2015). A novel prevention mechanism for sybil attack in online social network. In *2nd World Symposium on Web Applications and Networking (WSWAN)* (pp. 1–6). IEEE.
30. Sahoo, S. R., & Gupta, B. (2018). Security issues and challenges in online social networks. In *User Perspective. Computer and Cyber Security: Principles, Algorithm, Applications, and Perspectives* (pp. 591–606). CRC Press.
31. Blackshaw, D., & Cross, C. (2015). Improving police responses to online fraud. *Policing: A Journal of Policy and Practice, 9*(2), 119–128.
32. Heidemann, J., Klier, M., & Probst, F. (2012). Online social networks: A survey of a global phenomenon. *Computer Networks, 56*(18), 3866–3878.
33. Burns, S., & Roberts, L. (2013). Applying the theory of planned behavior to predicting online safety behavior. *Crime Prevention and Community Safety, 15*(1), 48–64.
34. Boshmaf, Y., Muslukhov, I., Beznosov, K., & Ripeanu, M. (2011). The socialbot network: When bots socialize for fame and money. In *Proceedings of the 27th annual Computer Security Applications Conference* (pp. 93–102). AGM.
35. Dewan, P., & Kumaraguru, P. (2015). Towards automatic real time identification of malicious posts on Facebook. In *13th Annual on Privacy, Security and Trust (PST)* (pp. 85–92). IEEE.

36. Hu, W., Gong, Z., Leong Hou, U., & Guo, J. (2015). Identifying influential user communities on the social network. *Enterprise Information Systems, 9*(7), 709–724.
37. Dakiche, N., Tayeb, F., Slimani, Y., & Benatchba, K. (2019). Tracking community evolution in social networks: A survey. *Information Processing & Management, 56*(3), 1084–1102.
38. Ionescu, L., Mirea, V., & Blajan, A. (2011). Fraud, corruption and cyber crime in a global digital network. *Economics, Management and Financial Markets, 6*(2), 373–380.
39. Huber, M., Kowalski, S., Nohlberg, M., & Tjoa, S. (2009). Towards automating social engineering using social networking sites. In *International Conference on Computational Science and Engineering* (pp. 117–124). IEEE.
40. Kayes, I., & Iamnitchi, A. (2017). Privacy and security in online social networks: A survey. *Online Social Networks and Media, 3*(1), 1–21.
41. Fire, M., Katz, G., & Elovici, Y. (2012). Strangers intrusion detection-detecting spammers and fake profiles in social networks based on topology anomalies. *Human Journal, 1*(1), 26–39.
42. Gupta, B. B., Sangaiah, A., Nedjah, N., Yamaguchi, S., Zhang, Z., & Sheng, M. (2018). Recent research in computational intelligence paradigms into security and privacy for online social networks. *Future Generation Computer System, 86*(1), 851–854.
43. Hayat, T. Z., Lesser, O., & Samuel-Azran, T. (2017). Gendered discourse patterns on online social networks: A social network analysis perspective. *Computers in Human Behavior, 77*(3), 132–139.
44. Bossler, A. M., & Holt, T. J. (2009). On-line activities, guardianship, and malware infection: An examination of routine activities theory. *International Journal of Cyber Criminology, 3*(1), 400–420.

Chapter 25
Artificial Intelligence: Revolutionizing Industrial Decisions

Ritu Singhal, Ishita Singhal, Varun Sharma, and Vikas Bansal

Abstract Artificial Intelligence has emerged as a powerful force in modern business, transforming how firms operate engage with customers, and make choices. The role of AI in data-centric approaches to industrial decisions, advancing technology, digitization, and business implications has been remarkable in the current market scenario and technological innovation. This chapter explores how AI technologies are transforming decision-making processes across industries, as well as the underlying techniques, benefits, problems, and future approaches. The chapter aims to provide a complete overview of AI's revolutionary impact on industrial decisions by examining use cases in real-time applications as well as research studies. This study investigates the wide range of AI applications in business, as well as the underlying technologies, benefits, problems, and opportunities. It seeks to provide a comprehensive acquaintance of how AI is transforming the corporate landscape through research studies and real-world examples. The study expands on the usage of cloud computing to support AI with large IT infrastructures for the industrial revolution.

Keywords Artificial Intelligence · Business · Decision-making · Business practices · Technological innovation · Cloud computing

25.1 Introduction

Artificial Intelligence (AI) has emerged as a transformative force in the industrial sector, driving significant changes in how businesses make decisions. By leveraging vast amounts of data, AI enables more accurate, efficient, and strategic

R. Singhal (✉) · V. Sharma
Indraprastha College of Women, Delhi, India
e-mail: toritu7@gmail.com

I. Singhal
Maharaja Surajmal Institute of Technology, I.P. University, Delhi, India

V. Bansal
Xebia IT Architects India Private Limited, Gurugram, Haryana, India

P. K. Kapur et al. (eds.), *Data-centric Approaches to Industrial Decisions*, Asset Analytics, https://doi.org/10.1007/978-981-96-7556-2_25

decision-making processes. AI in industry refers to applications of machine learning, natural language processing, computer vision, and other AI technologies to improve decision-making processes. AI systems can analyze vast amounts of data, recognize patterns, and provide actionable insights, thereby enabling industries to make more informed and timely decisions. The integration of AI into industrial decision-making has been driven by advancements in data processing capabilities, the proliferation of big data, and the development of sophisticated algorithms. This has led to the widespread adoption of AI across various sectors, from manufacturing to healthcare, finance, and beyond.

Incorporating AI into strategic business choices is a watershed moment in the corporate landscape. It has enormous potential for generating shareholder value, involves a reconsideration of traditional strategic planning methods, and encourages the growth of novel business models. As businesses traverse AI, the emphasis moves to influencing the power of AI to improve decision-making processes, boost corporate performance, and secure long-term commercial success in a world that is growing digitally.

25.1.1 Objectives

- To explore AI technologies and traditional approaches in the industry.
- To identify AI Technologies in a diverse range of industries.
- To study how AI Revolutionizes Industrial Decisions
- To analyze the benefits and challenges of implementing AI applications in industrial decision-making.
- To examine future trends and integration of emerging technology like cloud computing with AI for industry.

The chapter is organized as follows: Sect. 25.2 focuses on the Background of Artificial Intelligence in Business. Section 25.3 discusses AI in the Business Context. The comparison of AI with Traditional Approaches in the Industry is covered in Sect. 25.4. Section 25.5 discusses how AI revolutionizes Industrial Decisions. Section 25.6 showcases the Challenges of Implementing AI for Real-Time Applications in Industry. Section 25.7 discusses the Synergy of AI and Cloud Computing in the Industrial Sector. Finally, Sect. 25.8 concludes this article with future scope. Kindly provide the proper section citation for this chapter. I have updates proper section citation and editing section numbers to maintain better flow. kindly let me know if i need to edit again.

25.2 Background of AI in the Business Context

The emergence of artificial intelligence in business has been characterized by vast influence and continual innovation. Throughout the last decade, AI has progressed from a revolutionary technology notion to a critical component in strategic business analysis, transforming how organizations tackle problems and possibilities in the information age. The concept of AI was first proposed by computer science pioneers in the middle of the twentieth century when they started to investigate the idea of creating robots that could mimic human intellect. Early research aimed to create computational models and algorithms that could solve problems, recognize patterns, and engage in logical reasoning. AI has developed over the years from a theoretical concept to a useful technology with important applications in a variety of industries, including business.

25.2.1 Early Developments and Theoretical Foundations

The concept of AI emerged in the 1950s, motivated by technological developments and a desire to construct machines that could think and learn as humans. The Dartmouth Conference in 1956 officially established artificial intelligence as a subject of research. Researchers include John McCarthy, Marvin Minsky, Allen Newell, and Herbert A. Simon, who established the fundamental theories that would govern AI development for decades to come.

During this time, AI research was primarily concerned with symbolic AI, which involves employing symbols and rules to express knowledge and solve problems. Early AI systems were restricted in capabilities and required significant human intervention to function. Despite these limitations, the early work of the 1950s and 1960s laid the theoretical foundations for AI's future uses in business.

25.2.2 The Rise of Machine Learning and Expert Systems

The 1980s and 1990s saw a shift in AI research towards expert systems and machine learning. During the 1980s, expert systems became popular. These AI programs were created to imitate the decision-making abilities of human specialists in specialized fields. Financial analysis, medical diagnosis, and process control used expert systems in businesses such as finance, healthcare, and manufacturing.

Machine learning, a branch of AI that focuses on building algorithms that allow computers to learn from and predict data, gained popularity in the 1990s. Improvements in computer power, data availability, and algorithm development fueled the growth of machine learning. This era created the framework for modern machine-learning algorithms, which are today built into AI applications across a variety of industries.

25.2.3 The Emergence of Big Data and Modern AI

The early 2000s witnessed the start of a new age in AI, propelled by an explosion of digital data and advances in computational capacity. The availability of large amounts of data, along with strong algorithms and greater processing resources, allowed AI to make advancements in machine learning, particularly deep learning. Deep learning is a kind of machine learning that uses neural networks with multiple layers to learn complicated patterns and representations from big datasets.

In the commercial world, the ability to analyze and extract insights from massive volumes of data has become a game changer. AI-powered analytics, predictive modeling, and automation have transformed finance, healthcare, retail, manufacturing industries, and logistics. Businesses may now use AI to manage operations, improve customer satisfaction, and gain an edge in the marketplace.

25.2.4 AI's Integration into Business Processes

Today, AI is deeply integrated into various business processes, driving efficiency, innovation, and growth. Some of the applications of AI in business are Customer Relationship Management (CRM), Supply Chain Management, Financial Services, Marketing and Sales, and Human Resources Management through data-driven insights and automation.

25.3 Artificial Intelligence in the Business Context

In the business context, AI refers to the deployment of advanced computational algorithms and models that enable machines to perform tasks that would typically require human intelligence. These tasks range from simple automation to complex problem-solving and decision-making processes. AI in business is designed to enhance operational efficiency, improve accuracy, drive innovation, and provide valuable insights, ultimately contributing to a company's strategic goals and competitive advantage.

25.3.1 Categories of AI in the Business Context

AI in business can be categorized based on various dimensions, including its capabilities and functionalities.

Based on Capabilities

Narrow AI (Weak AI): AI systems that are designed to handle specific tasks, are highly specialized, and do not possess general intelligence or versatility beyond their designated function. Some of the examples are voice assistants Siri and Alexa, recommendation systems, Chatbots, etc.

General AI (Strong AI): Hypothetical AI that can understand, learn, and apply knowledge across a wide range of tasks, much like a human and capable of performing any intellectual task that a human can, exhibiting general intelligence. Currently are the subject of ongoing research.

Super Intelligent AI: An advanced level of intelligence that surpasses the smartest human minds as exhibits superior cognitive abilities, problem-solving skills, and creativity beyond the smartest human beings. This is a theoretical concept.

Based on Functionalities

Reactive Machines: AI systems that respond to specific stimuli without using memory or learning from past experiences. They have limited functionality and operate solely based on predefined rules. Examples include IBM's Deep Blue chess-playing computer and basic customer service Chatbots.

Limited Memory: AI systems that can use past experiences to inform current decisions and are capable of handling more complex tasks by utilizing historical data. Examples are autonomous vehicles (self-driving cars), predictive maintenance systems, etc.

Theory of Mind: AI systems that understand human emotions, beliefs, and social interactions. Though not yet achieved, aims to simulate a human-like understanding of social dynamics.

Self-aware AI: AI that possesses self-awareness and consciousness. This is a theoretical stage where AI systems possess self-awareness and consciousness. Currently theoretical, would involve machines having a sense of self and autonomy.

25.4 Artificial Intelligence in Industry

AI technology has become a cornerstone of modern industry, driving innovation, efficiency, and growth. AI encompasses various technologies, such as machine learning, natural language processing, computer vision, and robotics, each offering unique capabilities that enhance industrial processes.

25.4.1 Artificial Intelligence Verses Traditional Approaches in the Industry

AI technologies are transforming industries by offering new capabilities and efficiencies that traditional approaches struggle to match. AI technologies offer transformative advantages over conventional techniques in various industries. By automating data processing, enhancing decision-making, improving customer interactions, and increasing production efficiency, AI provides significant benefits that traditional methods cannot match [1–5]. AI drives innovation and cost savings, giving companies a competitive edge in today's fast-paced market.

The comparison in Table 25.1 highlights the differences between AI-driven and traditional methods in different aspects across various industrial sectors.

Tables 25.2, 25.3, and 25.4 comparisons highlight how AI technologies offer significant improvements over traditional approaches across various industries like healthcare [6], manufacturing, and retail industries. AI enhances accuracy, efficiency, and personalization, leading to better outcomes in healthcare, manufacturing, and retail industries. As AI continues to evolve, its adoption will become increasingly crucial for businesses seeking to maintain a competitive edge and achieve operational excellence.

25.4.2 Key Technologies of AI in Industry

AI encompasses a wide range of computational technologies that enable machines to perform tasks that typically require human intelligence. Each AI technology brings unique strengths and features, contributing to the transformative potential of AI across various domains and industries [7]. Key AI technologies include:

Machine Learning (ML)

By examining data patterns, machines can learn and adapt based on previous experiences in ML technologies.

In the industrial automation arena, heterogeneous data and its analysis receive much attention from academia and industry, because it has the potential to improve or enable operations like diagnostics, predictive maintenance, and condition monitoring. Kamm et al. [8] examine existing approaches for the machine-learning based analysis of heterogeneous data in the industrial automation domain.

Arefin et al. [9] investigated the multifaceted patterns of buyer behavior in the UK retail business using advanced data analysis with 495,478 consumers. They used different supervised machine-learning models, with XGB Classifier and ET achieving the highest accuracy in estimating customer lifetime value. This study sets a new standard in commercial analytics by providing a scalable and productive technique for future studies aimed at using data analytics to understand and forecast customer behavior in various business organizations.

Table 25.1 Comparison between AI-driven and traditional approaches followed in the industry

Aspect	Traditional approaches	AI technologies
Data processing	Manual data analysis and processing, often slow, and error-prone Limited capacity; difficult to handle large volumes	Automates data processing, reducing the need for manual intervention, and minimizing errors Capable of analyzing vast datasets quickly and accurately, providing real-time insights
Decision-making	Based on historical data and human intuition, limited by cognitive biases. Slower decision-making process due to manual analysis and deliberation	Data-driven decisions, leveraging predictive analytics and pattern recognition for more accuracy Enables real-time decision-making by quickly analyzing data and identifying patterns
Customer interaction	Human customer service representatives, limited availability, and consistency. Marketing and consumer engagement techniques rely on extensive segmentation and manual analysis	AI Chatbots and virtual assistants offer 24- hour support; Personalized marketing and recommendation systems analyze individual customer data to tailor interactions
Production efficiency	Manual labor and traditional machinery are prone to human error and downtime Maintenance is often reactive and issues are addressed only after they occur	AI-driven automation and robotics enable continuous production with high precision and reliability Predictive maintenance uses machine learning to foresee equipment failures, reducing downtime
Innovation	relies on human invention and trial-and-error approaches, is occasionally costly, and time-consuming; Limited by the cognitive capabilities and biases of human workers	AI-enhanced creativity, using AI algorithms like Generative Adversarial Networks (GANs) to generate new ideas and optimize innovation processes; Accelerates the innovation process by optimizing research and development efforts
Cost management	Higher operational costs due to manual processes and inefficiencies	Reduced operational costs through automation and optimization of resources Predictive analytics improve inventory management and reduce waste, leading to cost savings

Table 25.2 Healthcare Industry

Aspect	Traditional approach	AI approach
Diagnosis process	Manual analysis of medical records, lab results, and imaging	AI algorithms analyze medical images and patient data rapidly and accurately
Example	Manual examination of X-rays and MRIs	Google's DeepMind AI system for diagnosing eye diseases
Pros/Cons	Expertise-based, personalized care; Time-consuming; Prone to human error	Rapid, accurate, and consistent diagnoses; Reduces human error; Increases diagnostic capacity

Table 25.3 Manufacturing Industry

Aspect	Traditional approach	AI approach
Production	Relies on human labor and traditional mechanization, leading to inefficiencies	Use of AI-driven robots for continuous, precise production processes
Maintenance	Reactive, addressing equipment issues only after they occur	Predictive maintenance using machine-learning models to forecast equipment failures
Example	Human labor and traditional mechanization with frequent downtime	General Electric (GE) AI-based predictive maintenance for industrial machines
Pros/Cons	Human oversight; Prone to inefficiencies and errors; Higher maintenance costs	Reduces downtime and maintenance costs Enhances production efficiency and precision Continuous operation

Machine learning (ML) has had a significant impact on the manufacturing industry as part of the Industry 4.0 concept. The Industry 4.0 paradigm emphasizes the use of smart sensors, gadgets, and machines to create intelligent manufacturing plants that continually collect production data. The authors discuss the ability of machine-learning techniques to provide predictive insights has enabled the detection of complex manufacturing patterns and paves the way for an intelligent decision support system in a variety of manufacturing tasks such as intelligent inspection, predictive maintenance, quality enhancement, optimization, supply chain management, and task scheduling in manufacturing and production systems [10]. The paper thoroughly reviews ML technology applications (ANN) in biodiesel research for function approximation, optimization, and control problems [11].

A comprehensive investigation of ML algorithms used in operation management for production planning and control, as well as defect analysis by Bertolini et al. [12], suggest that ML will become ubiquitous in many sectors of operation management.

Table 25.4 Retail industry

Aspect	Traditional approach	AI approach
Customer interaction	Broad marketing strategies based on manual segmentation	Personalized marketing and customer interaction strategies using AI-driven recommendation systems
Product recommendations	Generic recommendations may not align well with individual preferences	Recommendations are customized to each consumer's activities and inclinations
Example	Generic product recommendations	Amazon's AI-powered recommendation system
Pros/Cons	Basic segmentation; Lower customer engagement; Less effective targeting	Personalized customer interactions; Higher engagement and sales; Improved satisfaction and loyalty

These researches indicate ML as one of the core enablers in AI to advance a traditional manufacturing system to the Industry 4.0 level.

Sircar et al. [13] conducted a comprehensive state-of-the-art review of machine learning to address challenges and issues in data processing and handling for the oil and gas industry, demonstrating that these AI techniques can be used for data processing and visualization in various sectors of the upstream oil and gas industries.

Deep Learning (DL)

It utilizes multi-layered neural networks for high accuracy in complex pattern recognition tasks, capable of handling unstructured data.

Hamouda et al. [14] recommend a novel privacy-preserving secure framework using blockchain-enabled federated learning to handle cyber threats and data privacy issues in the growing reliance on industry 4.0/5.0 with improved privacy, verifiability, and transparency.

Zipfel et al. [15] explore the quality of the European car manufacturing Industry as a case study and assess fault detection through unsupervised deep neural network models namely Skip-GANomaly, PaDiM, and PatchCore.

Natural Language Processing (NLP)

It makes it possible for systems to comprehend and react to human speech. It facilitates human–computer interaction and processes text data for tasks like language translation and sentiment analysis.

The review work explores multiple factors of the construction industry empowered by Text mining (TM) and natural language processing (NLP) and highlights their importance for automation to minimize manual tasks [16].

Computer Vision

It allows machines to interpret visual data from the world, crucial for tasks like image analysis and autonomous driving.

Xiong et al. [17] designed a framework using Deep-learning-based and computer vision approaches, based on a real use case from the packaging industry including an automated quality control (AQC) architecture with a computer vision-based training model to reduce waste.

Robotics

It involves the design and use of robots to perform tasks autonomously or semi-autonomously and enhances productivity and precision. Mia and Shuford [18] navigate the challenges and exploit the ongoing progression of robotics and AI algorithms, identifying this combination as influential in revealing the full potential in defining the future landscape of Industry 4.0.

Expert Systems

It emulates human expert decision-making, providing specialized knowledge and consistent decision-making capabilities. Wang et al. [19] propose a decision-making expert system EconG© driven by monitored power data that includes signal conditioning, peaks-valleys feature exaction, and compression approaches. It leads to reduce the storage volume of real-time monitored data to 6.5% of the source. The authors claim that deeper implementation of EconG© will help in responding quickly to explosive demands and improve intellectualization, sustainability, and flexibility for the 4th industrial revolution.

Reinforcement Learning
Learns optimal strategies through trial-and-error, effective in dynamic environments, with reward-based learning.

The paper uses the digital twin method, with the data-driven tool, simulates the model, and obtains the final tool wear monitoring values through particle wave fusion algorithm. The authors claim that this adaptive optimization method can delay tool wear in industrial production, and help improve production efficiency in intelligent manufacturing (Abou et al. [20].

Generative Adversarial Networks (GANs)
It generates realistic synthetic data, enhancing creativity and innovation with generator and discriminator networks. The study presents a survey of the most recent trends in GAN applications and projects them into industrial use cases, conducts tests with synthetic images, and examines the majority of Using picture anomalies and GAN breakdowns, optimum approaches for training are proposed. It demonstrates GAN's breakthrough in computer vision applications by capturing a dataset's distribution and producing high-quality similar images [21].

Table 25.5 provides an overview of various AI technologies, their definitions, assets, and key features.

25.4.3 Importance of AI Technologies in Industry

AI technologies have become crucial for the industrial sector, offering numerous benefits that significantly enhance operations, productivity, and competitiveness. The following points highlight the importance of AI technologies in various industrial applications:

Efficiency and Productivity
AI technologies automate repetitive tasks, reducing the need for human intervention and significantly increasing productivity. AI-driven robots, for example, can operate continuously without becoming tired in the production sector, increasing productivity.

Improved Decision-Making
Massive data sets are evaluated by AI systems to produce information that can be put to use. In finance, AI models predict market trends and assess risks more accurately than traditional methods, enabling better investment decisions [22, 23].

Table 25.5 AI technologies, their assets, and key features

AI technology	Definition	Assets	Key features
Machine learning (ML)	A branch of AI that enables machines to acquire knowledge from their experiences and get better despite any direct programming	Ability to handle large datasets; Improves over time with more data; Versatile applications	Supervised, unsupervised, and reinforcement learning; Predictive analytics; Model training and evaluation
Deep learning (DL)	A type of machine learning that analyzes complicated trends in datasets by employing multi-layered neural network models	High accuracy in tasks like image and speech recognition; Capable of handling unstructured data	Convolutional neural networks (CNNs); Recurrent neural networks (RNNs); Autoencoders
Natural language processing (NLP)	AI enables machines to understand, interpret, and respond to human language	Facilitates human–computer interaction; Processes large volumes of text data; Language translation	Sentiment analysis; Language generation and understanding; Speech recognition and synthesis
Computer vision	AI makes it possible for machines to comprehend and decide based on visual data	Exceptional precision in identifying and detecting objects; Automates visual inspection tasks	Image and video analysis; Facial recognition; Optical character recognition (OCR)
Robotics	AI is applied to design and use robots for performing tasks autonomously or semi-autonomously	Enhances productivity and precision; Capable of performing dangerous tasks	Autonomous navigation; Manipulation and grasping; Human–robot interaction
Expert systems	AI systems that replicate human decision-making skills in specific domains	Provides specialized knowledge; Consistent decision-making	Rule-based reasoning; Knowledge base; Inference engine

(continued)

Table 25.5 (continued)

AI technology	Definition	Assets	Key features
Reinforcement learning	A kind of ML where entities are rewarded or penalized for how they act	Effective in dynamic and complex environments; Learns optimal strategies through trial-and-error	Reward-based learning; Policy optimization; Exploration and exploitation
Generative adversarial networks (GANs)	A class of AI algorithms that use two neural networks to generate new, synthetic data	Generates realistic data; Enhances creativity and innovation	Generator and discriminator networks; Data augmentation; Synthetic data creation

Enhanced Customer Experience
AI-powered Chatbots and recommendation systems personalize customer interactions, providing instant support and tailored suggestions, which improve customer satisfaction and loyalty.

Innovation and Creativity
AI technologies like Generative Adversarial Networks (GANs) enable the creation of new and innovative products. In entertainment, AI generates music, art, and even entire movie scripts, pushing the boundaries of creativity.

Safety and Quality
AI enhances safety through applications such as autonomous vehicles, which reduce the risk of human error. In healthcare, AI improves diagnostic accuracy, leading to better patient outcomes.

Cost Reduction
By automating tasks and optimizing operations, AI helps reduce operational costs. In logistics, AI optimizes routes and improves supply chain efficiency, lowering transportation costs.

Scalability
AI solutions with cloud computing can easily scale to handle increasing amounts of data and complexity, making them suitable for large enterprises and fast-growing businesses.

Competitive Advantage
Companies that adopt AI technologies gain a competitive edge by offering superior products and services, improving operational efficiency, and making data-driven decisions.

AI technologies are transforming entire industries through increased productivity, superior decision-making, and stimulation of creativity. Compared to traditional approaches, AI offers superior data processing capabilities, more accurate and timely decisions, enhanced customer experiences, and significant cost reductions. As industries continue to adopt AI technologies, they gain a competitive edge, enabling them to thrive in an increasingly data-driven and automated world. Embracing AI is not just a technological advancement but a strategic imperative for future success.

The next section explores the profound impact of AI on industrial decision-making, examining its applications, benefits, and challenges.

25.5 AI Revolutionizes Industrial Decisions

AI is at the forefront of a new industrial revolution, transforming decision-making processes across various sectors. From manufacturing to healthcare, finance to retail, AI technologies are enhancing efficiency, accuracy, and strategic planning. The integration of AI into industrial decision-making has been driven by advancements in data processing capabilities, the proliferation of big data, and the development of sophisticated algorithms. These advancements have enabled AI to analyze vast amounts of data, recognize patterns, and provide actionable insights, thereby revolutionizing the way industries operate.

25.5.1 *Applications of AI in Industrial Decision-Making*

AI has become a transformative force across various sectors, with its impact on industrial decision-making being particularly profound. The integration of AI in industries revolutionizes how decisions are made, enhancing efficiency, accuracy, and innovation. Table 25.6 explores the diverse applications of AI in industrial decision-making, highlighting its benefits and the challenges it presents.

The significance of AI technology for the industrial sector is undeniable. From enhancing operational efficiency and predictive maintenance to improving quality control and supply chain optimization, AI drives significant improvements across various aspects of industrial operations. By leveraging AI technologies, industries can achieve higher productivity, better decision-making, increased safety, and a competitive edge in the market. As AI continues to evolve, its impact on the industrial sector will only grow, making it a critical component of modern industrial strategy.

25.6 Challenges for Implementing AI in Real-Time Applications in Industry

The applications of AI in industrial decision-making offer numerous benefits, including increased efficiency, reduced costs, improved quality control, and enhanced predictive capabilities but it also comes with a set of challenges that need to be addressed to ensure successful implementation. Some of the key challenges are:

High Initial Setup Costs
Implementing AI solutions often requires significant investment in hardware, software, and infrastructure. High initial costs can be a barrier for small and medium-sized enterprises (SMEs) and even large companies that are cautious about Return on investment (ROI). For example, the cost of deploying AI-powered computer vision systems for automated quality inspection can be prohibitively expensive for smaller manufacturers.

Data Quality and Quantity
AI models require large volumes of high-quality data for training and continuous learning. Inadequate or poor-quality data can lead to inaccurate predictions and decisions, undermining the effectiveness of AI applications. Predictive maintenance

Table 25.6 Applications of AI, real-time examples, benefits, and the challenges

Application	Description	Examples	Benefits	Challenges	References
Automated quality inspection	Uses AI-powered computer vision to inspect products for defects and anomalies	Siemens: AI-based visual inspection in manufacturing processes	Reduces human error; Increases inspection speed; Ensures consistent quality	High initial setup cost; Requires extensive training data	Siemens [24] Gopalakrishnan et al. [25] Linda [26]
Predictive maintenance	AI models analyze equipment data to predict when maintenance is needed, preventing breakdowns	General Electric (GE): for industrial machinery predictive maintenance Nestlé: Predictive Maintenance	Reduces downtime; Extends equipment lifespan; Lowers maintenance costs	Implementation complexity Potential data privacy issues	Cobus [27] Wulan Pangestika [28]
Process optimization	AI algorithms optimize production processes to ensure consistent product quality	Procter & Gamble (P&G): Optimizing manufacturing processes with AI	Enhances process efficiency; Minimizes waste; Improves product consistency	Dependence on quality and volume of data; Requires continuous monitoring and updating	Lisa [29] Nicholas [30] Thor [31]
Anomaly detection	Identifies unusual patterns in production data that may indicate quality issues	Tesla: Real-time anomaly detection in production lines	Early detection of quality problems; Prevents defective products from reaching customers	False positives can lead to unnecessary halts; Requires robust data infrastructure	AI & Robotics [32] Elhafsi et al. [33] Artem [34]

(continued)

Table 25.6 (continued)

Application	Description	Examples	Benefits	Challenges	References
Supply chain quality control	AI monitors and evaluates the quality of raw materials and components from suppliers, optimizing inventory management, logistics, and demand forecasting	Boeing: Assessing the quality of parts from suppliers with AI Central banks: AI and the economy	Ensures quality of incoming materials; Reduces risk of defects Improves supplier management	High integration costs; Complex supplier data integration	Lindsay Bjerregaard [35]; Boeing[50]; BIS Annual Economic Report [36]
Robotic process automation (RPA)	Automated robots carry out precise, tedious quality assurance tasks	Foxconn: AI-driven robotics for assurance checks on production lines	Increases efficiency; Reduces labor costs; Ensures consistent inspection standards	High initial investment; Maintenance and upgrades can be costly	Foxconn Robotics; Madison Huang [37] Freddie Pierce [38]

(continued)

Table 25.6 (continued)

Application	Description	Examples	Benefits	Challenges	References
Real-time monitoring	Continuous monitoring of production processes using AI to detect and address quality issues promptly	Nestlé: Real-time quality monitoring in food production	Immediate response to quality issues; Maintains high product standards; Reduces waste	Requires continuous data flow; High computational resource requirements	Innovation & Technology [39] Wulan Pangestika [28]
Machine learning models	Develops models to predict and prevent quality issues based on historical production data	BMW with AWS: Predicting potential quality issues in vehicle manufacturing	Proactive quality management; Reduces defect rates; Improves decision-making	Complexity in model development; High dependency on historical data accuracy	Marc et al. [40]

systems rely on historical and real-time data from equipment sensors; any gaps or inaccuracies in this data can compromise their reliability.

Complexity of Implementation
Integrating AI into existing industrial processes can be technically complex and may require significant changes to workflows and systems. This complexity can lead to longer implementation times and higher costs, as well as potential disruptions to ongoing operations. For example, implementing AI for process optimization may necessitate reengineering existing production lines and retraining staff.

Data Privacy and Security
Collecting and processing large amounts of data, including sensitive operational data, raises concerns about data privacy and security. Ensuring compliance with data protection regulations and safeguarding against cyber threats can be challenging and resource-intensive. Industries dealing with proprietary manufacturing processes must ensure that AI systems handling their data do not expose them to intellectual property theft.

Scalability Issues
Scaling AI solutions from pilot projects to full-scale deployment across all operations can be difficult. Ensuring that AI applications perform consistently and reliably at scale requires robust infrastructure and continuous optimization. For example, a predictive maintenance system tested on a small subset of machines may not perform as expected when deployed across an entire factory.

Interoperability with Existing Systems
AI systems need to seamlessly integrate with existing industrial systems and technologies. Lack of interoperability can lead to fragmented processes and data silos, reducing the effectiveness of AI applications. For example, an AI-driven anomaly detection system must be compatible with the existing manufacturing execution system (MES) to provide real-time alerts.

Lack of Skilled Workforce
Establishing and upholding AI solutions necessitates a staff with deep expertise across AI as well as statistics analytics. A shortage of skilled professionals can hinder the successful deployment and ongoing operation of AI applications. Many industries struggle to find qualified data scientists and AI specialists to develop and maintain AI systems.

Change Management and Resistance
Adopting AI technologies often involves significant changes in organizational processes and culture. Resistance from employees and management can slow down or derail AI implementation efforts. For example, workers accustomed to traditional inspection methods may resist transitioning to AI-based automated systems due to fear of job loss or distrust in new technologies.

Ethical and Bias Concerns
AI systems can inadvertently introduce biases based on the data they are trained on, leading to unfair or unethical outcomes. Addressing bias and ensuring ethical AI use is crucial to maintaining trust and achieving fair decision-making. For example, an AI system used for quality control must be trained on diverse and representative data to avoid biased decisions that favor certain product types over others.

Maintenance and Upgrades
AI systems require regular updates and maintenance to adapt to changing conditions and ensure continuous improvement. Ongoing maintenance and upgrades can be resource-intensive and require dedicated personnel. Predictive maintenance systems need regular updates to their algorithms based on new equipment data and operational changes.

Requires Extensive Training Data
The effectiveness of AI models is highly dependent on the quality and quantity of the data they are trained on. Obtaining and curating this data can be resource-intensive.

Dependence on Quality and Volume of Data
AI systems need accurate and comprehensive data to function correctly. Incomplete or poor-quality data can lead to inaccurate predictions and decisions.

Requires Continuous Monitoring and Updating
AI models need to be regularly updated with new data to maintain their accuracy and relevance. This requires ongoing monitoring and maintenance efforts.

While the integration of AI in industrial decision-making offers significant advantages, addressing these challenges is essential for successful implementation. Companies need to invest in the right infrastructure, ensure data quality, manage change effectively, and build a skilled workforce to fully leverage the potential of AI technologies. By overcoming these obstacles, industries can enhance their operational efficiency, improve product quality, and achieve greater competitiveness in the market. Cloud Computing is another emerging technology whose integration may resolve many AI challenges like Scalability, economy in infrastructure cost, reduction in Maintenance and Upgrade tasks, etc.

The next section highlights the advantages of their combination for industrial growth.

25.7 The Synergy of AI and Cloud Computing in the Industrial Sector

The integration of AI with Cloud Computing is catalyzing a paradigm shift in the industrial sector. This powerful combination is enhancing operational efficiency, driving innovation, and transforming decision-making processes. As organizations look to use data to gain a competitive edge, the combination of AI and cloud technology has proven to be a substantial game changer.

25.7.1 The Role of AI in the Industrial Sector

As described in the beginning, AI encompasses a range of technologies, including machine learning, computer vision, natural language processing, and robotics, which enable machines to learn, adapt, and make decisions. In the industrial sector, AI is utilized to Optimize Operations, Predictive maintenance timely maintenance and minimizing downtime, Quality Control, Supply Chain Management, Autonomous Systems (autonomous vehicles, drones, and robots) for enhancing operational efficiency, and many more.

25.7.2 The Role of Cloud Computing in the Industrial Sector

Cloud computing provides scalable, on-demand access to computing resources, including storage, processing power, applications, databases, networking, analytics, and intelligence over the internet. It marks a revolution in how corporations utilize and organize services. Features like scalability, Feasibility, cost-effectiveness, availability, and security make it the ultimate place to support AI for improved industrial think tanks. Cloud can help in the maintenance, updates, scalability, agility, and security of the underlying infrastructure [51,41, 42]. Emerging technologies like AI, Blockchain, and IoT rely heavily on the cloud for their computational power and scalability [43].

Cloud Resource Provisioning (CRP) is dissimilar from the conventional market for renting infrastructure in place of selling it. These resources are made available to customers as virtual machines (VMs) with various configurations comprised of CPU, memory, storage, and bandwidth using a pay-per-use business model for various industries including educational organizations for economical deals and 24/7 accessibility anywhere [54,44–46]. Fairness and QoS become crucial criteria for the whole assessment. Many algorithms put forth prioritize the genuine provider over others for the truthfulness and fair process with QoS [44, 45, 47]. CRP also helps in reducing e-waste generation and management of IT resources with energy efficiency [53,44, 48].

Its role in the industrial sector includes:

Scalability: Cloud computing offers a scalable infrastructure that can be adjusted based on demand, ensuring cost-efficiency.

Data Storage and Processing: The cloud supports large-scale data storage and processing, enabling real-time analytics and decision-making.

Collaboration and Accessibility: Cloud platforms facilitate collaboration and accessibility, allowing teams to access data and applications from anywhere.

Disaster Recovery: Cloud services offer robust disaster recovery solutions, ensuring business continuity in case of disruptions.

Cost-Effectiveness: The pay-as-you-go model reduces capital expenditures, allowing industries to invest in innovation.

25.7.3 The Collaboration of AI and Cloud Computing

The integration of AI with cloud computing amplifies their strengths, creating a synergistic effect that revolutionizes industrial operations. Key benefits include:

Enhanced Data Analytics

Cloud computing provides the necessary infrastructure to store and process massive amounts of data generated by industrial operations. AI leverages this data to extract actionable insights, enabling informed decision-making. This synergy enhances predictive analytics, real-time monitoring, and optimization of industrial processes. Siemens utilizes cloud-based AI to analyze data from its manufacturing plants worldwide, optimizing production schedules, reducing downtime, and improving overall efficiency [25].

Scalable AI Solutions

The cloud's scalability allows industries to deploy AI solutions without the need for extensive on-premises infrastructure. Industries can scale their AI applications up or down based on demand, ensuring flexibility and cost-efficiency. General Electric (GE) uses cloud-based AI to scale its predictive maintenance solutions across its global fleet of jet engines, ensuring optimal performance and reduced maintenance costs.

Improved Collaboration and Innovation

Cloud platforms facilitate collaboration by providing a centralized repository for data and AI models. This enhances cross-functional teamwork and accelerates innovation. Teams can collaborate on AI projects, share insights, and develop new solutions more efficiently. Johnson & Johnson leverages cloud-based AI to facilitate collaboration among its research and development teams, accelerating the discovery and development of new medical devices and pharmaceuticals.

Real-Time Decision-Making
The combination of AI and cloud computing enables real-time data processing and decision-making. Industries can respond quickly to changing conditions, optimize processes on the fly, and improve operational agility [49]. Tesla's manufacturing plants use cloud-based AI to monitor production lines in real-time, adjusting parameters and making decisions autonomously to ensure quality and efficiency[52].

Cost-Effective Deployment
Cloud computing's pay-as-you-go model reduces the need for significant upfront investments in hardware and infrastructure. Industries can deploy AI solutions cost-effectively, paying only for the resources they use. A small manufacturing firm can use cloud-based AI services to optimize its production processes without the need for large capital expenditures, leveling the playing field with larger competitors.

25.7.4 Challenges and Considerations

While the integration of AI and cloud computing offers significant benefits, there are challenges and considerations to address including Integration with Legacy Systems, Reliability and Downtime, Privacy, etc.

The integration of AI with cloud computing is revolutionizing the industrial sector, driving enhanced efficiency, innovation, and data-driven decision-making.

25.8 Conclusion

AI is revolutionizing industrial decision-making by enhancing efficiency, accuracy, and strategic planning. Though certain hurdles hinder its execution, the rewards vastly exceed the negative aspects. As AI technologies continue to evolve, their impact on industrial decisions will only grow, making it imperative for companies to embrace AI and stay ahead of the curve. By understanding and addressing the challenges, industries can harness the full potential of AI to achieve sustained growth and competitive advantage. By leveraging the scalability, flexibility, and collaborative capabilities of the cloud, industries can deploy powerful AI solutions cost-effectively and efficiently. Despite the challenges, the benefits of this integration far outweigh the drawbacks, positioning industries to thrive in an increasingly competitive and data-driven world. As AI and cloud technologies continue to evolve, their combined impact on industrial operations will only grow, making it imperative for industries to embrace and integrate these technologies to stay ahead of the curve.

References

1. Elahi, M., Afolaranmi, S. O., Martinez Lastra, J. L., & Perez Garcia, J. A. (2023). A comprehensive literature review of the applications of AI techniques through the lifecycle of industrial equipment. *Discover Artificial Intelligence, 3*(1), 43.
2. Gupta, S., & Biswas, S. (2023). The confluence of smart computing and traditional businesses to foster productivity, profitability, and prosperity. In *Artificial Intelligence Techniques in Human Resource Management* (pp. 87–112). Apple Academic Press.
3. Hirsch-Kreinsen, H. (2023). Artificial intelligence: A "promising technology". *AI & SOCIETY*, 1–12.
4. Jan, Z., Ahamed, F., Mayer, W., Patel, N., Grossmann, G., Stumptner, M., & Kuusk, A. (2023). Artificial intelligence for industry 4.0: Systematic review of applications, challenges, and opportunities. *Expert Systems with Applications, 216*, 119456.
5. Moro-Visconti, R., Cruz Rambaud, S., & López Pascual, J. (2023). Artificial intelligence-driven scalability and its impact on the sustainability and valuation of traditional firms. *Humanities and Social Sciences Communications, 10*(1), 1–14.
6. Mohanty, S. N., Chatterjee, J. M., Jain, S., Elngar, A. A., & Gupta, P. (Eds.). (2020). *Recommender system with machine learning and artificial intelligence: Practical tools and applications in medical, agricultural and other industries.* John Wiley & Sons.
7. AI for Social Good. (2023). *Understanding AI: Differentiating what is AI and what is not AI.* AI for Social Good. https://aiforsocialgood.ca/blog/understanding-ai-differentiating-what-is-ai-and-what-is-not-ai
8. Kamm, S., Veekati, S. S., Müller, T., Jazdi, N., & Weyrich, M. (2023). A survey on machine learning based analysis of heterogeneous data in industrial automation. *Computers in Industry, 149*, Article 103930.
9. Arefin, S., Parvez, R., Ahmed, T., Ahsan, M., Sumaiya, F., Jahin, F., & Hasan, M. (2024). Retail industry analytics: Unraveling consumer behavior through RFM segmentation and machine learning. In *2024 IEEE International Conference on Electro Information Technology (eIT)* (pp. 545–551). IEEE.
10. Rai, R., Tiwari, M. K., Ivanov, D., & Dolgui, A. (2021). Machine learning in manufacturing and industry 4.0 applications. *International Journal of Production Research, 59*(16), 4773–4778.
11. Aghbashlo, M., Peng, W., Tabatabaei, M., Kalogirou, S. A., Soltanian, S., Hosseinzadeh-Bandbafha, H., & Lam, S. S. (2021). Machine learning technology in biodiesel research: A review. *Progress in Energy and Combustion Science, 85*, Article 100904.
12. Bertolini, M., Mezzogori, D., Neroni, M., & Zammori, F. (2021). Machine Learning for industrial applications: A comprehensive literature review. *Expert Systems with Applications, 175*, Article 114820.
13. Sircar, A., Yadav, K., Rayavarapu, K., Bist, N., & Oza, H. (2021). Application of machine learning and artificial intelligence in oil and gas industry. *Petroleum Research, 6*(4), 379–391.
14. Hamouda, D., Ferrag, M. A., Benhamida, N., & Seridi, H. (2023). PPSS: A privacy-preserving secure framework using blockchain-enabled federated deep learning for industrial IoTs. *Pervasive and Mobile Computing, 88*, Article 101738.
15. Zipfel, J., Verworner, F., Fischer, M., Wieland, U., Kraus, M., & Zschech, P. (2023). Anomaly detection for industrial quality assurance: A comparative evaluation of unsupervised deep learning models. *Computers & Industrial Engineering, 177*, Article 109045.
16. Shamshiri, A., Ryu, K. R., & Park, J. Y. (2024). Text mining and natural language processing in construction. *Automation in Construction, 158*, Article 105200.
17. Xiong, F., Kühl, N., & Stauder, M. (2024). Designing a computer-vision-based artifact for automated quality control: A case study in the food industry. *Flexible Services and Manufacturing Journal, 36*, 1–28.
18. Mia, M. R., & Shuford, J. (2024). Exploring the synergy of artificial intelligence and robotics in industry 4.0 applications. *Journal of Artificial Intelligence General science (JAIGS), 1*(1).

19. Wang, J., Tian, Y., Hu, X., Fan, Z., Han, J., & Liu, Y. (2024). Development of grinding intelligent monitoring and big data-driven decision making expert system towards high efficiency and low energy consumption: Experimental approach. *Journal of Intelligent Manufacturing, 35*(3), 1013–1035.
20. Abou Akar, C., Abdel Massih, R., Yaghi, A., Khalil, J., Kamradt, M., & Makhoul, A. (2024). Generative adversarial network applications in industry 4.0: A review. *International Journal of Computer Vision, 132*(6), 2195–2254.
21. Gao, Z., & Li, L. (2024). Adaptive optimization of cutting parameters in milling industry considering dynamic tool wear in intelligent manufacturing driven by reinforcement learning. *The International Journal of Advanced Manufacturing Technology*, 1–10.
22. Deepthi. B, Gupta, P., Rai, P., & Arora, H. (2022). Assessing the dynamics of AI driven technologies in Indian banking and financial sector. *Vision*, 09722629221087371.
23. Yadav, P., Gupta, P., Sijariya, R., & Sharma, Y. (2024). Artificial intelligence in risk management. In *Artificial Intelligence for Risk Mitigation in the Financial Industry*, (pp. 1–25).
24. Siemens. (2023). https://references.siemens.com/en/reference/scalable-visual-quality-inspection-?id=37527
25. Gopalakrishnan, S., Hartman, N. W., & Sangid, M. D. (2023). A digital engineering framework to facilitate automated data exchange between geometric inspection and structural analysis. *Advances in Engineering Software, 183*, Article 103498.
26. Linda Wilson. (2024). *Siemens acquires automated visual inspection startup.* Inspekto markets an AI-enabled visual inspection in-a-box system. https://www.vision-systems.com/factory/manufacturing/article/14305479/siemens-acquires-automated-visual-inspection-startup.
27. Cobus van Heerden. (2020). *GE Digital: smarter manufacturing with AI & analytics.* https://manufacturingdigital.com/technology/ge-digital-smarter-manufacturing-ai-and-analytics
28. Wulan, P. (2024) *Nestle: Transforming with AI and predictive maintenance.* April 30, 2024, https://discoveryshift.com/2024/04/30/nestle-transforming-with-ai-and-predictive-maintenance/
29. Lisa, J. (2024). *P&G leans into AI for dynamic routing and sourcing optimization.* 1/23/2024 (https://consumergoods.com/pg-leans-ai-dynamic-routing-and-sourcing-optimization)
30. Nicholas DeNittis. (2024). Artificial Intelligence at Procter & Gamble February 2, 2024, (https://emerj.com/ai-sector-overviews/artificial-intelligence-at-procter-gamble/)
31. Thor Olavsrud. (2022). *P&G turns to AI to create digital manufacturing of the future.* 30 Sep 2022 https://www.cio.com/article/408351/pg-turns-to-ai-to-create-digital-manufacturing-of-the-future.html
32. AI & Robotics. (2024). https://www.tesla.com/AI
33. Elhafsi, A., Sinha, R., Agia, C., Schmerling, E., Nesnas, I. A., & Pavone, M. (2023). Semantic anomaly detection with large language models. *Autonomous Robots, 47*(8), 1035–1055.
34. Artem Grotov. (2020). *Self-driving anomaly detection.* February 21, 2020 (https://www.stackstate.com/blog/self-driving-anomaly-detection/).
35. Lindsay, B. (2019). *Boeing selects AI startup to enhance parts inspections.* 16 December 2019 (https://aviationweek.com/mro/boeing-selects-ai-startup-enhance-parts-inspections)
36. BIS Annual Economic Report. (2024). *Artificial intelligence and the economy: Implications for central banks.* https://www.bis.org/publ/arpdf/ar2024e3.htm
37. Madison, H. (2024). *Foxconn trains robots, streamlines assembly with NVIDIA AI and omniverse.* 2 June 2024 (https://blogs.nvidia.com/blog/foxconn-digital-twin-ai/).
38. Freddie Pierce. (2020). *One million Foxconn robots outsource human work.* May 18, 2020 https://supplychaindigital.com/digital-supply-chain/one-million-foxconn-robots-outsource-human-work
39. Innovation & Technology. (2024). (https://nestleyouthentrepreneurship.com/artificial-intelligence-and-data-science-support-innovation)
40. Marc, N., Aubrey, O., Jonathan-Edwin, A., Markus, F., & Mohan Gowda, P., (2024). *Accelerating industrialization of machine learning at BMW group using the machine learning operations (MLOps) solution.*

41. Buyya, R., Yeo, C. S., & Venugopal, S. (2008). Market-oriented cloud computing: Vision, hype, and reality for delivering IT services as computing utilities. In *2008 10th IEEE International Conference on high performance computing and communications*, (pp. 5–13). https://doi.org/10.1109/HPCC.2008.172
42. Singhal, R., Sonia, & Singhal, A. (2016b). Rising future of agile software development using cloud computing: A study using cloud computing in different phases of an agile method-CRUM. *International Journal of Computer Technology & Applications, 7*(2), 240–246.
43. Singhal, R., Sharma, V., Singhal, I., & Bansal, V. (2024). Blockchain-enabled auction for cloud resource provisioning: a survey on trust and economy. *International Journal of System Assurance Engineering and Management, 15*, 1–21.
44. Singhal, R., & Singhal, A. (2019a). A combinatorial fair economical double auction resource allocation model. *In International Conference on Advances in Computing and Data Sciences* (pp. 272–281). Springer, Singapore. https://doi.org/10.1007/978-981-13-9942-8_26
45. Singhal, R., & Singhal, A. (2019b). A combinatorial economical double auction resource allocation model (CEDARA). In *International Conference on Machine Learning, Big Data, Cloud and Parallel Computing (COMITCon)* (pp. 497–502). IEEE. https://doi.org/10.1109/COMITCon.2019.8862456
46. Singhal, R., Singhal, A., & Sonia. (2016a) Towards a generic e-cloud architecture for Universities. *IJWA, 8*, 36–43.
47. Singhal, R., & Singhal, A. (2021). A feedback-based combinatorial fair economical double auction resource allocation model for cloud computing. *Future Generation Computer Systems, 115*, 780–797. https://doi.org/10.1016/j.future.2020.09.022
48. Katal, A., Dahiya, S., & Choudhury, T. (2023). Energy efficiency in cloud computing data centers: A survey on software technologies. *Cluster Computing, 26*(3), 1845–1875.
49. Singhal, R. (2024). *Introduction to cloud computing (AWS),* Authors Press.
50. Boeing https://www.boeingsuppliers.com/quality.html
51. Buyya, R., Ranjan, R., & Calheiros, R. N. (2010). Intercloud: Utility-oriented federation of cloud computing environments for scaling of application services. In *International Conference on Algorithms and Architectures for Parallel Processing*, (pp. 13–31). Springer. https://doi.org/10.1007/978-3-642-13119-6_2
52. Jonathan M. Gitlin. (2024). *Elon Musk says Tesla is an AI company now.* 8/6/2024 (https://arstechnica.com/cars/2024/08/elon-musk-says-tesla-is-an-ai-company-now-heres-how-plausible-that-is/)
53. Singhal, R., Singhal, A., Sharma, V. and Saachi, (2019a) E-Cloud: A solution towards e-waste management for educational institutions. *International Journal of Recent Technology and Engineering (IJRTE), 8*(2S6), pp. 964–972. https://doi.org/10.35940/ijrte.B1184.0782S619
54. Singhal, R., Singhal, A., Bhatnagar, M., & Malhotra, N. (2019b) Design of an audio repository for blind and visually impaired: A case study. In *Advanced Computing and Communication Technologies*, (pp. 77–85). Springer, Singapore. https://doi.org/10.1007/978-981-13-0680-8_8

MIX
Papier aus verantwortungsvollen Quellen
Paper from responsible sources
FSC® C105338

If you have any concerns about our products,
you can contact us on
ProductSafety@springernature.com

In case Publisher is established outside the EU,
the EU authorized representative is:
Springer Nature Customer Service Center GmbH
Europaplatz 3, 69115 Heidelberg, Germany

Printed by Libri Plureos GmbH
in Hamburg, Germany